An Introduction to Economics, 4th Edition

Concepts for Students of Agriculture and the Rural Sector

09

An Introduction to Economics, 4th Edition

Concepts for Students of Agriculture and the Rural Sector

Berkeley Hill

Emeritus Professor of Policy Analysis
University of London (Imperial College London)

www.cabi.org

CABI is a trading name of CAB International

CABI	CABI
Nosworthy Way	38 Chauncy Street
Wallingford	Suite 1002
Oxfordshire OX10 8DE	Boston, MA 02111
UK	US
Tel: +44 (0)1491 832111	Tel: +1 800 552 3083 (toll free)
Fax: +44 (0)1491 833508	E-mail: cabi-nao@cabi.org
E-mail: info@cabi.org	
Website: www.cabi.org	

A catalogue record for this book is available from the British Library, London, UK.

10 0740313 5

Library of Congress Cataloging-in-Publication Data

Hill, Berkeley.
  An introduction to economics : concepts for students of agriculture and the rural sector / Berkeley Hill. -- Fourth edition.
    pages cm
  Includes bibliographical references and index.
  ISBN 978-1-78064-475-2 (alk. paper)
1. Economics. 2. Agriculture--Economic aspects. I. Title.

  HB171.5.H63 2014
  330.024'631--dc23

                        2014006674

ISBN-13: 978 1 78064 475 2

Commissioning editor: Claire Parfitt
Editorial assistant: Alexandra Lainsbury
Production editor: Tracy Head

Typeset by SPi, Pondicherry, India.
Printed and bound by Gutenberg Press, Tarxien, Malta.

Contents

Preface   vii

Acknowledgements   ix

1   What is Economics?   1

2   Explaining the Behaviour of Individuals: Theory of Consumer Choice   11

3   Demand and Supply: the Price Mechanism in a Market Economy   26

4   Markets and Competition   51

5   Production Economics: Theory of the Firm   72

6   Factors of Production and their Rewards: Theory of Distribution   115

7   Market Failure: Some Problems of Using the Market to Allocate Resources   138

8   Macroeconomics: the Workings of the Whole Economy   150

9   International Trade   172

10   Government Policy for Agriculture and Rural Areas   191

Appendix 1 Essay Questions   215

Appendix 2 Suggested Answers and Explanations for the Exercises following each Chapter   219

References   237

Suggested Further Reading   239

Index   241

Preface

This text aims to provide a simple but effective introduction to economics for students of agriculture, the rural sector and related topics in universities and colleges.

While only a small minority of these students will have studied economics as a subject at school, all are soon made aware that most of the fundamental problems facing agriculture and rural areas are of an economic nature or have economic implications. A rapid introduction to the concepts of the discipline is required, yet economics generally forms only part of their total studies, so time is at a premium. Several decades of teaching without access to a basic text of suitable complexity and content prompted the writing of this outline. It focuses on the parts of economic theory most relevant to this group of students as a solution to their particular educational problem.

This introduction describes universal economic principles. It is not a book on the economics of agriculture but one of general economics, illustrated primarily by examples drawn from farming, the food industry and the rural sector. Many books concentrating on the economics of agriculture, agricultural policy, farm management and rural resource use assume in the reader some basic knowledge of economics. This text is intended to provide that knowledge and pave the way to the various specialist areas. At the same time, the broad base of theory contained here enables these specialist studies to be seen in better perspective. For example, managers of individual farm businesses will find it increasingly relevant, in a changing world, to be aware of trade theory and the arguments put forward for allowing more international competition, of why inflation needs to be controlled, and of the concept of externalities (such as pollution) and how they can be taken into account.

Despite this emphasis on the broad approach to the discipline, students studying agriculturally related topics will discover that some facets of economics are of greater importance to them than others. Of particular value are price theory and (because farm management is often a prime interest) production economics. In most introductory texts intended for general use the balance does not favour these areas; instead, attention is switched to a description of the wider economy and its institutions. Intermediate university texts, while usually containing the necessary material, are often found too daunting by the novice. In tailoring an introduction especially for students of agriculture and the rural sector, it has been possible to produce a balance which it is hoped will best serve their needs in terms of content and complexity.

When preparing the second edition in 1990 I took the opportunity of updating the text. The first edition had been orientated to the UK. Some of the examples needed changing after a decade (the 1980s) in which market forces and private enterprise had been allowed to take an increasing role in the way that the economy was organized. However, the main change was the addition of a chapter on government policy and agriculture. This did not compromise the general approach of concentrating on economic principles, but recognized the importance to students of understanding the policy process. The impact of government intervention on the agricultural industry, land use and the rural environment is highly significant and pervasive. The inclusion of a description of the agricultural policy aims of the European Union (EU) and the ways in which these were turned into practical programmes was felt to represent a significant improvement to the text. Several reviewers of the first edition shared this opinion.

The third edition (2006) took this broadening a step further. The EU had been enlarged to include countries that formerly used systems of planning to solve economic problems, and agriculture was increasingly viewed in a global context. A new section on technological change was added to the chapter on production economics because innovations were (and remain) important in explaining the income pressure felt by farm operators and the changing structure of agriculture as farmers attempt to cope (by increasing the size of their business, diversifying or adapting in other ways, including quitting farming).

This fourth edition has had to acknowledge that since 2006 many further changes have taken place in the environment in which agriculture in Europe operates. Not only has the EU been enlarged to 28 Member States, but the main method of support of agriculture has shifted away from intervention in the markets for agricultural products and subsidies linked to production and towards a system of decoupled direct income payments. In the broader economy there has been a major hiatus in European banking, widespread economic recession and, in many countries, high levels of unemployment. Financial crises have been seen in some Member States that had adopted the euro. These developments carry implications not only for the chapter on agricultural policy but in many other places in the text. In the process of revision, the opportunity has been taken to make numerous other updates and small improvements including, where practical, adopting gender-neutral language, though readers should be warned that farmers are still often referred to as male, as in the real world they overwhelmingly are.

While the economic principles underlying the problems of agriculture and the policies that address them are universally appropriate and enduring, the scope has been broadening. Agriculture is now recognized as only one activity that is undertaken by many people who operate farm businesses. The old tradition of seeing farming as a distinct and separated activity is outdated, and even an introductory text must recognize the inter-relationship between what happens on farms and the other economic endeavours in which their operators engage. This applies not only to farms run as family businesses (sole proprietorships and partnerships) but also to those set up as companies (and thus having separate legal status).

Farmers and landowners produce environmental goods and services in addition to their traditional food and fibre products, and the economic explanations for this multifunctional nature and the basis on which they are rewarded for these non-market outputs have been incorporated into the text. So too has the recognition that agriculture is increasingly seen as only one aspect of what goes on in rural areas (a small and declining economic one but a major environmental one), and that it plays an instrumental role in policies directed at the countryside in general and in contributing to sustainability.

In any introduction there are always dangers stemming from oversimplification and exclusion. Hopefully, serious misconceptions are not propagated here. Students who find the treatment incomplete can turn to the many intermediate texts that are available. The general standpoint taken to the subject here is that of the free enterprise market economy. I am aware that no fundamental criticism of capitalism has been offered. Also, perhaps more attention has been shown to perfect competition than might be currently fashionable. However, it should be recalled that most of the students for whom this text is intended will find themselves operating in economies with agricultural industries that are capitalist in base, market dominated and where perfect competition perhaps comes nearest to reality.

At the ends of chapters are exercises that make use of the preceding material. The completion of these exercises forms an integral part of the teaching function of this text. At the end of the book will be found extended answers to the questions posed in the exercises, a list of essay questions and suggested further reading.

Acknowledgements

Colleagues at the former Wye College (University of London) to whom I owe thanks are the late Professor D.K. Britton and the late John Medland, both of whom on occasion taught introductory courses jointly with me. I am especially grateful to Professor Alison Burrell for reading the typescript of the first edition while a staff member at Wye and for her many helpful suggestions following hours of discussion. Errors of an economic nature and infelicities in presentation are, of course, fully my responsibility.

Past students at the Royal Agricultural University, Wye College and, following Wye's merger in 2000, Imperial College (Wye campus) would no doubt find much of the text familiar. Their feedback was extremely helpful in developing the contents of this book. The Wye campus is now closed and its teaching rooms currently unused, but what started as notes for a series of first-year lectures there is clearly finding a continuing use in other institutions. This forms an unexpected and welcome coda to a career spent enjoyably with generations of lively-minded young people.

This fourth edition is dedicated to my grandson Joe, who found an earlier edition of some value in his first encounters with economics.

1 What is Economics?

Introduction – the Essence of Economics

Economics is concerned with choice at all levels of society: choice by individuals, by firms, by local and central governments, and by international organizations. Economics is the study of why choices are necessary and how they are made, a study generally undertaken with the aim of improving in some way the outcome of choices.

At the level of the individual, a person leaving school and faced with a variety of alternative college or university courses has to select just one, knowing that the selection will heavily influence how the rest of his or her life is spent. Having made this major choice, whether to study economics or law or whatever, the student while at college must allocate time between the study of the chosen subject and leisure. If the objective is to pass the examinations and yet enjoy some sport, choice must be made on how time is spent in order to achieve academic success and still have pleasure from leisure. It will soon become clear that more hours spent playing football inevitably means fewer hours are available for work. Achieving a good balance between work and non-work is something that is of increasing concern in modern society.

Businesses, from family firms to multinational companies, can use their manpower, capital and premises or land in a variety of ways and it is the task of their managements to choose to allocate the firm's resources in the best ways, which usually means most profitably. Taking an agricultural example, a farmer can perhaps grow wheat on his farm and have no other enterprises, or grow grass and keep dairy cows, or maybe he can have a mixture of grass and wheat. It is up to the farmer to choose how to allocate his land and other resources, such as his labour force, between the alternative enterprises, bearing in mind what he is attempting to achieve from his farming. With a farm of a given size, using more land for cereals means that less is available for other crops.

The government, or state, acting on behalf of society, raises money through taxation or borrowing and spends it on defence, roads, health care and social services, pensions, education, and in many other ways. With a budget of a certain size, if the government wishes to spend more on the health service it will have to cut back on something else – say, welfare benefits. Just as with individuals, the government has to choose how to allocate its resources between many possible ways of spending, and by spending money on one thing it loses the opportunity of spending it on another.

The environment is another area where choice by society is involved. For example, society has sometimes to choose between either building a new airport in an area important for nature conservation or having the wildlife but also suffering congestion at existing airports. By enjoying the convenience of an extra airport, the opportunity of enjoying the unspoilt environment is lost.

Opportunity Cost

Whenever choice is involved, taking one alternative automatically means that all the others have to be left, or forgone. If the choice is made to build an airport on a piece of land, the opportunity to use that land for any other purpose is ruled out – for a nature reserve, or for farming, or for a golf course, etc.

The cost to society of having an airport may be thought of as what has to be gone without by having the airport. If the land is currently used for farming, being judged the best way to use it before the airport was proposed, then this activity has to be forgone when the airport is built. Costs are commonly thought of in terms of money, but money in this context is really only a useful common denominator in which the various items lost and gained can be expressed and compared. We should note that some of the important costs of having an airport, such as the extra noise involved and the aesthetic loss of unspoiled countryside and reduced biodiversity, are very difficult to express in money terms and may well get left out of the calculation, making nonsense of any precise monetary figure that may be arrived at.

In the example, the land could be used for a whole range of purposes, but only one at a time. If it is used as an airport, the cost of using it as such is really the best alternative use forgone, in this case for farming – this is termed the **opportunity cost** of using the land as an airport.

> The opportunity cost of any choice is the best alternative that is forgone by making that choice.

Another example is given by the student who can spend an evening drinking, or going to a cinema, or buying a book and studying it. If the pub is chosen, but the cinema would be next on the list of enjoyments, the opportunity cost of the night in the pub is the visit to the cinema.

Scarcity

Choice becomes necessary because the resources which individuals and society have at their disposal are insufficient to satisfy all their wants simultaneously and completely. As individuals we often come up against money, or rather spending power, as a scarce resource, so we have to choose how we use it. Governments frequently have limits to the budgets they can spend on competing projects. Time is another scarce resource, and it seems increasingly important to make best use of time as one becomes older.

Some commodities are not economically scarce: we do not normally have to choose how to allocate oxygen in the air – there is sufficient for each of us to use as much as we want for all purposes. It has no opportunity cost – we can use more of it by lighting a fire without having to breathe less. Such commodities are called **free goods**. A distinction should be made between scarcity of occurrence and economic scarcity. Discarded bedsteads that sometimes adorn ditches and lay-bys are not scarce in the economic sense because they are plentiful *in relation to the desire or want for them*. Essentially, no one has any use for them.

> A commodity is economically scarce when insufficient is available to satisfy all wants completely and the question of its allocation arises.

What is an Economic Problem?

An economic problem is said to occur whenever choice between alternatives presents itself. The government, faced with decisions on how to allocate its revenue between education, defence, housing or social services, is facing an economic problem. So is the farmer allocating his labour force over a range of jobs. So is the fat lady who has to choose between a range of slimming diets on which to spend her money and time. The objective of her choice is clear, at least on first examination: it is to lose weight. What she has to choose is how to achieve it. This leads us on to consider what lies behind any choice – its objective or final goal.

Objectives of choice

The woman in the above example appears to have an easily identifiable objective or goal – that of losing weight. However, by questioning more closely, we may find that the taste of her diet is important too. She therefore has two simultaneous goals: weight loss plus a diet that is not unpalatable. A third goal might be cheapness of the diet, because lower food costs mean that she can spend more on other goods and services that give her satisfaction. The woman's eventual choice of diet will be the result of her trying to achieve a whole range of goals simultaneously. It is difficult for an outsider to list all the goals and even harder to ascribe to each goal some measure of its relative importance. Yet the woman must have all these goals balanced, probably subconsciously, and her final choice will be her way of best achieving her objectives. Her choice will be rational and not just by chance. If presented with the same range of diets at some other time, as long as her objectives had not changed we would expect her to make the same, or a very similar, choice. While it is difficult to specify her objectives, it is clear that she has some. Economists say that the individual is attempting to maximize her *satisfaction* (or **utility**) by choosing as she does and make the assumption that the object of any choice is to get the greatest satisfaction which can be attained from the range of available alternatives.

The objectives of society in any choice are equally complex. Some of the objectives which are followed simultaneously and which are common to most countries, though with different balances that reflect circumstances, are:

- national security (external defence and internal policing, etc.);
- a healthy population;

- a high level of employment;
- a fair distribution of income and wealth;
- housing for all;
- stable prices;
- protection of the environment (biodiversity, landscape quality, prevention of water and air pollution, etc.);
- mitigation of and adaption to global warming;
- a high level of education for the population;
- sustained economic growth;
- promotion of harmonious international trade;
- individual liberty and the individual's control of his or her own environment; and
- democratic participation in major policy decisions.

Some of these objectives conflict with each other, and politicians will find themselves 'trading-off' one objective against another. For example, to curb inflation it may be necessary to create a pool of unemployment. To prevent atmospheric pollution it may be necessary to control factories more closely and force them to produce their goods in a cleaner and more costly way; this will push up prices, hamper international competitiveness and may retard economic growth. Society has, in its choices, to balance all its competing objectives according to their various weights of importance. In a democracy these weights are indicated primarily in the way that people vote at elections for parties or individuals who present different packages of objectives in their election manifestos. The political system should put in power the parties and individuals who, broadly speaking, reflect the wishes of society.

The objectives of firms are generally easier to define. Their prime objective is to make profit by providing customers with what they want, but other objectives might be the minimization of risk to the business, or to enhance the prestige of the company. In agriculture a farmer might consider profit to be relatively unimportant as a goal once a level has been attained which gives him an adequate living standard. He may then be more interested in doing what gives him pleasure – such as building up a pedigree herd of cows, or spending 3 days a week at local markets – and will arrange his farm with these non-profit motives in mind.

While objectives are generally difficult to define precisely, it is clear that individuals, society and firms all have them and that their behaviour is directed towards achieving them. Economics is concerned with how these objectives are approached.

We have dealt with scarcity of resources, choice and objectives. They can be combined into a definition of **economics**.

> Economics is the study of how individuals and society choose to allocate scarce resources between alternative uses in the pursuit of given objectives.

The Mechanism of Allocating Scarce Resources

In terms of how countries allocate their scarce resources, history suggests that two chief types of society can be distinguished: (i) the society with a planned economic system (sometimes called a **command economy**); and (ii) the society with a free-market economic system (or **market economy**). Problems of the allocation of the resources which can produce goods and services (i.e. land and natural resources, the labour force and managerial ability, and capital such as machines, buildings, power stations, etc.) occur in both systems. However, the ways in which the choices are made differ.

For the purposes of comparison of the two approaches it is useful to consider each in its most extreme form, i.e. a totally planned system of production and resource allocation contrasted with one relying entirely on the market with no central coordination. This approach, which uses 'models' of each system, is an attempt to capture the essential elements of a real-world situation without the complexities that would be encountered if we were to examine the details of the allocative process in real countries. The use of models permeates economic theory and will be encountered in this text when we consider the behaviour of individual consumers, of firms, of the whole economy and in a range of other contexts. Although models are often criticized as being 'unrealistic', by containing the most important elements of a situation and disregarding the rest, they may give us a far better insight into real-world situations and the likely outcome of actions or happenings than if we attempted to grapple with the complexities of the real world without their use.

Basically, in any economic system decisions have to be made on three questions:

1. *What* goods and services are produced. For example, should supersonic passenger aircraft be produced, or should the manpower and other resources involved be used to produce, say, more television sets?

2. *How much* of each good and service should be produced?

3. *Who should get* what is produced?

A country with a centrally planned economic system might answer these questions by using a committee, perhaps called the Central Planning Authority. This could make estimates of how many shoes, coats, television sets were needed by its population and then send directives to its factories telling them what and how much to produce. Assuming the Central Planning Authority also knew what resources were available in the country, it would probably also tell the factories how they were to produce their products. The nation's resources would thus be allocated in such a way that what needed to be produced (in the judgement of the planners) *was* produced. The presence of multiple sets of detailed instructions is why this approach can be labelled the 'command economy'. Naturally such an economic system could only work where the state owned the vast majority of the land and capital (machines and buildings, etc.) and where the inhabitants did what they are instructed to do. Furthermore, the Central Planning Authority would need to have available a vast range of frequently updated information on consumer preferences, the available resources and the technical details of production on which to base its planning decisions. Acquiring all this information would itself involves costs and absorb resources.

In the other extreme type of society, that of the unhindered free-market economic system, there is no central planning of what, how much and for whom production takes place. There is thus no centrally planned allocation of the nation's resources, and the need to assemble all the information necessary for planning is avoided. The resources are owned by individuals, not the state as in the centrally planned economy. What is produced, and how much, is determined by consumers going to shops and buying goods. Surpluses of certain goods soon disappear because, if shopkeepers cannot sell, say, a type of shoe and have to drop its price, they will not order any more from the factories, so production will be cut back. On the other hand, if not enough of a good is being produced, say, not enough large flat-screen television sets or smartphones, so that consumers are waiting for delivery, shopkeepers will badger the factories for supplies. Consumers' wants are

rapidly indicated to the producers, and they re-allocate their labour force, capital, etc. so that what the consumer wants and is able to pay for is produced. Resource allocation hence is a reflection of the wants of consumers: if consumers want a lot of smartphones, resources will be switched to producing more of them.

No country in the real world is either completely centrally planned or operates a completely unhindered free market system. In the 1990s many of the examples of central planning in Europe, in states controlled by Communist parties, abandoned their attempts at command economies in favour of market systems (and a simultaneous reinstatement of forms of democracy) (see Box 1.1). Countries such as the UK, France and Sweden have had elements of planning within their basically capitalist market systems, and the role given to it has risen and fallen over time. However, the problem of resource allocation exists in all types of society; only the mechanism of making the choice differs. The study of economics – the study of how individuals and society choose to allocate scarce resources between alternative uses in the pursuit of given objectives – is equally appropriate.

The Scientific Approach to Economics

Taking a scientific approach involves working out explanations (**theories**) of how we think things work, making predictions (**hypotheses**) based on these theories, and testing them against evidence to see if they are valid. As well as improving our knowledge of the world about us, a successful theory is useful in that it enables the outcome of various occurrences to be foreseen. Thus, by using economic theory, a government may be able to predict the sort of impact on national income that might follow from an increase in the price of oil brought about by a shortage of international supply or a rise in the tax placed on it.

The scientific approach to any subject follows the same basic pattern. The scientist becomes aware, through observation of her subject, of a pattern of circumstances. This might be that the roots of a plant tend to grow downwards even if the plant is turned upside down (termed 'geotropism'). Or it might be that consumers buy a greater quantity of petrol when its price is reduced. In an attempt to explain or to account for these observations, the

Box 1.1. The rise and fall of planned economic systems in Europe.

Letting market forces operate in a completely unrestrained way can produce some results that seem unfair and distasteful to many people. The market mechanism contains imperfections that are now widely recognized (see Chapter 7 'Market Failure'). The 19th-century industrialization of countries such as the UK led to some conditions developing, particularly in large towns and in the factories in which much of the population there worked, that caused social commentators to question the dominant role of the market. Karl Marx (1818–1883) was the most prominent of these; from 1849 he lived in London and witnessed early Victorian England's factories with their dangerous working conditions, long working weeks, child labour and employees suffering general poverty and bad housing. Marx was highly critical of the system in which the market was the driving force and of the extremes of income and wealth to which it gave rise. He saw the factories as destroying the natural willingness and ability of people to express themselves through their work (as craftsmen) by making them machine operators doing repetitive tasks in which they took no satisfaction other than to earn money. Competition between industrialists seemed destined to make conditions worse as each factory tried to undercut its rivals and, where possible, replace workers with more machinery. Could not a better system be devised in which society produced what each person needed and in which each person contributed according to their ability? Marx did not leave a detailed blueprint for the alternative system, though planning was implied.

Marx predicted that, ultimately, the increased misery caused by poverty, unemployment and lack of satisfaction with work would lead to the breakdown of the economy and society in a great revolution in which the 'proletariat' (literally those whose only thing they possessed to sell was their labour) would overthrow the capitalists (the owners of factories, etc.). However, the revolution did not come in the relatively industrialized UK with its tradition of democracy. Instead, it came in Russia in 1917, fundamentally still an agricultural country and one used to autocratic rule by the Czar. The leaders of the post-revolutionary USSR gave priority to building up their industrial sectors by using state planning bodies and a series of 5-year plans that involved moving resources out of agriculture, including the forced movement of labour and the extraction of capital by depressing the rewards from agriculture. Land was nationalized and farms were amalgamated into large-scale state farms or collective farms, designed to reap advantages of size. Much of this was done in a brutal way. Nevertheless, this planned approach resulted in rapid industrial growth which enabled the USSR to cope with the demands made on it by the Second World War and its part in the defeat of Nazi Germany.

In the tussle for territory and influence after the war, many countries in Eastern Europe came under the control of the USSR, run by the Communist Party, and adopted its system of central economic planning and one-party government. Eastern Germany (FDR), Poland, Hungary, Czechoslovakia, Estonia, Lithuania, Latvia, Bulgaria and Romania all had local strong Communist leaders and essentially 'command' economies. Some also 'socialized' their agricultures into large-scale units. However, planning was imperfect and led to waste, inefficiencies and poor innovation; targets were not set in rational ways and managers often had incentives that were perverse. The system proved to be too complex to produce the sort of economic growth that was experienced in the market economies of Western Europe and North America, though gradually some of their features were introduced to help matters (such as financial incentives that reflected the profitability of enterprises). By the 1980s a combination of political pressure for greater democratic freedoms, the loosened grip of the Communist Party and its autocratic leaders, and the failures of the planning system led to a collapse of the USSR and independence by its former satellite countries in a series of (mostly) non-violent revolutions, starting in 1989. The former planned economies of Central and Eastern Europe adopted market systems and, after a period of painful adjustment, were sufficiently advanced (in both economic terms and politically) to join the European Union (EU) in 2005 and 2007 (Bosnia in 2013).

Today there are very few countries that still adhere to the old 'command economy' model, with its associated lack of political freedom. North Korea remains the prime example.

scientist develops theories of what might be causing these behaviour patterns. Whether or not the theory is valid can only be established by a process of testing hypotheses that are developed from the theory. A hypothesis is a 'positive', or testable, statement about a certain sequence of events with the conditions under which the statement is supposed to apply clearly stated. A positive statement says what was, is, or will be and it can be tested by an examination of the facts. Normative statements, however, are value judgements. For example, 'old age pensions should provide a reasonable standard of living' is a normative statement. It may be morally right, but it is untestable. Economics as a science is concerned with positive statements, although this will not prevent economists from holding opinions, just like any other members of society, on what they consider to be ethically right or wrong.

A scientific hypothesis, then, is a positive statement about a certain set of conditions developed by the economic scientist from her observation of the world about her and her (provisional) explanation of how the various parts work together. For example, from existing information on the way that consumers spend their money, the relationship between the price of petrol and the quantity sold might be expressed by the following hypothesis: 'The price of petrol and the daily quantity sold are related so that, if the price of petrol is reduced, a greater quantity is sold per day, and if the price is increased a smaller quantity is sold, all other things being equal'. The last part, 'all things being equal' (sometimes given in the Latin form *ceteris paribus*) is very important; it means that other influences on the volume of petrol sold, such as incomes of car owners (as people become richer they tend to spend more on petrol), or the other costs of running a car, do not vary. If they did, it would not be possible to tell whether the change in petrol sales had been the result of the change in petrol price *or* changes in the other influences.

A shorthand way of writing the hypothesis might be:

$$D_{petrol} = f(P_{petrol})$$

Demand for petrol is a function of (or depends on) the price of petrol. This is called a 'functional relationship'.

If the price of petrol falls and the quantity sold *does* increase, the facts are in line with the hypothesis and thus support the theory. If the quantity sold does not increase, or decreases, the facts are not in line with the hypothesis (it is 'rejected') and thus do not support the theory. Thus the theory (the understanding of what is going on) must be modified in light of any new observations, or scrapped in favour of a theory that explains the facts better.

The process of theory formation and confirmation is shown in Fig. 1.1. If this diagram is understood, the basic scientific approach to any subject will have been grasped.

Problems Faced by Scientific Economists

Branches of science differ in the ways in which the predictions from theories can be tested. Two main groups are: (i) those sciences where controlled experiments are possible; and (ii) those where they

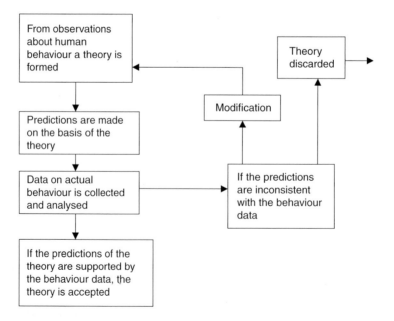

Fig. 1.1. Theory formation and verification.

are not. Physics, chemistry and most biological subjects fall in the first category, economics and sociology in the second.

If in physics a scientist were attempting to show that as the volume occupied by a gas is reduced its pressure increases, he could in his laboratory control the other influence on pressure (i.e. temperature). If he controlled the temperature of his experiment carefully he would get a consistent result however often he repeated his experiment, although small errors might creep in because of imperfections in his apparatus.

A biologist, wishing to verify the theory that increased nitrogen in the soil causes plants to grow more rapidly, might put a number of plants in a growth cabinet in which temperature, humidity, light, etc. could be controlled. Some plants would then be given nitrogen and others, often called the 'controls', would not. The scientist would then look for any differences in growth between the plants with and without nitrogen. Not all plants with nitrogen would grow at the same rate – some 'biological variation' would occur – but the average, or 'mean' growth of plants with nitrogen would be above the average growth of plants without if the theory were valid.

Both the physicist and the biologist in the examples have attempted to eliminate outside influences on their experiments by conducting them in carefully controlled conditions. The economist cannot do this. The basic material of her science is human behaviour, and people cannot be kept in laboratories. If the economist wishes to test her theory about the relationship between the price of petrol and the quantity sold, she cannot control the other factors which determine the quantity sold – such as the incomes of purchasers, or the prices of the other goods on which they could spend their money. Moreover, the economist cannot change the price of petrol – unlike the biologist who could put nitrogen on his plants or the physicist who compressed his gas. She has to wait until the price of petrol in the real world changes and then measure the shifts in quantity sold. Market surveys *do* partly solve this problem by asking consumers how much they *would* buy if prices were changed; however, they might not actually do what they say they would do. The economist also has to measure all the other influences on sales and, with the mass of data collected, use complex statistical techniques to analyse the observations and unravel the relationships. Statistics is a highly valuable tool to the economist. The lack of laboratory-like facilities for experiments is therefore a major but not insuperable problem faced in the evaluation of economic theories.

A further problem faced by social scientists is the variability of human behaviour. This is another, but probably more extreme, version of the biological variation pointed out earlier when the plants-and-nitrogen experiment was discussed. For example, a fall in the price of petrol might induce some people to buy much more, some a little more, some no more, and some less (they might think there was something wrong with it). It is difficult, then, to predict what the response of individuals will be and economic theories are therefore either concerned with the behaviour of groups of people collectively or with the 'typical' or 'representative' person who takes the form of a group average. This is because, in a group, the odd things that one individual may do are compensated for by the equally odd but opposite things other people may do, an effect sometimes called the **Law of Large Numbers**. Economic theories which apply collectively cannot be disproved by the behaviour of individuals; such theories must be tested for a group or **sample** of adequate size to allow for the odd behaviour of individuals.

Path dependency

Another complication to be faced is that, unlike laboratory conditions, in the real world decisions are almost always influenced by earlier decisions or happenings, and these will vary between individuals, firms and societies and are difficult to 'control' experimentally. In addition, outcomes will often reflect the way in which decisions are made; within organizations and society how decisions are reached is itself a choice. Decisions are therefore **path dependent**. In its broadest form this is expressed as 'history matters', a dictum that applies to almost all economic decisions. At a lower level, while the business choices of a farmer at any one time will reflect past decisions, they could also shift if he took a son or daughter as a formal partner who shares in management. And, as will be seen in Chapter 10, current decisions in the EU's Common Agricultural Policy are constrained by decisions made previously. Furthermore, the fact that the EU politicians making the choices are responsible for the agricultures in their home countries probably results in a differently shaped policy from that which would emerge if the decisions were taken by ministers of finance or of social welfare and poverty. Where we are depends to some extent on the path by which we got here.

Cause-and-effect relationships

Because economists are usually denied laboratory-type experiments it is dangerously easy to draw premature conclusions from data. For example, it can be observed that the number of churches and the number of public houses in towns are positively correlated. A premature conclusion might be that the existence of churches caused people to turn to drink; this is probably not justified. Further information on town sizes would show that the linking factor explaining what appears to be an association between the number of pubs and the number of churches is population size. Incomplete information, therefore, can be misleading and associations do not always mean that cause-and-effect relationships exist. It may be tempting to grasp at associations to support a case made for political reasons. For example, a rising number of immigrants is sometimes pointed to when trying to explain greater costs of national health care, but more detailed and careful analysis is likely to show that immigrants are on average lighter users of the National Health Service than UK-born people, and that it is the ageing of the population that is the key explanatory factor.

If a theory is shown to be valid, will it always be valid?

The short answer is 'no'. Economic theories may be shown to hold under a certain set of circumstances, but other circumstances may arise which invalidate them. As time passes and knowledge of how the world operates increases, new theories may be developed that explain situations better, so the old theories have to be discarded. This is happening continually in medicine, for example as we understand the workings of the human body and what causes changes to occur; the better understanding of conditions caused by psychological factors and of genetic makeup have displaced many older theories.

In economics the understanding of the basic subject under examination – the behaviour of people in their making of choice as individuals or in groups – is somewhat more stable. However, things do change. For example, economists are much more aware of how environmental and health issues form part of the objectives of people and need to be built into explanatory models. At farm level there has been a growth in the theory of explaining the behaviour of agricultural households that encompasses their activities as producers in agriculture, participants in local labour markets (as wage earners), as consumers and as enjoyers of leisure. Farmers are also affected by family circumstances, including their stage in the life cycle and the presence or absence of successors to whom the business can be passed down. In public choice, theory has developed in the way in which organizations decide their objectives and how pressure groups try to influence them.

At a more mundane level, technical change has invalidated many simple relationships. For example, in the late 19th century anyone looking for a reasonable indicator of the prosperity of a farm might have looked at the number of draught horses on it; a valid hypothesis under the technical circumstances then common might have been that the number of cart-horses on farms in England was directly related to the income of the farmer, all other things being equal. Farmers with big incomes with surpluses to invest tended to buy more horses as their businesses grew. However, with the introduction of the tractor this relationship changed, so that in the earlier part of the 20th century, when both tractors and horses were to be found, the farmers with higher incomes tended to have fewer horses (but invested in more tractors). The theory and related assumptions on which the hypothesis was based was not wrong for the earlier period but needed modifying, or superseding by a better theory, when conditions changed.

Conclusion

Economics is the study of how individuals and society choose to allocate scarce resources between alternative uses in the pursuit of given objectives. Economics is a science that studies human behaviour. It is therefore termed a *social* science, and has to apply a 'positive' scientific approach without the benefit of controlled experiments. Nevertheless, it is possible to formulate theories and test hypotheses which help to explain the fundamental economic problem of the allocation of scarce resources.

The first area of economics to be studied contains the theories which try to explain how individuals choose to allocate their purchasing power (or the scarce resource of the money available to spend); they are known as **Theories of Consumer Choice**.

Exercise on Material in Chapter 1

Answer the questions in the spaces provided. If you are unable to answer a question, note the particular difficulty and pass on to the next question. Answers and explanations are given in Appendix 2.

1.1. 'Economics is the study of maximizing production.' What essentials are missing from this definition?
   (a)
   (b)
   (c)

1.2. Briefly describe circumstances where the following commodities are (a) scarce, in the economic sense of the word, and (b) not scarce.

Air	(a)
	(b)
Water	(a)
	(b)
Sand	(a)
	(b)

1.3. (a) Is the following statement normative or positive? 'If the Government raises Income Tax, consumers will have less money to spend.'

   (b) What is the difference between a positive and a normative statement?

1.4. Opportunity cost is ....
....

1.5. A student can spend an evening either studying or playing table-tennis. What is the opportunity cost of:
   (a) his evening's table-tennis?

   (b) his evening's study?

1.6. A woman has the choice of two jobs, with no other alternatives. One pays £10,000/year, but is close to her home; the other pays £11,000/year but will involve daily travel of 1 h each way in the firm's bus, which is provided free. What is the opportunity cost of her taking:
   (a) the £11,000 job?

   (b) the £10,000 job?

1.7. A farmer can plant either wheat or potatoes in a field. The reward for growing either crop is the difference between the receipts from the crop less the costs of growing it. Let this difference be called the 'net revenue' for each crop. The farmer decides to grow wheat; what is the opportunity cost of his growing this crop?

1.8. Arrange the following steps in the sequence appropriate to scientific method (some branching with options may be involved).

Testing predictions against observations

Formation of hypothesis

Predictions made from hypothesis

Observations are consistent with predictions

Hypothesis accepted

Hypothesis rejected

Observations not consistent with predictions

Hypothesis modified or new hypothesis developed

1.9. Setting up a hypothesis
   (a) You read the following statement in the farming press: 'We are sure that farmers on dairy farms of 100–150 ha in the County of Devon who have invested in buildings over the last 10 years get lower net incomes now than those who did not.' Having in mind the stages shown in 1.8, what process could be followed to show whether this statement was valid?

   (b) If the hypothesis is accepted, and present income is found to be inversely related to investment over the last 10 years, have you proved that investment in farm buildings causes low incomes?

1.10. Comment on what is implied by the following statistical associations.

(a) Crime statistics show that the level of most crimes has been falling in recent years. Because local authorities have been requested to cut back on public expenditure, the number of uniformed police officers has also been falling. How does this fit with the claim often made by politicians that, in response to complaints by the public, they will cut crime by increasing the number of police?

(b) Tobacco firms oppose the move by politicians to make cigarette packages less attractive. They often claim that there is no causal relationship between the amount they spend on packaging and the number of people who smoke. What additional information would you request before supporting their opposition to making packaging plain?

(c) Some people advocate the partial extermination (culling) of badgers because growth in their numbers seems to have a relationship with the rising incidence of tuberculosis (TB) among cattle. Does this prove that badgers should be culled in the interest of cattle health? What evidence would you propose to collect to establish whether a culling policy should be implemented?

2 Explaining the Behaviour of Individuals: Theory of Consumer Choice

Introduction – the Concept of Utility

The economic systems we have been describing exist so that the goods and services we want for living can be provided. In satisfying these wants the goods and services generate satisfaction in people; to use the technical term of economics we say that they generate **utility**. An example of utility is as follows: we have a want for clothes, because we require clothes to keep us warm in cold weather and society also demands that bodies be covered to some extent in public. Articles such as shorts, jeans, etc. have the power to satisfy this want and are thus said to generate utility.

Articles can generate utility without being necessities for living; they need not be 'useful' things. Oil paintings give utility because they satisfy the wants of collectors who happen to like them without the paintings being of 'use' in any practical sense. New clothes may be wanted by fashion-conscious shoppers simply because they conform to what is now being worn by role models and therefore generate utility in a way that unfashionable goods cannot, not because they are better in terms of their ability to protect from the elements. Another example is addictive drugs that do their consumers no good (often the opposite) other than by meeting a craving.

Utility is different from **demand** because demand implies an ability to pay for a commodity. For example, for an impoverished man, a private yacht may generate considerable utility in its ability to satisfy his desire for pleasure, but this desire will not be reflected in orders to yacht builders. The poor man will not express any effective demand because he cannot pay for a yacht. However, give him a windfall from a lottery, so that he has money to spend, and the boatyard may well find that the demand for its products rising by at least one cash sale.

Before proceeding further it must be pointed out that in practice utility cannot be measured in the same precise and universal way that heat can be measured in calories or sound in decibels. If such a workable cardinal unit were available the explanation of human behaviour could well be much easier and public and private choices much more straightforward. However, just because actual utility or satisfaction is difficult or impossible to quantify in absolute units it does not mean that it is useless as a concept for explaining human behaviour. Neither love nor hate can be measured exactly, yet their existence is widely accepted and people's actions are affected by them.

For the moment let us assume that the utility generated within a person when a good or service satisfies a want is, at least in theory, measurable. Later we will return to the complications caused by the absence of a practical unit of measurement. Utility has several important characteristics.

1. *A commodity generates different utilities for different people, and for the same person at different times.* For example, a pack of hounds gives no utility to a person who is completely against blood sports while the same pack might be a source of great utility to someone keen on hunting. Sun creams generate little utility for people with dark skin types, but may give considerable utility to a fair-skinned European visitor to a hot country. An example of utilities changing with time is provided by thick woollen overcoats, which give little utility in a Russian heat wave but generate great utility for their owners in the depths of a Moscow winter. Discussion of utility must therefore take account of the time at which the good is used or consumed.

2. *For the same person different commodities may satisfy the same want to different extents. They thus generate different amounts of utility and it is possible to rank them in order of utility (or desirability).* For example, a car owner may desire a fuel to make its petrol engine run. She may be faced with a variety of possible fuels, such as good quality petrol, poorer petrol with a lower octane rating, vintage tractor fuel (TVO), household heating oil (paraffin) and industrial alcohol. Figure 2.1 shows that £1-worth of each of

© B. Hill 2014. *An Introduction to Economics*, 4th Edn: *Concepts for Students of Agriculture and the Rural Sector* (B. Hill)

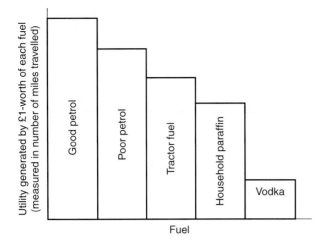

Fuel

Fig. 2.1. Utility from fuels (hypothetical data).

these fuels generates a different amount of utility. Some make it go well, while others hardly allow it to function at all. Given the choice, the motorist will select the fuel giving the highest utility – good petrol.

3. *The utility generated by a unit of a commodity varies according to how much a consumer has of that commodity already.* In general, the more of a good a person is already using to satisfy a want (i.e. the more he is already consuming) the less will be the utility generated by additional units of the good. Take the example of clothes. A man may want a dark suit so that he can go properly dressed to formal social events, job interviews and other occasions where such clothing is expected. He acquires a suit which gives him considerable satisfaction. However, he knows that occasionally it will be inconvenient to arrange the cleaning of the single suit, so he acquires a second one, thereby avoiding this inconvenience and gaining some additional satisfaction The two suits satisfy his wants for for-mal clothes to a greater extent than does one suit (i.e. the two generate more utility than one) but the utility arising from the second suit is much less than that from the first. Table 2.1 shows the amount of utility arising from different numbers of formal suits which this man might possess and the utility coming from the last suit purchased. While the *total* utility increases as the number of suits he possesses increases, it can be seen that the utility generated by each *additional* suit dimin-ishes. The total utility schedule and the schedule showing the utility coming from the last suit pur-chased are shown in graphical form by Figs 2.2 and 2.3.

Table 2.1. Utility generated by suits – total and marginal (hypothetical data).

No. of suits	Total utility	Utility generated by last suit[a]
0	0	–
1	10	10
2	15	5
3	17.5	2.5
4	18.75	1.25

[a]Utility generated by the last suit is termed marginal utility. Utility generated by the nth suit = Total utility of n suits – Total utility of $n-1$ suits.

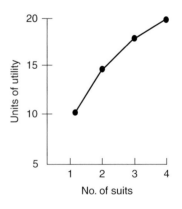

Fig. 2.2. Total utility.

The Margin

The concept of the **margin** is very important in economics and is encountered in many facets of the subject. We can illustrate the concept of the margin using the formal suits example. If the man does not

---

stop at buying two suits but buys a third, that third or additional suit can be regarded as the *marginal* suit. In more general terms, the last unit acquired is called the *marginal* unit. So instead of saying that the utility derived from additional suits as shown by Fig. 2.3 decreases, we can say that the utility generated by the marginal suit decreases as the number of suits the man possesses increases. The utility coming from this marginal suit is called **marginal utility**.

This reduction in marginal utility with increases in the number of units already being consumed is so commonly experienced that it has been formulated into the **Law of Diminishing Marginal Utility**. This says that *the utility of additional units of a commodity to any consumer decreases as the quantity of that commodity already being consumed increases.* Other examples of this law might be: (i) once one possesses a fountain pen or wristwatch,

the satisfaction given by additional pens or wristwatches is less than that given by the first; and (ii) if attempting to play tennis, the first tennis ball and perhaps the second tennis ball will generate high utility, but additional tennis balls will give less extra satisfaction until the stage may be reached where the twenty-fifth or so ball may give no additional satisfaction at all. It may even get in the way and be considered worse than nothing – it could generate negative utility or 'disutility'.

Examples can be cited where the utility from marginal units first increases and then decreases. This does not necessarily invalidate the Law of Diminishing Marginal Utility as a generality, but calls for a slight modification. If a person has a motor car with no wheels (perhaps they had all been stolen), the marginal utility from wheels will probably increase until the fourth wheel is acquired, as this enables the car to be driven. However, after this point has been reached, the utility given by each additional wheel may decline. In such cases it is said that marginal utility declines once the **origin** has been attained. In the case of the motor car the 'origin' would correspond to four wheels (see Fig. 2.4).

Free Goods

If a consumer is offered an unlimited quantity of an article at no cost, how much of this free good will he help himself to? How many mince pies will a boy eat at Christmas if he is offered a pantryful by his grandmother? The consumer, or the boy, will obviously continue to help himself until the satisfaction given by the last unit (or pie) taken has

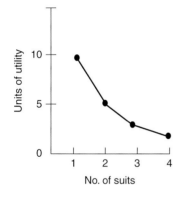

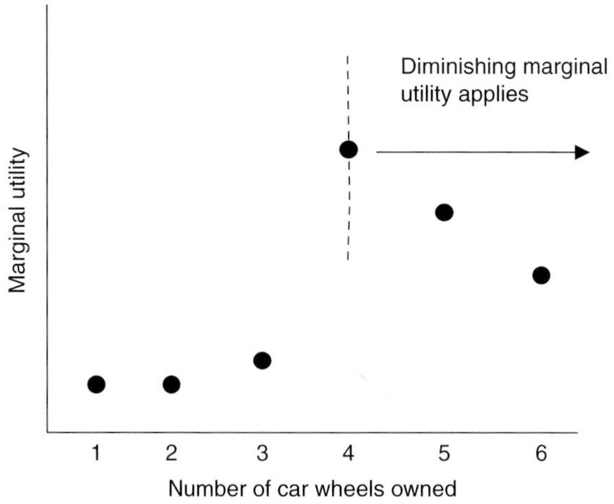

Fig. 2.3. Utility generated by the last suit.

Fig. 2.4. Marginal utility of a good used in sets, e.g. car wheels (data hypothetical).

diminished to nothing. In other words, the quantity of a free good which will be taken is the quantity where marginal utility is zero. Note that to the grandmother the mince pies are unlikely to be free.

Few commodities in life are free goods, that is, they are so plentiful in relation to the wants for them that no allocation is necessary and so they do not have a price, e.g. fresh air in most circumstances, and salt water at the seaside. Most commodities are economically scarce and, in market economies, have a price attached to them. To acquire them a consumer faces a cost and consequently he or she has to *choose* how to spend their limited funds in the ways to give them most satisfaction. This is, of course, a typical economic problem of allocating a scarce resource (purchasing power) between alternative uses (the range of goods and services open to the consumer) in pursuit of a given objective (to maximize personal satisfaction). Economists attempt to explain how consumers do this using **Theories of Consumer Choice**.

But before leaving the subject of free goods, it is worth raising the example of goods that are free to the individual but not to society at large. For example, the individual fisherman who goes out with his boat to take fish from the sea will behave rather like the boy – he is likely to keep catching fish without thought as to what is happening to fish stocks. Once he has got a boat and travelled to the fishing area the fish represent a 'free good'. But for society in general fish stocks are a scarce resource and it is necessary to make positive decisions about how they are to be used. It is likely that, in order to ensure that supplies of fish will continue in the future, some restraint in fishing is needed in the short term so that stocks are maintained at a sustainable level; fishing quotas are a useful tool in achieving this, or restricting the number of days boats can be at sea. This contrast between the individual's perspective and that of society will be revisited when we examine the problems of using the market mechanism to allocate resources in a later chapter.

Theories of Consumer Choice

The consumer is of fundamental interest to the economist because it is to satisfy the demands of consumers, directly or indirectly, that production takes place in market economies. Paradoxically, the objectives which lie behind the behaviour of consumers and the choices they make are among the most elusive because they are complex and, for the most part, unmeasurable. In the face of these difficulties economists have developed several theories to explain consumer behaviour, two of which will be described briefly here: (i) Utility Theory; and (ii) Indifference Theory. Both make the reasonable assumption that the objective the consumer has in mind is to get the greatest amount of satisfaction possible from the limited amount of purchasing power he or she possesses. We begin with Utility Theory that, while simple in concept, contains some difficulties which the second approach, using indifference curve analysis, overcomes.

Utility Theory of Consumer Choice

This theory imagines that it is possible for a consumer to draw up in their own mind schedules of total and marginal utilities for each of the goods that can be purchased, although these schedules may be totally subconscious. By subconscious reference to these schedules the consumer can allocate expenditure in the most pleasurable way.

Let us take a simplified example of a man who can spend his weekly income on only two commodities, bread and cigarettes. For convenience we will assume that one loaf of bread costs 10p and one packet of cigarettes also costs 10p (the fact that such figures are completely unrealistic is irrelevant when illustrating a principle). Hypothetical total and marginal utility schedules for bread and tobacco are shown in Table 2.2. In both cases the marginal utility can be seen to decline as the person's consumption of the commodity increases in line with the Law of Diminishing Marginal Utility.

Assume that initially the man has 80p to spend and that he buys six loaves of bread and two packets of cigarettes. This will give him a total utility of 450 + 70 = 520 units of utility. Is he spending his 80p in the best possible way? To answer this question we must look at the marginal utility schedules. The sixth loaf of bread bought gives 50 units of utility and the second packet of cigarettes generates 31 units of utility. If he were *not* to buy the second packet of cigarettes but instead bought one extra loaf of bread, i.e. a seventh loaf of bread, he would lose 31 units of utility from tobacco but gain 40 units of utility from bread. Thus total utility would increase by 9 units. Seven loaves of bread and one packet of cigarettes is a better choice than six loaves of bread and two packets of cigarettes. Total utility would be 490 + 39 = 529 units.

Table 2.2. Total and marginal utilities of bread and cigarettes (data hypothetical).

No. units[a] consumed/week	Bread		Cigarettes	
	Total utility	Marginal utility	Total utility	Marginal utility
1	100	100	39	39
2	190	90	70	31
3	270	80	92	22
4	340	70	105	13
5	400	60		
6	450	50		
7	490	40		
8	520	30		
9	540	20		
10	550	10		

[a]The unit of bread is the loaf; the unit of cigarettes is the packet.

Should he transfer his expenditure on cigarettes entirely to bread? If he sacrificed his one remaining packet of cigarettes, 39 units of utility would be lost but the eighth loaf of bread would bring 30 units. On balance 9 units of utility would be lost by transferring the money from a first packet of cigarettes to the eighth loaf of bread. In the circumstances, then, the most satisfactory allocation of his expenditure is on seven loaves of bread and one packet of cigarettes.

This example is of relatively large units of expenditure, i.e. 10p per unit with a total expenditure of 80p. If we could buy bread by the slice and individual cigarettes, so that the quantity which could be bought would go up in very small steps, we would find that the best allocation of the consumer's money would be achieved when the last penny spent on bread gave the same amount of satisfaction as the last penny spent on cigarettes. Put another way, at the optimum allocation of a person's spending power the marginal pennyworth of bread will give the same utility as the marginal pennyworth of cigarettes. The principle can be extended, so that consumers should adjust purchases so that the last penny spent in *any* direction yields the same (additional) satisfaction, thereby achieving the greatest amount of total satisfaction that can be attained with a given level of expenditure. This is an example of the **Principle of Equimarginal Returns** which is encountered in many branches of economics, as will become evident from later chapters.

Where goods cannot be easily broken down and bought by the pennyworth, it can be shown that a consumer achieves maximum satisfaction from his or her expenditure when the ratio of the marginal utilities of the last unit of each good is the same as the ratio of their prices. For example, if the price of beer per pint is twice that of a pint of milk, a consumer, in order to achieve maximum satisfaction, will have to adjust his or her purchases so that the last pint of beer generates twice the utility of the last pint of milk.

Objections to the Utility Theory

The Utility Theory of Consumer Choice is open to criticism on the grounds that satisfaction and utility are very difficult to measure. No unit of utility has yet been devised comparable with, say, the calorie as a unit of heat. Calories are definable in an exact way, and apply universally to heat generated by sunlight, oil, combustion, friction, etc. In contrast, it is impossible to construct reliable utility schedules for individuals, let alone groups of people, if no practicable unit of measurement exists. Despite the apparent obviousness of the Utility Theory for explaining consumer choice, it is untestable and therefore cannot be validated. To meet these objections an Indifference Theory has been developed, the formulation and testing of which does not necessitate measuring satisfaction or utility in absolute units.

An Indifference Theory of Consumer Choice

Again we take a simple model in which a man has the choice of spending his resources on two commodities, in our case beer and milk. Table 2.3 shows the various combinations of beer and milk which give the same level of satisfaction to the consumer.

Five litres of beer and 2 l of milk consumed per week give the consumer the same amount of satisfaction as 4 l of beer and 3 l of milk, or 2 l of beer and 8 l of milk. The consumer is indifferent to which combination he consumes. This sort of schedule can actually be constructed (in contrast with the utility schedules given earlier which must remain hypothetical) in a simple experiment that starts by giving a consumer 5 l of beer and 2 l of milk and then proceeds by taking beer away from him, bargaining with milk to compensate. Thus in this approach we are not attempting to measure the consumer's level of satisfaction as Utility Theory implies; we are simply assuming that a consumer knows how much extra of one commodity is required to compensate for the loss of another commodity – in other words, to keep his level of satisfaction the same. Given that people will normally be used to making choices in everyday life, this is a realistic assumption. A parallel might be experienced by a River Authority; it is impossible in practice for them to measure the volume of water in their river in gallons (or litres), but it is easy for them to watch its level. Gains from heavy rainfall can be compensated for by opening sluices and drought conditions compensated for by restricting pumping for irrigation so that the river level is kept constant. As a consequence, the volume of water in the river is also kept constant although the precise gallonage is not known.

You will notice from Table 2.3 that as we reduce the weekly consumption of beer, the consumer demands greater and greater quantities of milk to compensate. For example, when we reduce his consumption of beer from 5 l to 4 l, we are only required to give him one extra litre of milk to compensate. However, if we wished to reduce his quantity of beer from 2 l to 1 l, he would demand an extra 6 l of milk to compensate

(i.e. to increase his weekly consumption from 8 l of milk to 14 l). The number of units of one good (milk) required to compensate a consumer for the loss of 1 unit of the other (beer), i.e. to maintain the same level of satisfaction, is called the **marginal rate of substitution (MRS)**. It increases as the consumer's consumption of beer decreases, and decreases as his consumption of beer increases. The MRS can also be measured by the number of units of one good (litres of milk) that a consumer is willing to *give up* to gain 1 unit of the other (litre of beer).

Another indifference schedule could be constructed by giving the consumer at the start 5 l of beer and 4 l of milk. This would mean that he was initially on a higher level of satisfaction than with the previous combinations because the quantity of milk is greater. A schedule such as Table 2.4 might result. Obviously any combination in Table 2.4 is superior to any combination in Table 2.3. Plotting these schedules gives us the **indifference curves** we see in Fig. 2.5. Indifference curves further away from the origin imply higher levels of satisfaction.

The following characteristics of indifference curves should be noted:

- Indifference curves cannot cross.
- Indifference curves normally have a negative slope.
- Indifference curves are normally convex to the origin.
- Assuming that the goods on the axes are readily divisible, an additional indifference curve can always be drawn between two other indifference curves. For example, we could break down the litres and deal with millilitres.

When attempting to explain consumer behaviour using indifference curves, we construct a simple model by assuming that the consumer has a set of

Table 2.3. Combinations of beer and milk which give the same level of satisfaction.[a]

Litres of beer	Litres of milk
5	2
4	3
3	5
2	8
1	14

Table 2.4. Combinations of beer and milk which give the same level of satisfaction.

Litres of beer	Litres of milk
5	4
4	5
3	7
2	11
1	20

[a]Although only whole-number combinations are given for simplicity, it is assumed that both commodities are available in fractions of litres so that intermediate combinations involving part litres could be shown in an expanded schedule, i.e. 1.6 l of beer and 2.5 l of milk could appear in Table 2.3.

them for the two commodities, he has a fixed income – all of which is spent on the two commodities, and that prices are fixed (at least to start with). Figure 2.6 shows four indifference curves belonging to a consumer for beer and milk.

The optimum way in which the consumer can spend his income can be illustrated by superimposing on to the indifference curve map a line called the **budget line** or **iso-expenditure line** (Fig. 2.6). The budget line shows all the combinations of beer and

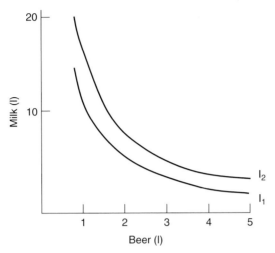

NB: An indifference curve shows the combinations of two goods which give the same level of satisfaction. Curves further from the origin correspond to higher levels of satisfaction.

Fig. 2.5. Indifference curves.

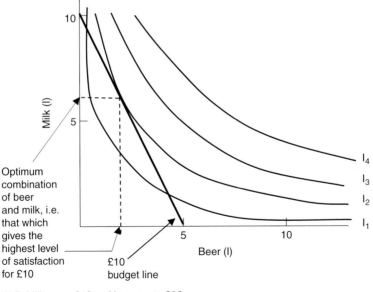

Optimum combination of beer and milk, i.e. that which gives the highest level of satisfaction for £10

£10 budget line

N.B. Milk costs £1/l and beer costs £2/l

Fig. 2.6. Indifference curves and budget line.

Box 2.1. The marginal rate of substitution (MRS).

By convention, the MRS is estimated with respect to changes by 1 unit in the consumer's holding of the good on the horizontal axis of indifference curve maps. In the example chosen, beer is put on the horizontal axis and the MRS measures the varying numbers of litres of milk required to compensate for the loss (or gain) of 1 l of beer. Because of this convention, the MRS of goods at any point on an indifference curve is numerically the same as the slope of the curve. The slope of the indifference curve, and hence the MRS, declines as we move along the horizontal axis away from the origin. To be more mathematically precise, the slope (and MRS) should be said to *increase* as we move away from the origin along the horizontal axis. This is because the slope (and MRS) is a negative quantity and a figure of –5, which might be found near the origin, is mathematically a smaller quantity than a figure of –1, which might be found further from the origin. However, this effect of the negative sign is usually omitted and the MRS is said to decrease with distance from the origin.

milk that could be purchased by the consumer at a given level of income and prices. If we assume that the income of the consumer is £10/week and that milk costs £1/l and beer costs £2/l (it does not matter when establishing the principle that these are unrealistic figures), the budget line can be drawn by connecting the quantities on each axis that can be purchased for £10. Thus, this line connects 10 l of milk to 5 l of beer because these are the quantities of goods which could be purchased weekly if *all* his money were spent on one good *or* the other. In addition, any combination of beer and milk lying on this line could be bought by the consumer. This is called an *iso-expenditure line* because the same amount of expenditure takes place at all points along it.

It can be seen in Fig. 2.6 that at two points the budget line cuts the lowest indifference curve, and the consumer could, if he wished, buy the combinations corresponding to either of these points and thereby experience the level of satisfaction corresponding with indifference curve I_1. The consumer can never buy a combination lying on indifference curves I_3 and I_4 because they lie to the right of his budget line. The levels of satisfaction corresponding to these curves are beyond the reach of his spending power.

However, his budget line *does* just touch indifference curve I_2 at one point. Here the budget line and indifference curve are tangential (i.e. their slopes are the same). The combination of milk and beer corresponding to this point is the optimum (or best) combination the consumer can obtain because it represents the highest level of satisfaction he can attain at the given level of income and prices. There is no point in choosing any combination lying on a lower indifference curve than I_2 because the satisfaction thereby achieved will also be lower. We assume that consumers always aim for the maximum

satisfaction available with their resources. Note that there are many suboptimal ways for the consumer to spend his income, but only one 'best' way.

Income effect

If the consumer's income is increased to £15/week, his budget line is shifted to the right and upwards, parallel to the original budget line. This is because, with his higher income, he will be able to buy more beer and/or more milk. This shift in the budget line may enable him to touch indifference curve I_3 or maybe even I_4, or some intermediate indifference curve which could be inserted between the existing curves. There is no guarantee that the ratio between the numbers of litres of beer and milk at the higher optima will be the same as at the optimum established on I_2, since this will depend on the shape of the higher indifference curves. The **income effect** is shown in Fig. 2.7.

Price effect

The effect of a change in price of either beer or milk will be to change the *slope* of the budget line. If the price of beer halves to £1/l, the £10 budget line will now connect 10 l of beer to 10 l of milk as the price of a litre of each will now be £1. This has been shown in Fig. 2.8.

The new budget line is tangential to I_4 and consequently a new optimum combination of beer and milk is established for the consumer. Note that when beer becomes cheaper and the slope of the budget line changes, the new optimum combination which the consumer can buy with his £10 includes a larger proportion of beer than before. When the price of beer falls, more of it is bought.

Box 2.2. MRS and ratio of prices.

At the optimum choice for the consumer the slopes of the indifference curve and budget line are the same. The slope of the indifference curve is the number of units of the good on the vertical axis (y) required to compensate for a change of 1 unit of the good on the horizontal axis (x), and is the same as the marginal rate of substitutions of y for x.

$$\text{Slope of indifference curve} = \frac{\text{Change in no. units of } y}{\text{Change in no. units of } x} = \text{MRS} \tag{i}$$

$$\text{Slope of budget line} = \frac{\text{Quantity of } y \text{ which can be bought for } £a}{\text{Quantity of } x \text{ which can be bought for } £a}$$

Quantity of either good which can be bought is inversely related to its price.

$$\text{Slope of budget line} = \frac{\text{Price per units of } x}{\text{Price per units of } y} \tag{ii}$$

At the optimum choice for the consumer combining (i) and (ii)

$$\text{MRS} = \frac{\text{Price of } x}{\text{Price of } y}$$

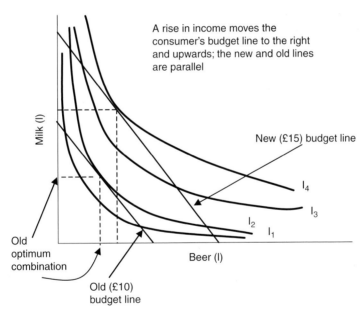

A rise in income moves the consumer's budget line to the right and upwards; the new and old lines are parallel

Fig. 2.7. Income effect.

This is a highly important link between price of a commodity and the amount which consumers are willing to purchase.

The relationship between price and quantity which consumers purchase is an important element of demand. If we were to keep the consumer's income and the price of milk constant by drawing a number of budget lines corresponding with a range of prices of beer, we could construct a table showing the quantities of beer which the rational consumer

---

would buy per week at different prices. Such a table is called a demand schedule. If the price of beer is plotted against the quantity that is bought per week, with price on the vertical axis, the result is a **demand curve**. This is shown in Fig. 2.9 (in which the demand curve happens to be a straight line). All other influences on the amount that is bought are assumed to be held constant; the term *ceteris paribus* is used to describe such circumstances. The curve shows the amount of a commodity which a rational consumer will purchase, faced with a choice between commodities and changes in the price of the commodity in question. Such curves will be met many times in later sections.

Demand, Marginal Utility and Consumer Surplus

A demand curve shows the quantities of a commodity which a consumer would buy at a

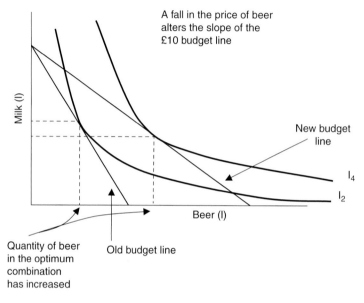

Fig. 2.8. Price effect.

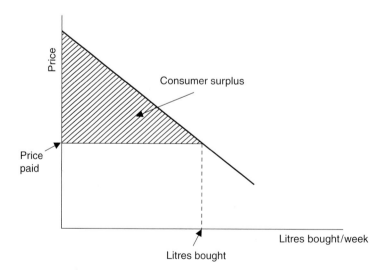

Fig. 2.9. Amount of beer bought at different prices of beer.

range of prices, *ceteris paribus*. But the relationship between price and the amount that consumers are willing to buy can be looked at from another angle, which links back to the earlier consideration of marginal utility. It also shows what the prices would have to be if the consumer were to be persuaded to buy given quantities.

If a rational consumer is offered an additional unit of a commodity, how much will he or she be willing to pay for it? Its purchase will mean diverting spending power from other goods and services, so some opportunity cost is involved. In these circumstances money is a useful common denominator in which the utility from spending alternatives can be expressed. Staying with the example of beer, if the consumer is willing to pay a high price for the extra pint we know that the amount of utility which can be derived from the additional pint is great (i.e. marginal utility is high). He is willing to forgo substantial amounts of other goods and services on which he could otherwise spend this money. Assuming that the Law of Diminishing Returns applies, we would expect the price which a consumer would offer for an additional pint would be smaller if he already has a large quantity. He might be prepared to pay £2.20 for the first pint, £1.50 for the second, and £1.00 for the third (assuming he bought them all together and that his rationality did not change as a result of buying and consuming the beer). Thus a demand curve can be interpreted as representing the diminishing marginal utility to a consumer as the amount already possessed increases, with the price standing as a proxy for the marginal utility.

If in practice a consumer can buy all his beer at the same price (£1 per pint) he will buy three pints. The marginal utility of a fourth pint is less than the utility to be derived from spending the money in other ways, so he does not purchase it. In a market economy individual consumers do not usually negotiate separately for each unit of a commodity they purchase; rather, they buy as much or as little as they wish at the going price. It follows that the price which the consumer would have been willing to pay for the first pint in our beer example is much greater than the price actually paid; he enjoys a surplus. Similarly, but to a lesser extent, he pays less than he would be willing to for the second pint. Adding all the differences between prices paid and those which the consumer

would be willing to pay for the successive units purchased (derived from marginal utilities and shown by the demand curve) gives the amount of total **consumer surplus**. On Fig. 2.9 it is indicated by the area between the price paid and the demand curve. If the price of beer falls, the amount of consumer surplus rises and if beer prices go up, consumer surplus is squeezed. This concept is often used when analysing the impact on consumers of agricultural policies which raise the price of food to consumers; a reduction in the amount of consumer surplus is interpreted as making consumers worse off.

'Lumpy' commodities

The indifference curves in Figs 2.5–2.8 were drawn smooth because it was assumed that between the whole-litre combinations of beer and milk shown in Figs 2.7 and 2.8, each schedule corresponding to a given level of satisfaction, any number of further combinations could have been inserted using part litres. If, however, only whole litres were available to the consumer, the 'curve' should not have been drawn smooth, but as a series of separated points, as in Fig. 2.10. Commodities which only come in relatively large units are termed 'lumpy'. Indifference Theory deals with the situation of 'lumpiness' better than Utility Theory. If a budget line touches the indifference 'curve' at a point, it can be seen that quite a large shift in angle of the budget line is required before a new optimum choice results (i.e. relative prices have to alter considerably before the consumer's best choice at one level of satisfaction is changed). If liquids were available to the consumer by the millilitre as opposed to by the litre, the indifference curve would be much smoother because of all the intermediate combinations, and the optimum allocation of spending on beer and milk would be consequently much more sensitive to relative changes in price.

The Stable and the Unstable Equilibrium

When a consumer has allocated their spending power in the way which gives them the greatest level of satisfaction attainable at the given level of income and prices, the consumer is said to be in **equilibrium**. Furthermore, it is a **stable** equilibrium because, if some temporary factor such

Box 2.3. When price per unit differs.

Where prices of units of goods are *not* the same, maximum satisfaction is obtained when the ratio of marginal utilities (MU) is the same as the ratio of prices.

1. Maximum satisfaction is obtained from a given level of expenditure on two goods when the last penny (or smallest unit of expenditure) spent on either good gives the same amount of satisfaction. This is an example of the Principle of Equimarginal Returns.

2. Where goods cannot be bought in units of one pennyworth, an *approximation* of the MUs of the last pennyworth is given by the following:

$$\text{MU of 1 penny spent on good A} = \frac{\text{MU of 1 unit of good A}}{\text{Price in pence of 1 unit of A}}$$

Similarly

$$\text{MU of 1 penny spent on good B} = \frac{\text{MU of 1 unit of good B}}{\text{Price in pence of 1 unit of B}}$$

Note that the price per unit of A and B need not be the same.

3. If, at maximum satisfaction

MU of 1 penny spent on good A = MU of 1 penny spent on good B

Then, at maximum satisfaction

$$\frac{\text{MU of 1 unit of good A}}{\text{Price in pence of 1 unit of A}} = \frac{\text{MU of 1 unit of good B}}{\text{Price in pence of 1 unit of B}}$$

This is usually written as

$$\frac{\text{MU of good A}}{\text{MU of good B}} = \frac{\text{Price of A}}{\text{Price of B}}$$

4. For example, if a jacket (good A) costs twice as much as a shirt (good B), a consumer would be spending his money in the most satisfying way if he got twice as much satisfaction out of his last jacket than out of his last shirt.

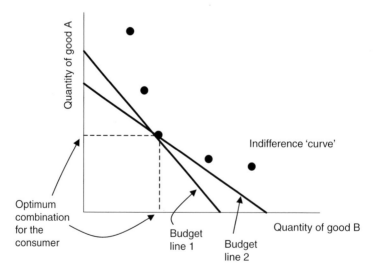

Fig. 2.10. 'Lumpy' commodities.

as rationing forces the consumer to buy some other combination of goods, when these factors are removed, there will be a tendency to return to the original combination, driven by the consumer's desire to maximize their satisfaction. An analogy is a ball in a basin; if displaced to the side, it will return by gravity to the lowest point to restore the status quo. These examples are of a stable equilibrium because the forces called into play by disturbing the stable state act to restore the stable state.

An example of an **unstable** equilibrium is a coin balanced on its edge. Undisturbed it will stand for ever, but a push will upset its balance and it will fall over because no forces exist which tend to restore the initial position.

Economics is closely concerned with equilibria. Possibly the most important example is the equilibrium which results when the demand for goods and services meets the supply of these goods and services. A balance between demand and supply is achieved through the price mechanism, and this is the subject of the next chapter.

Exercise on Material in Chapter 2

Answer the questions in the spaces provided. For some you will need to draw graphs on the squared areas, though you may prefer to use a separate piece of graph paper. Please note that for graphs drawn by hand some tolerance must be allowed between the answers you obtain and the 'correct' answers given in Appendix 2.

2.1. In which of the following situations (a) does bread generate utility? (b) is there a demand for bread? Indicate using Yes/No/Don't know.

	Situation	Utility	Demand
(i)	A starving man wanting a loaf of bread		
(ii)	A starving penniless man wanting a loaf costing £10		
(iii)	A starving wealthy man wanting a loaf costing £10		
(iv)	A starving man who has an aversion to bread (the aversion is total)		
(v)	A starving wealthy man who has an aversion to bread (again the aversion is total; resale is banned)		
(vi)	A man whose hunger has already been fully satisfied		

2.2. Given the following schedule of total utility for commodity X, derive the corresponding marginal utility schedule. Plot both schedules.

Units of X consumed per time period	Total utility	Marginal utility
1	9	
2	21	
3	35	
4	50	
5	65	
6	79	
7	91	
8	100	
9	105	
10	105	

Area for plotting the graph

Total utility

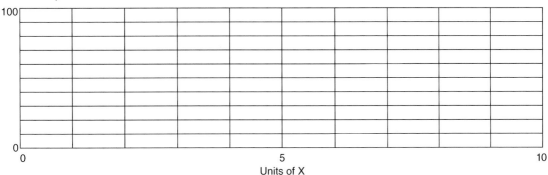

Units of X

Marginal utility

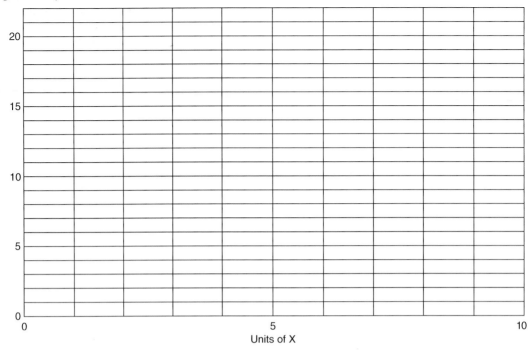

Units of X

2.3. The 'Law of Diminishing Marginal Utility' states that ....
....
....

2.4. Is the marginal utility schedule derived in 2.2 above consistent with the 'Law' stated in 2.3?

2.5. Give an example of a commodity that could have the characteristics shown by commodity X.

2.6. A consumer has £20/week to spend as he wishes on goods A and B. The prices of these commodities, the quantities he now buys and his (subjective) estimates of the utility provided by these quantities are as follows:

Goods	Price	Units currently bought/week	Total utility	Marginal utility
A	70p	20	500	30
B	50p	17	1000	20

Using the above information cross out the inappropriate alternatives in the following statement:

Assuming that the Law of Diminishing Marginal Utility applies to both goods A and B, for maximum satisfaction this consumer should buy *more/less/the same quantity* of A and *more/less/the same amount* of B. This adjustment of the purchasing pattern will make the ratio of *marginal utilities/total utilities closer to/more different from* the ratio of *prices/quantities bought*.

2.7. Below are some of the combinations of amounts of two commodities, X and Y, which correspond to points on three indifference curves for a consumer.

Indifference curve I_1		Indifference curve I_2		Indifference curve I_3	
X	Y	X	Y	X	Y
1	50	5	60	10	60
5	30	10	40	15	45
10	20	15	30	20	36
15	15.5	20	24	25	30
20	12	25	20	45	18
25	10	35	15	55	15
55	5	55	10	70	12
70	4	70	8		

(a) Draw these indifference curves (we assume that they are smooth).

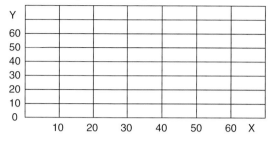

(b) Calculate the marginal rate of substitution on:
I₁ between X5 and XI0
I₂ between X15 and X20
I₃ between X20 and X25

(c) Draw in the consumer's iso-expenditure line (budget line) for each of the income and price levels quoted below, and determine the quantities of the two commodities he or she purchases in each of these situations.

Income per week (£)	Price of goods (each)		Quantity purchased per week	
	X	Y	X	Y
60	£2	£2		
88	£2	£2		
110	£2	£2		
110	£2	£3		
110	£2	£6		

2.8. From the final three rows of the table shown in 2.7(c) derive a demand schedule for commodity Y.

Price of Y	Quantity demanded
£2	
£3	
£6	

Plot this schedule as a demand curve, with price on the vertical axis.

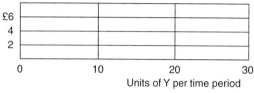

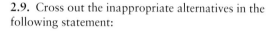

2.9. Cross out the inappropriate alternatives in the following statement:

A consumer is said to be in equilibrium when the *marginal/average/total utility* from his purchases is *minimized/maximized*. This also implies that the satisfaction he derives from his purchases is *minimized/maximized*. He allocates his spending, probably subconsciously, to this end by practising the Principle of *Diminishing Marginal Utility/Equimarginal Returns*. The consumer's equilibrium is termed *stable/unstable* since any departures from this position caused by external forces will call into play forces which tend to restore the equilibrium position.

3

Demand and Supply: the Price Mechanism in a Market Economy

Introduction

The study of the demand and supply of goods and services, and the way they interact, forms a fundamental part of economics. Indeed, the majority of economic problems we come across in everyday life can be explained, although perhaps not solved, by a careful examination of the demand and supply of goods or services. It has even been suggested that a parrot could be turned into a passable economist simply by teaching him to say the words 'demand and supply' in reply to all questions.

By way of introduction to this important area of study let us take an example from agriculture. At various times of the year some farmers want to buy barley while others are willing to sell barley – a demand and a supply both exist. Transactions occur at markets where buyers and sellers can meet each other. These used to be physical places, but nowadays this could equally well be Internet websites. Let us imagine that we can select one market on one day (say, in early December) and have the power to ask as many questions as we like. Furthermore, let us imagine that we can dictate what the price of barley is to be.

If we take all those who wish to buy barley and tell them that the price will be £100/t on that day – assume this is a high price for the particular season – very few farmers will wish to buy any and the quantity sold will be small. If we say that the price will be £90/t, more buyers will be interested. As the price is lowered, increasing interest will be shown – more farmers will wish to buy and each will tend to buy more. If we make a table of the quantity of barley we could sell at various prices we end up with a **demand schedule**.

If we moved on to suppliers of barley we would find that at a low price the quantity which they would be prepared to sell would be small. But if we offered a higher figure the quantities offered would increase – more sellers would want to sell and each would want to sell a greater quantity. Again we could draw up a table or schedule of prices and quantities of barley which suppliers would be prepared to supply at these prices.

We now have two schedules (see Table 3.1), one of demand and one of supply. If we set the price of barley at £60/t buyers will want 300 t but sellers will only be prepared to provide 50 t, so there will be a shortage at that price. If we set the price at £100/t suppliers will be prepared to sell 120 t, but only 20 t will be wanted. Glancing at the two schedules will show that there is only one price at which the quantity demanded and supplied exactly balances – £80/t, when 90 t will change hands. This is known as the **equilibrium price**.

> The equilibrium price in a market is that price at which the quantities that are willingly demanded and supplied in a given time period are equal.

If we stop dictating prices and give the buyers and sellers free access to one another we will find that, with all the haggling (or negotiating) that goes on, the actual price paid for barley that day will settle down at £80/t. If an auctioneer sold the barley his price would also settle at £80/t.

In our simple analysis the equilibrium price has been seen as the result of the interaction of the given supply and demand schedules for barley. However, had the weather turned severe early in the winter, the demand for barley might have been greater; this would have forced up the price in the market. Buyers would have been prepared to pay higher prices for the same quantity. If the preceding summer had been very favourable for cereal production, the supply of barley might have been greater, causing a slump in its market price. Had the price of wheat or maize been lower, livestock producers might have switched to this as a substitute for barley in their animal feed mixes, thereby producing a lower demand for barley and a lower price. Cereal dealers might have formed a 'ring' to fix prices and thus upset the free bargaining between buyers and sellers. Not all buyers and sellers are of equal astuteness, nor are they equally well informed, and prices of individual lots could be expected to vary around an average. Not all barley is of the same quality and this would also affect the price. Clearly this example as it stands is an over-simplification of what determines prices in the real world, but often only by reducing situations to their simplest elements can explanations be offered for many real-world phenomena. Once a basic model is understood, complications such as those listed above can then be incorporated.

Demand and supply curves

It is common practice, and an invaluable aid to comprehension, to express demand and supply schedules in graphical form. When plotting schedules it is conventional to place price on the vertical axis, the curves corresponding to the schedules already given are shown in Fig. 3.1. This is sometimes called the 'scissors graph' because of its shape; most demand curves slope downward from left to right – more of the commodity is demanded as price falls – whereas supply curves slope upwards from left to right – more is supplied as price rises. Where the two curves cross (intersect) is the equilibrium price at which the quantities demanded and supplied exactly balance.

The graph also illustrates the total value of the barley which changes hands; this value is the same as the total expenditure of all the buyers of barley taken together or all the revenue of all the sellers of barley taken together. If the equilibrium price was £80/t and the quantity supplied and demanded was 90 t, barley to the value of £7200 changed hands (£80 × 90 t). On the graph this is represented by the shaded rectangle which is subtended by the point of intersection (i.e. the rectangle bounded by lines drawn from the point of intersection at right angles to the axes). It is important to be able to identify this area, as we shall be referring to it later.

Because of the great practical importance of prices to all sections of the community and to the functioning of the market-based economies of the Western world, it is worth examining more closely the mechanism by which prices are arrived at. To do so we will examine separately: (i) the demand for goods; (ii) the supply of goods; and (iii) the ways in which supply and demand interact.

Table 3.1. Market intentions of buyers and seller (t barley/day).

Price per tonne	Quantity that buyers are willing to purchase	Quantity that suppliers are willing to sell
60	300	50
70	170	72
80	90	90
90	50	104
100	20	120

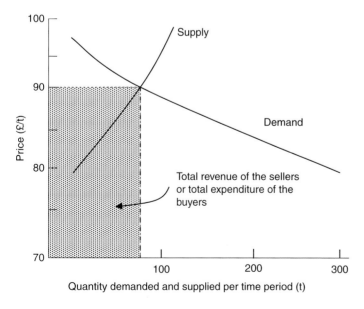

Fig. 3.1. Demand and supply curves for barley on a day in December.

The Theory of Demand

Demand for a commodity (such as eggs, an example of something that consumers might purchase regularly) occurs when people have a desire for the commodity coupled with the willingness and ability to purchase it. A starving man may want some eggs very badly. He may even physically need them to keep body and soul together, but unless he has the purchasing power in his pocket to buy those eggs he will no impact on the market price of eggs. People who die from lack of food in poor drought-struck countries do so because they lack the resources to buy from other parts of the world where it is plentiful. On the other hand, if a farmer has the financial resources which would enable him to buy a new car but he refuses to replace his old one for sentimental reasons, even though it is in an advanced state of decay, there is no demand from him for a car, in this case through lack of desire. Similarly, religious taboos may eliminate demand for beef or pork irrespective of the spending power of the people who adhere to the religions.

Household and market demand

Up to now we have concentrated on individual people and their demands. Often individuals exist in groups – households – the purchases of which are made collectively and reflect the wants and incomes of the various members. For example, a man and wife and (say) two children tend to buy things as a household rather than as four separate individuals. So we talk of **household demand**. When referring to the demand of a whole community for a commodity – such as the demand for milk in the UK – we use the term **market demand**. Increasingly markets are connected, and we talk of **global demand**.

If a household has a demand for eggs and buys, say, seven, the demand does not then evaporate, unless its tastes change and it 'goes off' eggs. Demand is a **flow**. Eggs are purchased time after time, and if we say a household's demand for eggs at a given price is seven eggs, we must add per day, or whatever the relevant time period is.

Factors affecting demand

If we continue with our example using eggs we can show that there are several factors affecting a household's demand for them. Perhaps the most obvious is their price: as the price of eggs rises one would expect fewer to be demanded. But the level of incomes of consumers also bears upon the demand for them: wealthy people may buy more than poorer people. Also what about foods that are eaten in place of eggs (breakfast cereals maybe) or eaten with eggs (such as bacon)? What happens to the demand for eggs when the prices of these vary?

The factors affecting household demand for a commodity are:

1. The price of the commodity itself (P_A).
2. The incomes of consumers (Y).[1]
3. The price of competitive (or substitute) goods and the price of complementary goods ($P_B, \dots, P_N$).
4. The tastes of consumers (T).

An abbreviated way of stating that the demand by consumers for good (A) is a function of, or depends on, these factors is:

$$D_A = f(P_A, Y, P_B, \dots P_N, T)$$

In addition, when considering what determines total or market demand for a commodity, two other factors must be recognized:

5. The size of the population.
6. The income pattern of the population.

In real life the demand for any commodity is determined by all the variables acting simultaneously. To help analyse how each bears upon demand we must simplify the real-world situation by imagining that all the variables remain constant except for one (that is, all other things are held equal, or *ceteris paribus*). By varying that one (say, consumer income) we can focus more clearly on its relationship to demand. This process can be repeated for all other variables singly.

How demand for a commodity varies with the price of that commodity

$$D_A = f(P_A, \text{other variables constant})$$

When the prices of most commodities go up, the quantities of them which consumers are willing to buy fall. Conversely, when prices are lowered, the quantities demanded increase. An explanation for this was offered in Chapter 2 on consumer choice. Plotting such a situation produces the familiar down-sloping demand curve.

Halving the prices of goods will usually increase the demand for them; for some goods a 50% fall in price will increase their demand enormously (think what would happen to the sale of Shell petrol if Shell alone cut its prices by that amount), but for other goods a similar percentage price cut would make very little difference to the quantity demanded (tap water bought by households is a good example). In the first case demand for that brand of petrol is said to be price sensitive or **elastic** and in the second case demand is not sensitive to price, or price **inelastic**. Note that the demand curve for the elastic demand good (Fig. 3.2) is much flatter than that for the inelastic demand good (Fig. 3.3). (It is not important in this instance to be specific about the units of quantity to which the prices relate as it is the percentage changes that are the focus of interest.)

It is possible to measure the degree of elasticity of demand – its responsiveness to price changes. The index used for this purpose is the **price elasticity of demand** (see equation at bottom of page).

Taking figures from the petrol demand curve: when the price falls from 10 to 8 (a fall or negative change of 2 units) the quantity demanded increases from 5 to 9 units (a rise or positive change of 4 units).

Percentage change in quantity demanded =
$4/5 \times 100 = 80\%$
Percentage change in price = $(-)2/10 \times 100 = (-)20\%$
Price elasticity of demand $(E_{Dp}) = 80\%/(-)20\% = (-)4$

To ensure that you know how the calculation was done, work out the price elasticity of demand for water using the figures on the graph. You will find that you get a result nearer zero. A coefficient of -1 means that an x per cent price change results in a change in the quantity demanded also of x per cent. A figure between 0 and -1 implies that the percentage change in quantity is smaller than the percentage price change (i.e. demand is relatively price inelastic) whereas a coefficient of more than -1 (say -2 or -3) implies that the percentage quantity change is greater than the percentage price change (i.e. demand is relatively price elastic).

It should be noted that these coefficients are accompanied by a negative sign because a movement

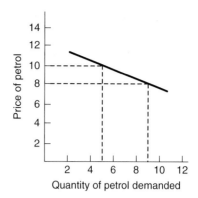

Fig. 3.2. Demand curve for petrol.

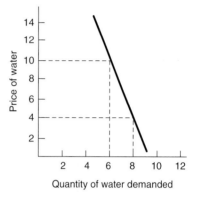

Fig. 3.3. Demand curve for water.

along a downward-sloping demand curve involves either a negative change in price (and a positive change in quantity) or a negative change in quantity (but a positive change in price). Frequently, however, the sign is omitted and the coefficient of elasticity is taken as the ratio of the proportionate changes.

What determines the sensitivity of the quantity demanded to price?

Several factors influence the sensitivity of the quantity demanded to price, reflected in the steepness of the demand curve:

$$\text{Price elasticity of demand } (E_{Dp}) = \frac{\text{Percentage change in quantity demanded}}{\text{Percentage change in price}}$$

1. *The presence of good substitutes.* Shell petrol is a good substitute for other brands of petrol in the eyes of most motorists, although individual drivers may have preferences. If the price of Shell were increased by a few pence, sales would fall greatly because motorists would switch to alternative petrol brands whose price had not increased and which were now cheaper. The demand for individual brands of petrol is therefore highly price elastic.

But take a second example, milk. While consumers may be prepared to switch between supermarket suppliers, for many purposes there is no effective substitute for fresh liquid milk (powdered whiteners do not taste the same in tea and coffee), so if the price were raised by a few pence, the total quantity bought would only decrease by a small amount. On the other hand, if the price were lowered, it is unlikely that much increase in sales of milk would occur since there will be little demand transferred to milk from substitutes. The demand for milk is said to be inelastic with respect to price changes because of the lack of good substitutes, and its demand curve is hence quite steep.

2. *The cost of the article in relation to household income.* The typical British household spends very little on salt – perhaps not more than a few pounds in a year. If the price of salt doubled, households would probably still buy the same quantity because in their budget it represented an insignificant weekly outlay even at the higher price. On the other hand, to a student with a car, the cost of fuel represents a considerable proportion of his available spending power, and he will cut down inessential travel (such as going home to see his parents) if the price of petrol rises.

3. *The essential or non-essential nature of a commodity.* It is difficult to get along without some commodities like basic foods, water, clothing and shelter. So if the prices of these go up, the quantity demanded will hardly change. On the other hand, it is quite likely that if the prices of essential goods go down, little extra will be taken by the consumer. To commuters, committed to a daily rail journey to work, travel costs will also appear as essential and thus price rises will not affect the number of season tickets sold, at least in the short term.

4. *Habits.* Well-established habits can make consumers' buying patterns insensitive to increases. Indeed, the tax on tobacco takes advantage of this. If putting a tax on cigarettes cut the number sold dramatically, little revenue would be raised. If a tax was already in place, raising it further would bring in more revenue only if there was a small effect on the quantity purchased. Habits can thus be exploited by governments. When examined closely many of our purchases are habitual.

Significance of price elasticity of demand (E_{Dp}) to agriculture

Food, is agriculture's main product, either directly for consumption by humans or animals or, increasingly, as the essential component of processed food products in various forms. Foodstuffs taken together have no substitute and are essential to life. On these counts alone we would expect the demand for food in relatively well-fed, high-income countries such as those that comprise the EU to be insensitive to price, giving rise to a steeply downsloping demand curve and a price elasticity of demand coefficient close to zero. Substitution between different types of food is of course possible, and demand for certain items – quality steak cuts for example – is quite responsive to price changes. But in general quantities demanded are insensitive to price, with price elasticities typically between −1 and 0. This is especially true of foods which are not expensive. Some examples are given in Table 3.2. For foods that are heavily processed and packaged, the price elasticity of the basic material provided by agriculture is generally much lower than for the product as it appears on the supermarket shelf, for reasons we will not explore here. The consequence is that farmers and horticulturalists find they face typically a price-inelastic demand for the products which characterize this industry.

Table 3.2. Examples of coefficients of price elasticities of demand (source: Defra, 2013).

Food	Price elasticity of demand (2001–2009)
Salmon	−0.7
Fresh vegetables	−1.0
Processed vegetables	−0.1
Potatoes	−0.5
Milk	−0.7
Cream	−0.5
Cheese	−0.6
Eggs	−0.6
Poultry	−0.9
Beef	−0.6
Pork	−0.8

The steeply downward-sloping demand curve which the agricultural industry faces for its products has important implications for the impact which increased levels of production can have on the revenue to the industry from selling its output, and hence on the incomes of farmers. To illustrate the implications we will use a simple model based on potatoes showing the impact on revenue to potato growers caused by: (i) the influence of weather on crop yields; and (ii) the spread of new higher-yielding varieties. This model contains the essential elements which apply to many agricultural products.

When favourable weather conditions cause potato yields to be heavier than expected, the effect on the supply curve of potatoes is to shift it to the right. At each price level growers would be willing to supply a greater quantity than in a year of normal yields (this is anticipating a more detailed study of Supply Theory, but the reader should be able to assimilate this idea already). This is illustrated in the graph in Fig. 3.4 which shows the supply curve for potatoes in a normal year and in a heavy-yield year. The supply curve for the year of favourable weather intersects the demand curve at a much lower price. Note that the point of intersection has moved along the demand curve, which itself has not shifted. At this lower price consumers are only willing to take a little more than they were at the old, normal-year price. Although farmers sell a somewhat greater quantity, this increase is not sufficient to compensate them for the much larger

(in percentage terms) price fall. The result is that the revenue which the industry receives (quantity sold × price per tonne, represented in Fig. 3.4 by the size of the stippled rectangles) is less in the year of heavy yields and with greater tonnage sold than in the year of normal yields. This fall in revenue which accompanies increased production is hugely significant, as will be explained below.

Conversely, a drought year with lower-than-normal yields would shift the supply curve of potatoes back to the left, greatly raising the price at which the supply and demand curves intersect. Revenue would increase because the fall in quantity sold would be more than compensated by the increased price. We have a situation, then, where the unpredictable influence of weather can cause outputs to vary, resulting in more-than-proportional changes in price, so that higher outputs are associated with lower revenues and vice versa.

Consider also what happens when a new variety of potato is developed which yields more heavily, so that for the same costs in terms of land, fertilizer and labour, a greater quantity can be produced. At any given price farmers will be willing to sell more than previously, with the effect that the supply curve moves to the right. This is similar to the favourable weather example above but differs in that this rightward movement is permanent whereas that resulting from good weather was temporary and could be reversed. The effect on the revenue which the farming industry receives from potato sales is, however, the same: the total value of potatoes

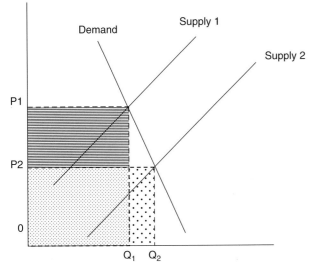

Fig. 3.4. Demand and supply curves for potatoes (hypothetical) at the industry level: the effect of favourable weather on the supply of potatoes.

sold is less after increasing production through adopting the new variety than before.

Recall that here we are dealing with a model which contains the essential features of a situation, thereby making it easier to comprehend and to make predictions, by omitting many of the real-world complications. For example, we have ignored the effect on total supply which imported potatoes might have in years of high prices and we have not considered the activities of government-backed organizations in restricting the supply on the market in order to prevent prices falling to very low levels, although the model provides an explanation for such collective action. For some commodities (e.g. milk) these complications may be of major importance in determining prices and revenues.

Nevertheless, the insensitivity of demand to changes in price brought about by shifts in the supply curve, which is a characteristic of the demand for many agricultural products, can be seen to be a source of many of the problems faced by the farming industry. These range from the instability in revenues from crops like plums and blackcurrants, where relatively minor yield variations can have major repercussions in terms of fruit prices and hence farm incomes, right up to fluctuations in the prices of internationally traded commodities like coffee. Low price-elasticity is also at the root of the long-run decline which can be observed in the prices of agricultural products and plays an important role in the cycles of low and high prices which characterize certain commodities such as pigmeat. Both of these phenomena will be re-examined later in detail.

Returning to our potato example, if increasing output reduces the agricultural industry's revenue from potatoes, the logical way for farmers acting together to raise their revenue is to reduce production, forcing the supply curve back to the left, pushing up prices more than proportionally. For reasons explained below, it is unlikely that farmers will come to such an arrangement voluntarily, and some sort of government involvement will be necessary. In reality, supply restriction has often been practised in agricultural policies as a way of keeping up prices and revenues for farmers. For example, the UK government used this mechanism for several commodities from the mid-1930s onwards, giving power to restrict supply to Marketing Boards (state-sponsored cooperatives) or similar regulatory bodies. This applied in particular to potatoes, milk, sugarbeet and hops. Supply restriction has

also been an important feature of the Common Agricultural Policy (CAP) of the EU (with its quotas on milk and sugar production, cereal land set-aside and so on). Although farming benefits through higher incomes the consumer has to pay higher prices. This subject is considered at greater length in Chapter 10.

Attempts by the whole industry to restrict output and raise prices will require that individual farmers are prevented from taking advantage of the higher prices by expanding, thereby undermining the collective restriction on output. There will be a tendency for this to happen because, while the demand curve faced by the whole farming industry is down-sloping, individual farmers do not face a downward-sloping demand curve for the products of their single businesses. For example, if a farmer uses a new variety of potatoes and has a bumper crop, he will be able to sell them all at the market price and will have a higher revenue than he would selling the fewer potatoes of the old variety. The quantity which he supplies to the market does not affect the price of potatoes because the quantity he has for sale is tiny compared with the total that is being sold. He faces an infinitely elastic (horizontal) demand curve for his products (Fig. 3.5).

However, if a sufficient number of farmers simultaneously adopt the new variety, their combined increase in output will push down the market price and hence be felt by the individual growers. Similarly, an agreement between producers to restrict output in order to raise prices and incomes can survive a few 'rogue' farmers who run counter to the agreement by expanding, but if the number of 'rogues' becomes too great the collective action will be frustrated. The tendency to break ranks is the reason why voluntary arrangements between farmers to restrict supply do not generally work. A degree of compulsion to comply is needed, which may take the form of a quota on the acreage of a crop grown (as formerly applied to potatoes) or quantity sold (milk) imposed by some central authority run by the government. Fines can be then imposed and enforced for exceeding the quota or permitted area of crop or numbers of animals.

Individual farmers will always be looking for new techniques, varieties, etc. which increase the quantity they can produce because this increases their revenue. But remember that increases in production act against the interests of the whole industry by greatly pushing down prices. Here is a great dilemma in agriculture – farmers individually will

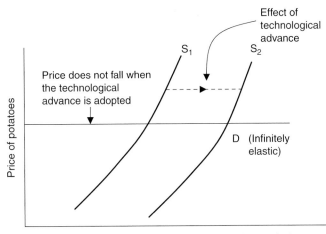

Effect of
technological
advance

S_1 S_2

Price does not fall when
the technological
advance is adopted

D (Infinitely
elastic)

Quantity demanded and supplied per time period

Fig. 3.5. Demand curve for the crop
(potatoes) of an individual farmer
(hypothetical).

be trying to expand but farmers acting collectively, or the government acting on their behalf, will be trying to limit expansion to maintain overall prices.

We must now move to another important factor affecting demand.

How demand for a commodity varies with the income of consumers

D_A = f(Y, all other variables constant)

If a household receives a 10% increase in its income it will increase the quantity of goods and services it buys, but it will not increase its expenditure on all commodities equally. Probably very little extra bread will be bought, maybe 21 loaves/month instead of 20 (a 5% increase), but three times as many bottles of wine might be purchased (a 200% increase). Changes in the quantity are related to changes in income by the **income elasticity of demand**. The formula used to estimate this elasticity for any commodity is given in the equation at the bottom of the page.

In the example of bread, when income rose by 10%, purchases of bread rose by 5%. Hence the income elasticity of demand for bread is 5%/10% = 0.5. Similarly, the income elasticity of demand for wine is 200%/10% = 20.

Bread has a low income elasticity of demand and the number of loaves bought is little affected by the change in income. On the other hand, the number of bottles of wine bought is greatly sensitive, and the income elasticity of demand is high. For some commodities an increase in income may result in less being bought, and for these income elasticity of demand would be negative. Margarine is such a commodity – as incomes rise consumers switch to other spreads. These goods with negative income elasticity of demand are called **inferior goods**, a technical term rather than a judgement on their quality.

Figures 3.6a and 3.6b show that the effect of a rise in income is to shift the demand curves for bread and wine to the right. The demand curve for bread is moved less than that of wine as the demand for bread is the less sensitive to income changes. Note the difference between this effect and that described in the previous section on price elasticity of demand (which does not involve a shift of the curve).

Income elasticity of demand: quantity and expenditure measures

The formula given for calculating income elasticity of demand uses a physical measure of the change in demand caused by a change in income (e.g. percentage change in the number of bottles of wine bought per week). Occasionally it is useful to

$$\text{Income elasticity of demand } (E_{Dy}) = \frac{\text{Percentage change in quantity demanded}}{\text{Percentage change in income}}$$

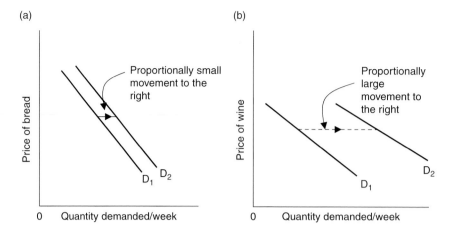

Fig. 3.6. The effect of a rise in income on the demand curves of (a) bread and (b) wine.

calculate the elasticity in terms of the percentage change in expenditure on a given item (see equation at bottom of page).

This formula will give the same figure as the previous formula except where people switch to a more expensive grade of good. For example, a man may still only buy one bottle of wine, but buy a more expensive wine; the quantity will not change but the expenditure will. Strictly speaking, we are not dealing with expenditure on the same good – cheap wines are not the same as dear wines – but in the real world it may be necessary in practice to lump a collection of dissimilar goods together for the purpose of measurement. Hence published data on income elasticities of demand will usually be labelled as being derived from 'quantity' or 'expenditure' measures.

Engel's Law

In our example of bread and wine the proportions of the household's expenditure which went on these two items changed when its income increased. Some numbers will make this clear. Say the first income was £10/week, of which £2 (20%) was spent on bread and 30p (3%) on wine. If income went up by £1 to £11 per week (a 10% rise) and the amount spent on bread increased by only an extra 10p (a 5% rise on the initial spending of £2),

bread would now account for £2.10 out of the £11 of income, or about 19%, a fall of 1 percentage point from the initial situation. Turning to wine, if three times the original amount was now spent on this commodity (90p in place of 30p), this would now account for 8% of income – a rise of 5 percentage points. The 19th century statistician Ernst Engel noticed that any additional income tended to be spent more on luxuries and non-essentials than on essentials and his observation, commonly known as 'Engel's Law', can be formulated as follows:

> The proportion of personal expenditure devoted to necessities decreases as income rises.[2]

This is represented diagrammatically in Fig. 3.7, which illustrates that expenditure on food and clothes forms a larger proportion of total expenditure of people with low incomes than of those with high incomes.

People's incomes differ and what will be a luxury to a poor person (e.g. butter) may seem very ordinary to a well-off man, so ordinary that, with further increases in income, he switches to low-calorie substitutes which please him better and he buys less butter. A generalized curve relating income to the quantity of a commodity bought can be constructed (Fig. 3.8). This curve can be divided into four phases:

$$\text{Income elasticity of expenditure on good A} = \frac{\text{Percentage change in expenditure on A}}{\text{Percentage change in consumer income}}$$

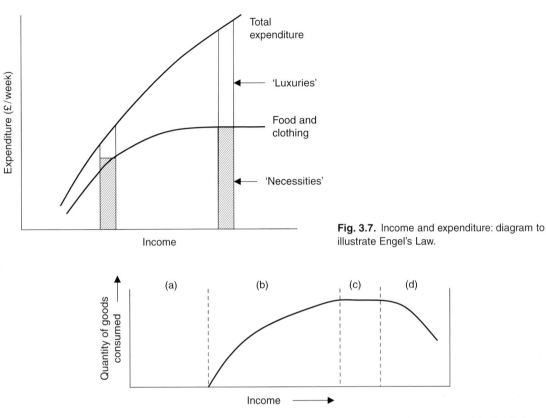

Fig. 3.7. Income and expenditure: diagram to illustrate Engel's Law.

Fig. 3.8. Generalized representation of the relationship between the quantity consumed of any good and the level of a consumer's income (for explanation of phases (a)–(d) see text).

- Phase (a) – income is so low that no butter is bought with increases in income. Income elasticity of demand for butter is zero.
- Phase (b) – with increases in income more butter is bought; at first a 1% change in income causes a large percentage increase in consumption. Income elasticity of demand is initially high, but falls as income increases.
- Phase (c) – increases in income cause no greater quantities of butter to be consumed. Income elasticity of demand is zero.
- Phase (d) – increases in income reduce the demand for butter which thus becomes an 'inferior good' with negative income elasticity. Not all goods necessarily become 'inferior' at high income levels.

The relationship in Fig. 3.8 is a general one and can apply to many commodities. For the typical UK inhabitant some goods will be in phase (d) (e.g. margarine – E_{Dy} negative), some in phase (c) (e.g. fish – E_{Dy} virtually zero), some in phase (b) (e.g. yoghurt – E_{Dy} positive), and some in phase (a) (e.g. a non-food example) luxury cars). The income elasticity of demand of all foods taken together is about +0.23, which means that food in general lies in the right-hand half of phase (b). Examples of the coefficients (expenditure and quantity purchased) for some UK foods are given in Table 3.3.

Significance of income elasticity of demand (E_{Dy}) to agriculture

Using the E_{Dy} of agricultural products we can show why agriculture is a declining industry in terms of the proportion of the nation's income that is spent on it, and consequently in terms of the proportion of the nation's stock of resources (principally manpower) it employs. This is a finding for virtually all countries as they grow economically over time, which implies that the income per head increases. It is particularly to be found among the more industrialized countries

Table 3.3. Examples of income elasticity of demand (UK coefficients) (source: MAFF, 1988).

Commodity	Income elasticity of demand (1986) of	
	Expenditure	Quantity purchased
Fresh potatoes	−0.20	−0.32
Margarine	−0.13	−0.26
Liquid milk	−0.10	−0.11
Bread	−0.04	−0.17
Processed cheese	0.14	0.09
Breakfast cereals	0.26	0.28
Carcass meat	0.35	0.27
Fresh green vegetables	0.43	0.17
Fresh fruit	0.70	0.62
Fruit juices	0.94	0.93
All food	0.23	

that form the members of the Organisation for Economic Co-operation and Development (OECD). The explanation is as follows.

We know that the E_{Dy} of food is low compared with that of the products of most other industries. A simple model can be constructed to show the relevance of this. Let us imagine an economy made up of only three people: (i) a farmer; (ii) a laundryman; and (iii) a consumer called Mr 'N', who buys eggs from the farmer and sends his shirts to the laundry. Mr 'N' has an income of £10/week, of which £1 is spent on eggs and £1 on laundry services. The farmer and laundryman therefore receive the same revenue. If we assume that half the revenue of each is kept as income (the other half going on costs of producing eggs or laundering clothes) then the farmer and laundryman will have the same amount of money remaining in their pockets; they will have income parity.

If Mr 'N''s income now goes up by 10%, he will want to buy more eggs and use the laundry more, but it is unlikely that his expenditure will increase on both in the same proportions. Knowing the income elasticities of demand (E_{Dy}), we can work out the increases in his expenditure (see the formula given earlier). Assume the E_{Dy} of eggs is 0.20 and the E_{Dy} of laundry services is 2.00. After Mr 'N''s rise in income the farmer will find that Mr 'N' will want £1.02-worth of eggs (the 10% income rise × the coefficient of income elasticity × the initial amount spent on

this commodity); this is 2% more than before. Mr N buys £1.20-worth of laundry services, 20% more than before. A greater expansion in demand has occurred for laundry services than for eggs. The revenue coming to these two producers will have expanded to different extents and, because half their revenue is retained as income, their incomes will have gone up by differing magnitudes. While in absolute terms both will be better off, the farmer will be relatively worse off because his income will have fallen behind that of the laundryman.

Despite the crudeness of the example, it is a fair representation of what happens in the real world. As a country becomes more developed and income per head of the population increases, although in absolute terms the agricultural industry gets richer, it falls behind other industries and hence relatively it becomes poorer. This is not the 'fault' of agriculture, or a sign of its technical inefficiency, but simply the result of the relative income elasticities of demand for its products and those of other industries.

The ever-widening gap between the income of agriculture and other industries results in the resources in agriculture (manpower, capital, etc.) earning lower returns than elsewhere. There are cries from farmers for subsidies to fill the gap, but this can only be a short-term solution as the gap is bound to widen as a country grows. The long-term solution is to encourage resources in agriculture – and this principally means labour and management – to transfer to other forms of production where higher rewards are available. This transfer will raise the average income of the resources left in agriculture and lower that of resources in other industries, so that parity is again approached. If the shift in resources is rapid, the gap in relative rewards will be small, but larger if they are reluctant to leave agriculture. At the same time, the reallocation of productive factors away from agriculture will reflect better the changing overall pattern of demand for goods and services which accompanies economic growth. This will be re-examined in subsequent chapters.

Income elasticities of demand are also highly relevant to making estimates of how the demand for food rises as countries undergo economic development. As Box 3.1 shows, rising incomes in poor countries have much more of an impact on food demand than the same proportional increase in rich countries. Estimates of the extra demand are important information in planning how to avoid national food shortages.

Box 3.1. Income elasticity of demand for food in developing countries.

The ability to feed the world population in the near future depends critically on the capacity of food supply to meet an increasing demand. As population rises, more people need to be fed, and as income grows more households' disposable income is available for food consumption. A crucial question is at what rate world food demand is expected to increase, especially in developing countries. The income elasticity coefficients of food demand are of great importance when assessing this. An International Comparisons Project (ICP), coordinated by the Economic Research Service of the United States Department of Agriculture, covers demand elasticities in more than 100 countries. Major findings from the 2005 ICP are that:

- As countries become more affluent, increases in income have a declining impact on food spending. For instance, a dollar increase in income would cause food expenditures in the Democratic Republic of Congo to increase by 63 cents, but only by 6 cents in the USA. In contrast, recreation expenditures in the Democratic Republic of Congo would not increase at all, while in the USA recreation expenditures would increase by 13 cents with an additional dollar of income. Note that these are responses to an absolute increase in income, not percentages of income changes, which is what coefficients of income elasticity of demand reflect.
- The income elasticity of demand for food varies greatly among countries and is highest among low-income countries, where it varies from 0.85 for the Democratic Republic of Congo to 0.71 for Armenia. It ranges between 0.71 and 0.57 for middle-income countries, and from 0.56 to 0.35 for high-income countries. The average income elasticity for low-income countries is 0.78, over 1.5 times the average for high-income countries.
- Within the food category there are differences. With affluence, the portion of additional food expenditures allocated to cereals and other staples decreases (and may even fall in absolute terms if they are considered as 'inferior goods') and that on animal protein rises. Consequently, the composition of the food basket changes. This leads to 'Bennet's Law', named after the first person to note that the proportion of calories that people obtain from starchy foods declines as their income increases.
- The own-price elasticities for the food subcategories also vary by affluence, in line with economic theory: low-income countries are more responsive to price changes compared with higher-income countries. For instance, the own-price elasticity value for breads and cereals ranges from –0.50 for the Democratic Republic of Congo to near zero for the USA.
- Overall, low-income countries are more responsive to changes in income and food prices and, therefore, make larger adjustments to their food consumption pattern when incomes and prices change.

By combining estimates of additional demand through population growth with that of increases coming from increases in income it is possible to make predictions of the anticipated rise in food demand.

Source: adapted from Muhammad *et al.* (2011).

It is now time to move to the third factor influencing demand for a commodity.

How demand for a commodity is affected by changes in price of competitive and complementary goods

$$D_A = f(P_B, \ldots P_N, \text{other variables constant})$$

It is necessary, when considering how the demand for a particular good is affected by the prices of other goods, to group them according to the relationship they hold in the minds of purchasers. The two main types of relationships are where the goods are seen as competitive or complementary.

Competitive goods

Competitive goods are those which are to some extent substitutes for each other, and they compete to satisfy the consumer's want. An example is butter and margarine. When the price of butter rises, consumers will buy more margarine. Figure 3.9 shows that a rise in the price of a competitive good shifts the demand curve of margarine (i.e. the curve showing quantities of margarine demanded at different prices of margarine) to the right. This happened in the UK in the 1970s when consumers saw butter prices rise when Britain joined the EU's CAP. Margarine manufacturers experienced an expansion in the demand for their product; they were also aware that CAP schemes to sell butter at subsidized prices to pensioners could reduce margarine demand.

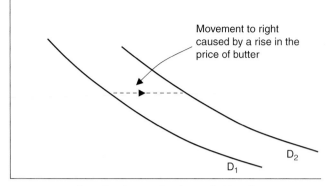

Movement to right caused by a rise in the price of butter

D_1 D_2

Quantity of margarine demanded/week

Fig. 3.9. Effect on the demand curve of a price rise of a competitive good.

The sensitivity of demand for margarine to the price of competitive goods, called the **cross elasticity of demand**, is given by the formula at the bottom of the page.

The E_{Dx} of competitive goods is positive, since a rise in price of one will cause more of the other to be bought. A rise in the price of the Ford Focus will result in an increase in demand for Volkswagen Golfs; they are strong competitors (or good substitutes) as family and business cars in European markets.

Complementary goods

Complementary goods are those which are usually used together. An example is oil and petrol for cars. If more petrol is bought in any one year, more oil is also bought because cars use both together. Such goods are sometimes called **joint demand** goods. If the price of petrol rises but that of oil is unaltered, not only will less petrol be bought but less oil too. Cross elasticities of joint demand goods are negative. The effect on the demand curve for oil of a rise in the price of petrol is to shift it to the left, as in Fig. 3.10.

There is a link between the cross elasticity of demand between two competitive goods and the price elasticities of demand of each good. Goods which are sensitive to prices of competitors will also tend to have shallow demand curves (i.e. elastic demand) because consumers will switch rapidly between competitors as one good becomes relatively cheaper or dearer than the other.

How demand is affected by tastes of consumers

$D = f(T, \text{other variables constant})$

Demand has been examined up to now by holding all the influences on demand, except one, constant, and observing the relationship between changing that variable and the quantity of goods consumers take from the market. But if we held constant the price of the good (P_A), the income of consumers (Y) and the prices of competitive and complementary goods ($P_B, \dots P_N$), there may still be changes in demand caused by changes in the tastes of consumers. An increase in demand caused by a swing in taste towards a commodity will shift the demand curve for that commodity to the right, so that more will be bought at any given price. This is shown in Fig. 3.11. The change in taste may be caused by a range of factors, such as by personal experience (e.g. the demand for pizzas in the UK has risen in part through people's experience of Italian food on holidays abroad), by advertising or by a desire to

Cross elasticity of demand (E_{DX}) for good A $= \dfrac{\text{Percentage change in quantity of good A demanded}}{\text{Percentage change in price of good B}}$

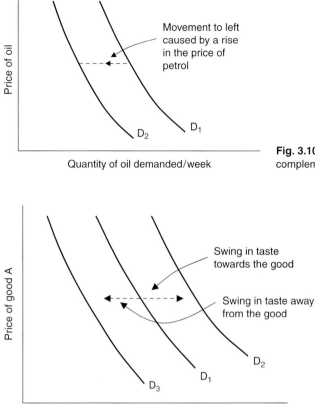

Fig. 3.10. Effect on the demand curve of a price rise of a complementary good.

Fig. 3.11. The effect of changes in taste on the demand curve for a good.

'keep up with the Joneses'. This status-seeking consumption shows that the quantity and quality of one consumer's purchases are partly determined by what other consumers do.

When dealing with relatively short periods of time, economists assume that the consumer tastes remain constant. If they did not, it would not be possible to conclude that, for example, the quantity of butter bought varied with the price of butter, with consumer income and prices of all other goods held constant. Fluctuations in demand could be explained by rapid changes of taste and, in some cases, are. Fashions for some types of food come and go, and in part advertisements from food companies attempt to stimulate demand for particular types of yoghurt, low-fat spreads and so on. Rapid taste change can come from health scares, both imagined and more real ones: for example publicity about salmonella contamination of eggs caused a substantial drop in UK demand during late 1988,

without much evidence that there was a new problem. Such events are unusual. When dealing with relatively long periods of time, however, changes in consumer tastes cannot be ignored.

Additional factors influencing market demand

At the beginning of this chapter on demand it was pointed out that two additional factors must be considered when analysing the demand by the whole population (market demand) for a good as opposed to the demand shown by individual households.

The first is *the size of the population*. A rise in the country's population will increase demand for goods even if average income per head and all prices remain the same. There are simply more people desiring to be clothed and fed, etc. Assuming that they possess purchasing power, the country's demand curve will be shifted to the right. In most developed countries populations are not growing much, and in some are declining. This helps explain the static demand for agricultural commodities there. In contrast, population growth in some developing countries is significant, though this may not be reflected in the market because of a lack of income.

The other factor to be considered is the *distribution of incomes* between households. A change in this distribution can take place without affecting average income per head, but it will cause the demand for some goods to increase and demand for others to fall. For example, if income tax is increased on unmarried people and decreased on married ones, leaving average spendable income unchanged, it could be expected that the demand for sports cars in the country as a whole would decline but the demand for family saloons would increase.

Shifts along the demand curve and shifts of the whole demand curve for a commodity

To conclude this section on demand, it is worth recapitulating on the ways the factors described above affect the demand curve. A demand curve shows the quantities of a commodity which consumers are willing and able to take from the market at a range of given prices of the commodity. If the price of the commodity is altered, because a shift in the supply curve causes the demand curve and the supply curve to intersect at a different level, more (or less) will be bought; this involves a movement along the demand curve. However, a change in price can result from a move of the whole demand curve while the supply curve remains static. The whole demand curve will be shifted to the left or right by: (i) a change in consumer income; (ii) a change in the prices of competitive or complementary goods; (iii) changes in taste; (iv) population changes; or (v) a change in the distribution of incomes between households. It is helpful to bear in mind when trying to explain movements in prices whether there is a shift along a demand curve, or whether the curve itself is moving, or whether both are occurring.

Theory of Supply

Having considered the demand for commodities in detail, the same approach will now be adopted to the supply of goods and services.

The meaning of supply, and factors that determine it

The supply of a commodity can be defined as the quantity that producers are willing and able to offer for sale in a given time period. Like demand, the supply of goods and services is a flow, and a time period is always implied (though not always stated specifically).

The supply of any good depends on five factors:

1. The price of the good (P_A).
2. The prices of other goods that firms could produce or do produce (P_B, ... P_N).
3. The prices of factors of production, i.e. the prices of the commodities which are used up by firms (F_A, ... F_M). These are supplied by other firms or individuals: examples are electrical power, raw materials, labour.
4. The state of technology (T).
5. The goals, or objectives, of firms (G).

In abbreviated form, supply can be shown as a function of these factors by the following:

$$S_A = f(P_A, P_B, ... P_N, F_A, ... F_M, T, G)$$

The relationships between changes in each of these five variables and changes in supply will now be explored in turn, with the other four being held constant. They will be met again in more detail in Chapter 5 when we consider the production decisions of individual farm businesses.

How the supply of a commodity changes with changes in that commodity's price

$$S_A = f(P_A, \text{other variables constant})$$

Normally, as the price of a good increases producers are willing to supply a greater quantity. This is shown by the supply curve in Fig. 3.12; it rises from left to right and hence has a positive slope. (Recall that a demand curve normally has a negative slope and slopes downward from left to right.) Taking the supply of eggs as an example, it is reasonable to assume that, if the price of eggs increased, more would be put on the market because existing egg producers would expand and some farmers not producing eggs would set up production. Why this happens will be developed in Chapter 5 'Production Economics'; at this stage it is appropriate to accept that producers behave in this way because greater profits can be made by producing more eggs at higher prices.

The upward-sloping supply curve can be regarded as what is normal. However, Fig. 3.13 shows a rather odd supply curve occasionally encountered, especially when producers have invested heavily in specialist buildings and equipment (i.e. they have high 'fixed costs'), such as is required in modern large-scale egg production. If prices are lowered to a certain point P, the quantity supplied contracts in the typical way, but if prices are lowered further, supply increases. This can be explained by producers trying to maintain their incomes by producing more as prices fall. For example, a farmer faced with a fall in the price of eggs might cram an increased number of birds into his hen house and aim for a higher output. Although profit per egg would fall with a fall in price, by achieving a higher output his overall profit may be maintained or even increased. This type of response, however, will only happen if there is a positive margin above variable costs (such as poultry feed) and should be treated as short term and exceptional.

The responsiveness of producers in terms of output to changes in the price of their product is measured by the **price elasticity of supply** (E_{Sp}). It is defined by the formula at the bottom of the page.

The word 'price' is often dropped from the title. Usually the elasticity of supply is a positive figure. (It is only negative in the exceptional circumstances of the 'reverse' supply curve, when the quantity supplied increases when prices fall.)

For some commodities the quantity supplied is extremely responsive to price changes. Such supply

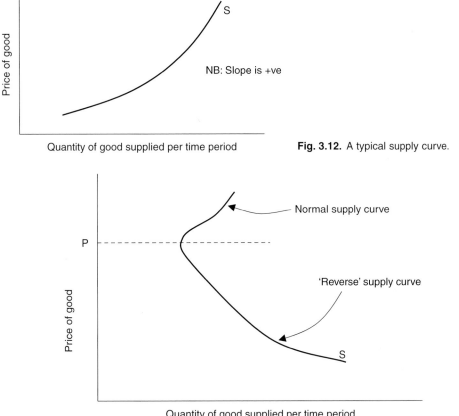

Fig. 3.12. A typical supply curve.

Fig. 3.13. Supply curve that is partially 'reverse'.

$$\text{Price elasticity of supply} = \frac{\text{Percentage change in quantity of good A supplied}}{\text{Percentage change in price of good A}}$$

is called **elastic** and the E_{Sp} would be high. Where changes in supply occur without any change in price being necessary, supply is called **infinitely elastic**. An example might be an ice cream seller at the seaside: within a given range of quantities they are willing to sell as much or as little as buyers want without changing the prices. At the other extreme, the quantity supplied is very unresponsive, possibly totally unresponsive, to price changes. An example is Cup Final tickets; once the arena has been sold out, however high the 'black market' price rises, no more seating spaces can be supplied. In such a situation, supply is completely inelastic.

Figure 3.14 shows what happens when a change in demand, caused perhaps by an increase in consumers' incomes, meets supply situations of infinitely

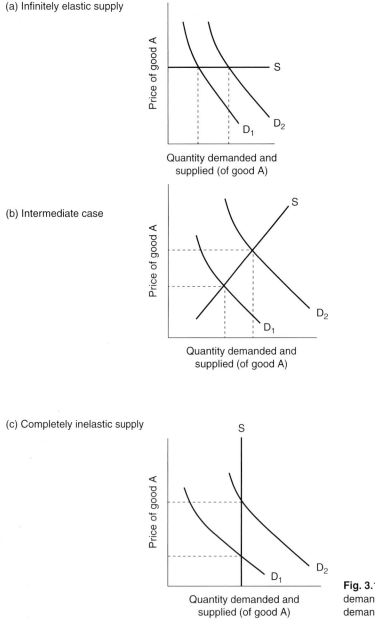

Fig. 3.14. The effects on price and quantity demanded and supplied when a shift in the demand curve occurs.

elastic supply (Fig. 3.14a), completely inelastic supply (Fig. 3.14c), and an intermediate case (Fig. 3.14b). With an infinitely elastic supply, no price increase occurs and the quantity which changes hands increases; with completely inelastic supply the price increases but no increase in quantity results, and in the intermediate case both the price and the quantity sold increase.

What influences the elasticity of supply?

Production is not an instantaneous process and it often takes a considerable time for producers to implement their decisions to change their output levels in response to price changes. Furthermore, in some cases, while it may be easy to cut back production, it may be difficult to expand so that the responsiveness to *price falls* may differ from the responsiveness to *price rises*. Hence when considering the factors which influence price/supply responsiveness it is necessary to specify whether expansion or contraction in the quantity supplied is being implied. The influences on the elasticity of supply are as follows:

1. *The time period under review and the length of the production cycle (chiefly applies to expanding the quantity supplied)*. It is obviously impossible to double the supply of barley from farms in 3 months if the price of barley jumps as the result of an increase in demand, so in the short term supply is inelastic. Nevertheless, farmers can increase their production of barley dramatically by planting more at the next spring, so in the longer term the supply is quite responsive to the price and the sharp short-term price rise may then be largely eroded. The elasticity of supply will depend on the time scale chosen and the length of the production cycle of the commodity under consideration, which is the time between instigating production and finally selling the product. The term **supply lag** is used to refer to the length of time before supply responds to a price signal.

Taking a second example, it is difficult to increase within a year the production of beef in response to price rises because about 2 years elapse between the conception of a calf and the slaughter of the finished animal. Generally the longer the production cycle the less elastic is the supply in a given period of time. Taking a third example, the supply of strawberries on any particular day during the picking season is very inelastic. Because they are highly

perishable, the fruit has to be sold whatever the price. The overall season's prices will be taken into account in farmers' decisions about production for next year and subsequently, so the long-term supply is much more sensitive to price levels.

2. *The cost structure of production (chiefly applies to reductions in the quantity supplied)*. Calves can be reared in several ways. They could be housed in expensive specialized buildings with efficient use of feedstuffs, or in cheap, rough straw shelters in which much food is wasted. The overall cost per calf may be the same because the differences in the costs of the buildings may be counteracted by the different costs of feedstuffs. However, the composition of the total cost differs and the behaviour of farmers when prices for reared calves fall will depend on which method of calf rearing they are using. The farmer with the cheap straw building may decide to stop production and burn his straw (i.e. with his cost structure dominated by feed costs his supply will be price sensitive) whereas the farmer with the specialized expensive buildings may decide to 'grin and bear' the low prices of his reared animals as long as they cover the cost of feedstuffs. Any margin left over will help pay for the expensive building with which he is encumbered. Because of his cost structure the output of reared calves will not be greatly affected by falling prices, and his supply will be inelastic; he may even try to expand production if prices fall (see the case of the 'reverse' supply curve given above).

Another example of the influence cost structure could be expected to have on the responsiveness of supply to falling prices is provided by farms who depend to different extents on hired labour. Many small-scale dairy farms in Europe, on which the workforce consists entirely of the members of the farmer's family, can keep on producing when prices fall (i.e. their supply of milk is price inelastic) because they have no hired labour to pay and are able to put up with lower profits simply by spending less and having a lower standard of living. On larger dairy farms with hired workers forming a bigger proportion of the labour force, wages have to be paid as a contractual obligation; reducing wages is not usually an option. Consequently a fall in the price of milk would be followed more quickly by a squeeze on profit left to the farmer and swifter actions, such as a switch to other enterprises. The general decline in the proportion of hired labour in the UK agricultural workforce which has been noted for many years would be expected to make supply of UK agricultural commodities less elastic, though in practice there may be larger influences moving in the other direction.

3. *The specific nature of the factors of production (chiefly applies to reductions in the quantity supplied).* The term **factor of production** is used to describe the goods and services which are used up in any production process. For example, the growing and sale of wheat requires: (i) land; (ii) seed; (iii) fertilizer; (iv) machinery; (v) labour to operate the machines; and (vi) someone to take the decisions of when to plough, when to plant, etc. These factors can be classified in various ways; one way is to group them into those that can only be used for a single or a few closely related lines of production – termed **specific factors** – and those which can be used in many lines – **non-specific factors**. Seed wheat is a specific factor – it can only produce wheat – whereas most types of land are relatively non-specific because a range of crops can be grown, or land can be covered with houses.

Where factors are specific to a particular form of production, that production will tend to continue with a fall in price of the product (i.e. supply will be inelastic as prices fall). This is because the factors may have no alternative use, or have only very low earnings in other uses. Earnings in an alternative use are termed **transfer earnings**, because they are what the factor would earn if it transferred. The transfer earnings of a factor are synonymous with the opportunity cost of keeping them in their present employment (see Chapter 2). An example of the effect of low transfer earnings is where a farmer is practically incapable of doing any other job except farming. His transfer earnings would be negligible so he continues farming even when prices drop to very low levels. Another example of a specific factor resulting in inelastic supply is that of the specialist calf-rearing house given earlier.

Elasticity of supply with rising and falling prices

Attention has drawn to the fact that that the supply of a commodity may have a different sensitivity to rising prices than to falling ones. This can produce the odd-looking double-supply curve in Fig. 3.15. This could well describe the situation experienced with products like pigmeat: as prices for pigmeat rise, farmers readily switch to pig production by building up their breeding stock and rapidly constructing new buildings. If prices level off and then fall back as a result of the increased supply reaching the market, farmers will be reluctant to scrap their new buildings and to slaughter their breeding animals. Supply response will be inelastic to price falls until the price drops to a very low level at which not even feed costs are covered, a point when many farmers decide they can no longer endure this level of price. Supply then contracts rapidly with a subsequent rise in the price for pigmeat. Such behaviour patterns by farmers are an important contributing factor to the cycles of low and high prices found in pigs, blackcurrants, etc.; these cycles will be described more fully later.

It is now necessary to move to the second factor determining the supply of a commodity.

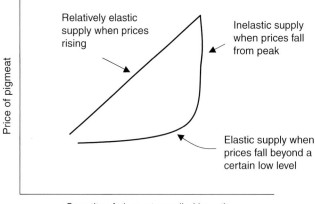

Fig. 3.15. Supply curve for pigmeat: supply under falling and rising prices (hypothetical data).

How the supply of a commodity depends on the prices of all the other commodities which are within the firm's production possibilities

$S_A = f(P_B, \ldots P_N,$ other variables constant)

Farms with arable land can usually vary the mix between different forms of cereal crop. If the price of wheat rises, but the price of barley remains the same, the supply of barley can be expected to fall because cereal farms will switch more land into growing wheat. Both crops compete for the farms' resources, and an increase in production of one crop, encouraged by a rise in its price, will necessitate a reduction in production of the other. Hence wheat and barley are termed **competing products** on the farms.

The sensitivity of the supply of barley to the price of wheat is measured by the **cross elasticity of supply** (see equation at bottom of page).

With competing products the cross elasticity of supply will be negative.

The effect on the supply curve of barley of a rise in the price of wheat is to shift it to the left (Fig. 3.16); at each price of barley farmers will supply less barley than previously because more acres will have been used for wheat and less for barley.

Some products exist whose production process is inseparable from that of some other good: an example is mutton and wool. If farmers are attracted into producing more wool by high prices, more sheepmeat will have to be produced, almost as a by-product. Such closely linked products are termed **joint products**; the cross elasticity of supply between such products is positive. The effect of a price rise of wool will be to shift the supply curve of sheepmeat to the right – more will be supplied at the same prices than before the rise in the price of wool (Fig. 3.17).

It is now appreciated that some of the joint products of agricultural production are not sold on the market but are none the less important to society. For example, the appearance of the landscape is often highly valued, particularly in relatively wealthy countries. How the countryside looks is often a by-product of the way it is used by agriculture (and forestry). Farming can also make contributions to the social life of rural areas. These environmental or social services are increasingly reasons why governments support their farmers. Their economic characteristics will be met again later in this book.

We have considered in the examples goods which are capable of being produced by farms (i.e. they lie within the farm's production possibilities). Prices of goods which are outside the farm's production possibilities will have no direct influence on the supply of those goods which are inside its possibilities. For example, the price of watches will have no influence on the supply of pigmeat, because pig farms are not equipped with the necessary machinery or worker skills to produce watches

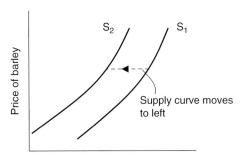

Quantity of barley supplied per time period

Fig. 3.16. The effect on the supply curve of barley of a rise in price of a competitive product (wheat).

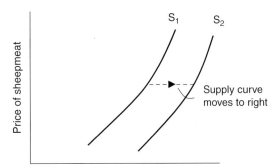

Quantity of sheepmeat supplied per time period

Fig. 3.17. The effect on the supply curve of sheepmeat of a rise in the price of a complementary product (wool).

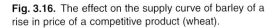

$$\text{Cross elasticity of supply} = \frac{\text{Percentage change in quantity supplied of good A}}{\text{Percentage change in price of good A}}$$

and watch factories are not equipped to rear pigs. The cross elasticity of supply between these two commodities would be zero.

How the supply of a commodity depends on the prices of factors of production

$$S_A = f(F_A, ...F_M, \text{other variables constant})$$

Factors of production are those goods and services which are used in the processes of production and include land, labour, capital assets (machinery, buildings, working capital) and business acumen. A more rigorous examination of factors of production will be considered in Chapter 6. At this stage it is sufficient to know that each must be paid (or rewarded) for the part it plays, in the form of rent, wages, interest and profit. The first three (rent, wages and interest) will need to be paid by a firm out of its revenue, and anything left over will be the profit for the firm's operators. If the cost of, say, labour rises without the price of the product increasing to compensate, the profit will be squeezed and reduced. The firm's operators may consider that this lower profit is insufficient to compensate them for the risks they are taking in production, and hence may pull out of production. The supply of the commodity reaching the market will fall and the supply curve will shift to the left.

The amount of profit that is just adequate to prevent a firm's operator (called the entrepreneur) from ceasing production is called **normal profit**. If the prices of factors fall, the profits of firms already producing will increase to greater-than-normal profits. This surplus is termed **supranormal profit**. These high profits will encourage new firms to enter this line of production, existing firms will expand, and supply will increase. The effect of a fall in the cost of factors is thus to shift the supply curve of the product to the right, implying that more will be supplied at the original product prices.

How the supply of a commodity depends on the state of technology

$$S_A = f(T, \text{other variables constant})$$

The significance of the state of technology, and advances in that technology, are of such great importance to the economics of agriculture that a major section of Chapter 5 is devoted to the subject later in this book. However, some account must be given here for the sake of completeness in examining the theory of demand and supply.

In agriculture there is a constant search for new strains of plants or breeds of animals which grow faster or yield more heavily than the existing varieties. Total costs to the farmer may rise – new cereal varieties may require heavier applications of fertilizer to realize their potential – but yields will rise too, so that the gap between total costs and total revenues, which we can loosely call profit, increases. Another type of technical advance is that which enables the same quantity of a good or service to be produced at a lower cost. An example might be the introduction of a feeding trough in a piggery that reduces the wastage of food so that the same quantity of pigs could be produced at a lower cost. Both examples mean that the average cost of producing a unit of output falls.

Technical advances that do not also result in an increase in output are comparatively rare. Indeed, many advances require an increase in output before their full benefit can be reaped. For example, a new high-powered tractor may only reduce the cost of cultivations if it is used as fully as possible.

The effect of an advance in technology is to shift the supply curve of a good to the right. Since the Second World War the supply curves of almost all crop and livestock products in European countries have moved persistently in this direction, overall by about 2–3% per annum but in some cases much faster.

Earlier in this chapter it was shown that the demand for many products of farming is relatively inelastic, giving them steeply down-sloping demand curves. An increase in supply caused by adopting the technical advance will depress the product's price so much that the amount of revenue to farmers as a group is actually reduced. (Revenue = quantity sold × price.) This was illustrated in Fig. 3.4. Nevertheless the spread of such advances cannot be halted. This is because the first farmers to adopt a new machine or variety of crop benefit most because they can increase their production while the price of their product is still high. Prices only start to fall when the bulk of 'middle-of-the-road' farmers take up the technical advance. The last, and most conservative, farmers find they are not only stuck with the old-fashioned way of producing but also with falling prices. They, too, are then forced to adapt, by which time prices may have fallen to a level at which the 'new' techniques may be no longer profitable. By this time, of course, the most progressive operators have probably moved on to a further advance in technology with yet lower costs and higher outputs. And so the treadmill of technological advance goes on turning. Clearly the secret of making a success of such a changing situation is to be able to recognize early what is a real technical advance and then to exploit it. Not all new machines or crop varieties prove themselves to be advances in technology. Some turn out to be technically interesting but of higher cost than existing farming practices.

How the supply of a commodity depends on the goals of firms

$S_A = f(G, \text{other variables constant})$

It is commonly assumed that the goal, or objective, of any firm is to maximize its profits. However, this is not always the case, and a change in the objectives of a firm can alter the supply curve independent of any change in product prices, factor costs or state of technology. Even when money profits are the only goal there may be a conflict between short-term and long-term profits. A butcher may decide to sell his beef at a lower price than he could get in a period of temporary beef scarcity in order to maintain the goodwill of his customers. Without that goodwill he might find his clients drift permanently to other butchers. Firms may be less interested in high profits than in low-risk profits. This is understandable when there is much borrowed capital; a young farmer with a large overdraft may prefer a dairy herd and a regular monthly milk cheque than to risk the overall more profitable but less reliable return from beef production. If the government intervenes to reduce the riskiness of farming, for example by guaranteeing a minimum price for agricultural products, farmers will alter their production patterns. They will specialize more and increase output as a result; they no longer need to take so many precautions to avoid risk.

Where the personal and business lives of the firm's operator are inextricably mixed, as in farming, the running of the farm may be geared as much to providing a certain lifestyle (a good environment, independence, and perhaps with sufficient spare time for attending local markets or enjoying field sports) as to money profits. This is particularly the case once a certain standard of living for the farmer has been attained. In addition, a farmer may enjoy growing a particular crop or class of livestock; some farmers regard themselves as 'sheep-men' and would be unhappy if sheep were banished from their farms in the interest of higher profits. Really, they are preferring to take part of their rewards from farming as non-monetary satisfaction. However, if these non-monetary goals are changed in favour of other non-monetary goals or in favour of money profits, the supply curves of the firm's products are bound to be affected and moved to the right or left depending on the individual circumstances. This often happens when a son takes over a farm from his father; their different set of aims and priorities will be reflected in the way the land is used and what the business produces.

The review of the factors bearing on supply is now complete. Having considered demand and supply separately, it is appropriate to examine in the next chapter how they react when they meet in markets.

Notes

[1] The letter Y is used here to denote income, rather than the more obvious I, because Y is the commonly accepted and long-established label for income when describing the workings of the entire economy (see Chapter 8 on macroeconomics).

[2] Engel's Law is commonly restricted to describe the fraction of income spent on food alone and originally took the form of an observation that the poorer the family, the greater the proportion of total expenditure it must use to procure food.

Exercise on Material in Chapter 3

Answer the questions in the spaces provided. For some you will need to draw graphs on the squared areas, though you may prefer to use a separate piece of graph paper. Answers and explanations are given in Appendix 2.

3.1. The following imaginary demand schedules for onions relate to the three different income groups in a community. Complete the total market demand schedule for the community.

Price per kilogram (£)	Demand (thousand kg/week)			
	Rich group	Middle group	Poor group	Total market demand
1.0	4	10	10	
0.9	5	12	12	
0.8	6	15	16	
0.7	6	18	23	
0.6	6	22	30	
0.5	5	26	40	
0.4	5	29	52	

3.2. Cross out the inappropriate alternatives in this statement:

If the price elasticity of demand for a good is equal to minus 3, then a 1% price fall will *raise/lower* the quantity *demanded/supplied* by 3%. The total revenue of the sellers of the good will *increase/decrease/remain unchanged* and the total expenditure of the buyers of the good will *increase/decrease/remain unchanged*.

3.3. What is the coefficient of income elasticity of demand for housing of a man who, when his income goes up from £2000 per annum to £2400 per annum, spends an extra 20% on housing?

3.4. Define Engel's Law.

Which of the following are compatible with Engel's Law?
   (a) As a country's income goes up, less is spent on essentials.
   (b) As a country's income goes up, more is spent on non-essentials.
   (c) As a country's income goes up, proportionally more is spent on non-essentials.

3.5. Cross out the inappropriate alternatives in this statement:

The elasticity of supply with respect to product price is estimated as the percentage change in quantity supplied per time period, *multiplied/divided* by the percentage change in *price/quantity* demanded. The elasticity of supply is generally *higher/lower* in the short run than in the long run, and this implies that the long-run supply curve is generally *more/less* steep

than the short-run curve. In addition, the elasticity of supply will be *higher/lower* if the producer has no alternative lines of production open to him; if the production cycle is short the elasticity of supply will be *greater/lesser* than if it is long.

3.6. List:
   (a) three events that would cause a shift in the demand curve for beef.

   (b) three events that would cause a shift of the supply curve of beef.

3.7. You are given the following information about the market for carrots:

Price (pence/kg)	Amount demanded per week (million kg)	Amount supplied per week (million kg)
9	30	62
8	35	60
7	41	57
6	45	53
5	49	49
4	53	45
3	57	41

   (a) Plot both schedules on the same graph.

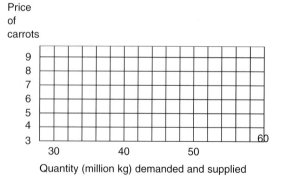

Quantity (million kg) demanded and supplied

   (b) What would be the equilibrium price?

   (c) What would be the effect of the government's fixing the market price at (i) 6p and (ii) 4p?

(d) The government decides to act as the sole buyer of carrots from producers and the sole seller of carrots to consumers. If it decides to fix the price to producers at 6p/kg –
  (i) What quantity will it need to purchase?

  (ii) What will be its total expenditure?

  (iii) What price will it need to charge consumers to clear this quantity?

  (iv) What will be its revenue from selling carrots?

  (v) What will be its net gain or loss from this transaction (ignoring administration costs)?

(e) Using the original schedules and curves, estimate the new equilibrium price after a change in consumer taste has caused demand to increase by 4 million kg at all prices.
New equilibrium price
What quantity would be sold at that price?

(f) Using the original schedules and curves, estimate the new equilibrium price after the government has placed a purchase tax of 1p/kg on carrots. (This has the effect or raising vertically the supply curve by 1p at all levels of supply.)
New equilibrium price
What quantity would be sold at that price?

(g) Using the original schedules and curves, estimate the new equilibrium price after the government has introduced a production grant of 2p/kg. (Producers after receiving the grant will be willing to supply at 7p the quantity which they were previously only willing to supply at 9p.)
New equilibrium price
What quantity would be sold at that price?

(h) What will be the effect on equilibrium price when an increase in consumers' incomes occurs if:

  (i) carrots have a negative income elasticity of demand?
  (ii) carrots have a positive income elasticity of demand?

3.8. You are given the following market demand schedule:

Price (units)	Quantity demanded (units)
10	0
8	2
6	4
4	6
2	8
0	10

(a) Plot this schedule on a graph, and measure the slope of this straight-line demand curve.
Slope

(b) What is the price elasticity of demand
  (i) from price 9.2 to price 8.8?
  (ii) from price 7.2 to price 6.8?
  (iii) from price 3.2 to price 2.8?

Is the following statement true or false?
'The price elasticity of a straight-line demand curve is constant along its whole length.' *True/False*

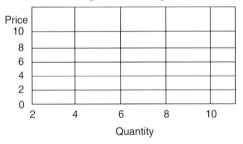

(c) The above estimates of elasticity refer to arcs of the demand curve. The elasticity at any one point can be estimated from the formula at the bottom of the page.

Using this formula, calculate the price elasticity of demand at:

  (i) price 2
  (ii) price 6
  (iii) price 8

$$\text{Price elasticity of demand at price X} = \frac{1}{\text{Slope of demand curve at price X}} \times \frac{\text{Price X}}{\text{Quality demanded at price X}}$$

Cross out the inappropriate alternatives in the following statement:

Where two demand curves of different slopes cross, price and quantity demanded at that price are the same for both curves. At the point of intersection the steeper curve possesses the *higher/lower* price elasticity of demand, and is therefore said to be the *less/more* elastic of the two.

3.9. A developing country has population of 10 million and consumes 1 million t of cereals/year at the start of a 10-year period. Assuming a population growth, over the 10-year period, of 10%, an income growth per head of 30% and an income elasticity of demand for cereals of 0.5, what is your estimate of the amount by which supply must expand if cereal prices are not to rise at the end of the decade?

4 Markets and Competition

Introduction

If people were entirely self-sufficient, producing the full range of food, clothes and other requisites from their own resources and the gifts of nature, as might a person marooned on a desert island, then a market in commodities would not exist. Once, however, complete self-sufficiency is left behind and specialization is indulged in for the greater output that it can generate, with farmers specializing in food production, builders in constructing houses, doctors practising medicine and so on, then the problem of disposing of surpluses in exchange for deficits is created. Farmers produce far more food than they and their families require and will wish to dispose of this excess in exchange for clothing, heating and other consumer goods. Similarly, doctors will not be always treating themselves and will wish to exchange their services for non-medical goods and services supplied by others.

By this process of specialization and exchange a higher living standard (implied by the ability to consume more goods and services) is enjoyed than would be possible under self-sufficiency (the formal proof of this is given in Chapter 9).

For most people specialization takes the form of selling their labour to firms in return for wages. These firms in turn specialize in the production of cars, carpets, crayons, etc. Such specialization is thus on an organized economy-wide basis, but the aim of increased output and, hence, enhanced living standards is the same. Working for the government in the armed forces or the civil service is just another form of specialization, although the products of defence or administration are less easily identifiable than is factory output.

The economy thus consists of supplies and demands of myriads of items which must be brought into contact if an exchange economy is to function. The mechanism by which supply and demand interact is called 'the market'. While it is true that traditionally these interactions between buyers and sellers often took place at specific locations – for example at cattle markets – the term is not now usually so restricted. The cereals market has widened with improvements in communications from local affairs held in 'corn exchanges' to as far as the telephone or e-mail can reach. National, international or even 'global' markets for major commodities (cereals, oil, gas, coal, shipping, etc.) are concepts easily contemplated, because the actions of buyers and sellers in, say, the USA can affect prices in Britain and throughout the world.

As we saw in the previous chapter, where a number of willing suppliers coexist with a number of willing buyers it is possible for a price to evolve at which the quantity that sellers are willing to put on the market is equal to the quantity desired from the market by the buyers. Prices can be agreed either between individuals or through an auction system. In a market for, say, store beef animals, if one farmer tries to secure a price higher than the general market level, no one will be interested, and to sell he will need to lower his asking price. Alternatively, if he asks less than the market price he will be bought up immediately, so there is no point in asking a lower price if he can sell at the going market rate. If farmers are unaware of prices elsewhere in the market, they may make transactions above or below the market price, but the better their knowledge of other transactions the smaller will be the price variations. A state of perfect knowledge would exist if all buyers and all sellers knew how much each buyer was prepared to buy and each seller was prepared to supply at each price, and only one market price would then exist.

In studying the interactions between demand and supply economists find the use of models of a number of different market arrangements of help in analysing real-world situations and in making predictions. Recall that a model tries to capture the essential elements in a real-world situation while discarding the rest in order to give greater insights into problems than would result from an approach which insisted on taking everything into account. An example is the **perfect market**, with 'perfect' being a technical description rather than a judgement of what is desirable. This model is of particular relevance to agriculture because many farm products are sold under market situations which contain features closely approaching those of the perfect market model. A perfect market exists when:

1. There are both a large number of buyers and a large number of sellers, such that the quantity bought or sold by each represents only an insignificantly small portion of the total.
2. There is a homogeneous product (i.e. the product from any one supplier is indistinguishable from that of any other – a fair example might be barley or milk).
3. There are no special factors (such as differences in transport costs) to cause any buyer to deal preferentially with any seller.
4. There are no prejudices in the minds of buyers in favour of or against any seller.
5. Buyers and sellers are free to stop and start buying from the market or supplying the market (i.e. there is unrestricted entry to or withdrawal from the market).
6. A state of 'perfect knowledge' exists.

In such a perfect market only one price would exist at any one time. Any attempts by buyers or sellers to deviate from this price would be thwarted and the price would return to its equilibrium. Both buyers and sellers would therefore be price takers in that they could not influence market price by withholding their sales or purchases and would have to accept the going price.

Functions of the Price System

In a perfect market the price system performs a number of important roles. *First*, in the short run it matches demand to supply. For example, while it is possible to buy fresh strawberries in the UK at any time of the year by importing them, during the British strawberry season the supply to the market is increased vastly, creating a new supply curve. In the short run this seasonal supply is very inelastic (strawberries must be sold irrespective of the price, or they will deteriorate). To clear the market of this abundance of fruit a fall in price is necessary; at lower prices consumers are willing to buy greater quantities and prices will fall to the level at which demand and supply are again matched. Figure 4.1

illustrates the situation. It assumes that the demand for strawberries is not affected by the season. In reality, this may not be entirely valid, and people's taste for strawberries may be somewhat greater in summer than in winter caused, perhaps, by what they see taking place at tennis tournaments (i.e. the demand curve may shift to reflect changing tastes).

Second, the price mechanism signals changes in consumer demands to producers. If consumers take, for example, an increasing liking to beef (the sort of thing that happens when incomes increase), the demand curve for beef will move to the right, resulting in higher beef prices. Beef producers will make high profits; they will attempt to expand production, and new producers will be attracted into beef production, eventually increasing the supply. New producers are attracted because existing producers will be making profits higher than those necessary to compensate them for the risks involved in beef production.

As was seen in Chapter 3, the lowest level of profit an entrepreneur finds acceptable to compensate him or her for the uncertainty involved in a particular type of enterprise is termed **normal profit**. The entrepreneur who persistently earns less than normal profits will eventually quit production. If more is earned more, this 'supra-normal' (or 'surplus', or 'excess') profit will attract other producers; their efforts will increase total supply and lower prices until all 'surplus' profit has been eliminated. Producers will then be earning only normal profits and no further new ones will set up in production.

Third, and closely linked with the previous point, the price system performs an allocative function. Productive resources, controlled by entrepreneurs, are attracted into producing those commodities which consumers demand. Changes in consumer demands result in a switching of resources to satisfy those changed demands, because they result in commodity price changes to which producers respond. If productive resources are capable of switching from product to product (say, if land is capable of rearing beef animals instead of growing cereals, and farm staff are as competent with livestock as with crops) resources will be so allocated that consumers will get the goods and services they demand, as indicated by the way they spend their money.

The *fourth* function of the price mechanism is to ensure that goods and services are produced in the most efficient manner. At this stage we can interpret

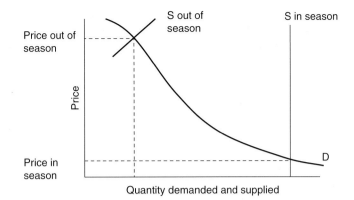

Fig. 4.1. Demand and supply for strawberries.

this as meaning at the lowest average cost. Competition between producers will allow the more efficient to undercut the lowest price at which the less efficient can operate, so the inefficient will be forced out of business. If one size of firm proves to be particularly efficient, then firms larger or smaller will be forced to adopt this size of operation or go out of business, and eventually all firms will be of this optimal size. The price system thus ensures that demands are satisfied at the lowest cost.

Imperfect Competition in Markets – Markets that are Not 'Perfect'

Although the price system plays an important role in allocating the available resources to best satisfy the demands of consumers, it is not a perfect allocator (for reasons that will become apparent). Consequently the best interest of society is not necessarily served by allowing producers to respond solely to the demand expressed by individuals, and a 'perfect market' will not necessarily produce an optimum allocation of the nation's productive resources. For example, an unhindered price mechanism would ensure that 'hard' addictive drugs would be readily available at minimum cost, being produced in the most efficient manner, whereas society judges (subjectively) that such addiction is neither good for the individual nor for the larger community. In an unhindered system environmental pollution by producers would impose heavy costs on the rest of society, and public goods (such as the defence forces or the community health system) would not be adequately provided. Government interference can give a preferable allocation. These 'external costs' and 'public goods'

will be considered in more detail in Chapter 7 when various aspects of market failure are reviewed.

At this stage we are concerned with the imperfections in the price system that occur when individual buyers or suppliers can influence prices in the market for whatever reason. On the supply side, a state of **perfect competition** is said to exist under the conditions laid down for a perfect market earlier (many producers each with an output insignificant when compared with total output, a homogeneous product, unrestricted entry and exit from the market, etc.). As soon as individual firms are in a position where they can influence price, **imperfect competition** is said to exist. A **complete monopoly**, where one firm controls all the output, is the opposite extreme to perfect competition. An example is the state-owned monopoly of the postal services to be found in many European countries. Where supply emanates from a few large firms, such as with farm fertilizers, an **oligopoly** exists. Often supply comes from a mixture of big firms and small firms, and the degree of control over price will then be related to the size of the firm.

On the demand side of the market a **complete monopsony** (note spelling) occurs when there is only one buyer. For half a century in England and Wales the Milk Marketing Board was virtually the only buyer of milk from farms in these countries and thus had considerable power in dictating the price received by farms. Similarly, an **oligopsony** is where only a few buyers exist – an example would be where there are only a few employers of a specialist skill in a locality.

The imperfections encountered in markets for farm products are not typical of those found in most non-agricultural sectors of the economy.

Many non-agricultural goods in developed countries are produced by large firms which are in monopoly positions or are part of an oligopoly, and the danger is that these firms will exploit their power over the market to charge the numerous but individually defenceless consumers more than they would have to pay under perfect competition. Farming is often faced by the opposite phenomenon: there are many producers of cereals, meat and milk, each of whom are insignificant when compared with total industrial output, but often farmers are dealing with large buyers, such as large supermarket companies which are now thought to have the power to determine the prices received by UK farmers for some commodities, especially if they decide to act together, because they buy a substantial proportion of the farming industry's output. In practice the monopsony is rarely complete.

From the general description of markets given above we turn to examine three models in more detail, first, perfect competition viewed from the levels of the individual producer and of the whole industry, secondly monopoly and thirdly monopsony.

The Individual Producer in Perfect Competition

The agricultural industry approximates to the perfect competition model. It consists of many farmers whose individual outputs form an insignificantly small proportion of total output and whose products are virtually identical. Under such conditions the individual producer cannot affect the market price for his product; the individual farmer is a 'price accepter' or 'price taker'. Furthermore, the same price per unit will be received whatever quantity the producer places on the market; it matters nothing whether the farm sells 1 t of barley, 10 t or 200 t, the price per tonne will be the same. (It is worth noting that transport costs are ignored in the model of perfect competition.)

The money received by a firm from selling its products is called its revenue. The revenue from selling a given quantity of output is termed **total revenue** (TR), while dividing total revenue by the quantity of output sold gives the **average revenue** (AR) (e.g. if the total revenue from selling 10 t of wheat is £900, the average revenue is £90/t). Average revenue is therefore the technical term for the price of the product. In the case of the farmer selling barley the total revenue received will be directly proportional to the

quantity sold. Doubling the quantity sold will double the total revenue, but the average revenue (price per tonne) will be constant with changing output. This is illustrated in Table 4.1.

Marginal revenue (MR) is the increase in total revenue to the producer gained by producing and selling one more unit of output (e.g. the MR of the fourth tonne of barley is the total revenue (TR) from 4 t minus the TR from selling 3 t).

$$MR_n = TR_n - TR_{n-1}$$

Table 4.1 and Fig. 4.2 show that each additional tonne produced increases the total revenue by the same amount (£50), which is the same as the price (average revenue, AR) of barley. To the individual producer under perfect competition, then, AR equals MR, and both are constant with increasing output. The AR schedule is also the demand schedule.

What level of output should the producer aim to produce and sell? A producer aiming to maximize profits will be seeking the largest possible difference between total revenue (TR) and total costs (TC), i.e. profit may be defined as:

Profit = TR – TC

The output at which this is achieved will generally not be the highest output possible because costs of production must also be taken into account. It is commonly experienced that, once a certain level of output has been achieved, the cost of producing additional units of output rises. An example is barley yields; each additional tonne of barley per hectare requires ever-larger applications of fertilizers, herbicides, etc. and hence the cost of each additional unit of yield is greater. The addition to total costs caused by producing the last (marginal) unit of output is called the **marginal cost** (MC) for that level of production.

Table 4.1. The revenue of a producer under perfect competition.

Tonnes of barley supplied to market	Price = average revenue (£)	Total revenue (£)	Marginal revenue (£)
1	50	50	50
2	50	100	50
3	50	150	50
4	50	200	50
5	50	250	50
6	50	300	50

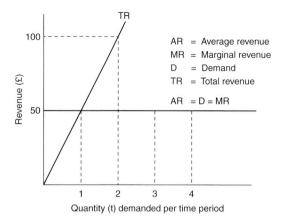

Fig. 4.2. Total, average and marginal revenues.

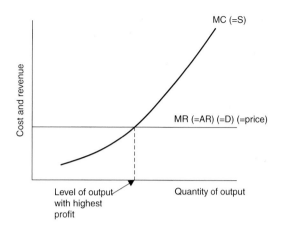

Fig. 4.3. Individual producer in perfect competition.

MC nth unit = Total cost of producing n units −
Total cost of producing n–1 units

If the addition to total revenue caused by producing, say, the fourth tonne of barley (i.e. its MR) exceeds the addition to total costs its production entails (i.e. MC), then it will pay the farmer to produce that extra unit of barley because he makes some profit from the unit. It will pay him to expand his output until the MC just reaches MR. Beyond this point, further expansion to gain the highest possible yields per hectare would reduce total profit, because each additional unit of output would cost more to produce than it brought back in extra revenue.

The highest profit will be made when MC = MR. We also know that, under perfect competition, MR = price of the product, so we can state that, under perfect competition, maximum profits will be reaped when output is adjusted so that MC = price of the product (see Fig. 4.3).

The Perfectly Competitive Industry

We move now from considering the market situation of an individual producer in a perfectly competitive industry to that of the whole industry. The supply curve for a whole industry, made up of many small producing units, can be estimated by adding together the quantities which each firm would be willing to supply at a given price, and repeating this process for a range of prices (this is known as the horizontal summation of their supply curves).

The demand curve for the whole industry is not horizontal, as was that faced by the individual firm. As has been pointed out earlier, an industry like

agriculture faces a downward-sloping demand curve for its product because, to encourage greater consumption, prices must fall. In the case of agriculture, the price fall must be relatively large. The downward-sloping demand curve cuts the industry's supply curve at an equilibrium price which the individual producers have to accept. Figure 4.4 should make this clear.

Competition between firms in a perfectly competitive industry will ensure that production is carried out in the most efficient way, as the following explanation shows. As well as his marginal costs of production described above, each producer will have a set of average costs (total costs divided by the quantity produced) which also vary with the level of output (e.g. the average cost of producing barley per tonne). Average and marginal costs relate to each other, as shown in the left-hand part of Fig. 4.4: AC is U-shaped, first falling as output is increased and then rising, and the rising MC curve cuts the AC curve at its lowest point (a relationship that will be explained in more detail in Chapter 5). To continue production, in the long run a producer must receive a price for the product (AR) so that all costs are covered, i.e. the lowest acceptable AR corresponds with the lowest AC that he can produce at.

Those inefficient firms whose lowest attainable average costs are not covered by the product's price will soon leave the industry. They are termed '**sub-marginal**' producers – and only the more efficient will be left in production (see Fig. 4.5). If one size of business happens to be more efficient than all other sizes, then firms will move to that optimum size. This will tend to increase the output of the industry,

Box 4.1. Demand and supply in relation to AR and MC.

In Fig. 4.3 the AR line has also been labelled D (for demand) and the MC curve S (for supply); some explanation of this is necessary. The AR curve, which in this example takes the form of a horizontal straight line, shows how the product price changes as the producer supplies different quantities to the market. But a demand curve also shows how much of the commodity buyers who collectively constitute the market are willing to buy at given prices (see Chapter 3). Clearly an AR curve and a demand curve are showing the same relationship approached from two directions.

The demand for barley in this example is infinitely elastic because the market will absorb very little or as much as the farmer wishes to supply, at the same price. Theoretically, if he were to lower his price an infinitely small amount below market price, demand for his product would increase by an infinitely large amount and, if he raised his price above market price, his sales would fall to zero.

Turning to the supply label, the MC curve of a firm in perfect competition is the same as its supply curve. Recall from Chapter 3 that a supply curve shows the quantities which producers are willing to put on the market at different prices. A price fall to our producer in perfect competition would mean a vertical shift downwards of the AR = MR curve, intersecting the MC curve at a lower quantity of output, and indicating that he should produce less. The MC curve thus traces the quantities which the profit-maximizing producer will produce and supply to the market at a range of prices, and this is what a supply curve shows. In Chapter 5 it will be demonstrated that MC can at first fall before the rising characteristic with increasing output, shown in Fig. 4.3, sets in. Strictly, then, the supply curve of a firm in perfect competition coincides only with the rising part of the MC curve that lies above its average variable cost (AVC; see Fig. 5.22 in Chapter 5).

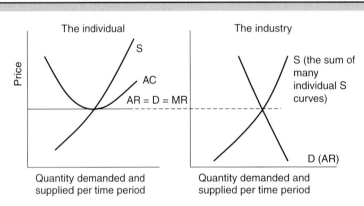

Fig. 4.4. The individual producer in a perfectly competitive industry and the whole industry.

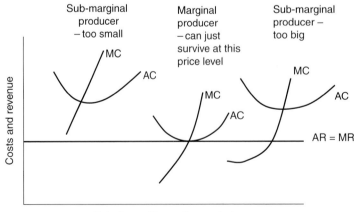

Fig. 4.5. Firms in perfect competition.

depress the market price and force those firms who refuse to adjust out of business. In the long run firms will find themselves producing at the lowest point of their average cost curves – the position indicated in Fig. 4.4 – when MR not only is equal to MC (the optimum output position for individual firms) but to AC as well. At this output each firm will be earning just normal profits (explained earlier as what is adequate to compensate its entrepreneur for the uncertainties of production, but no more).

A Firm Operating under Imperfect Competition

The next market model attempts to analyse the profit-maximizing behaviour of a monopolist in terms of the market situation he faces. Unlike the firm in a perfectly competitive industry, a firm operating under conditions other than perfect competition must face the fact that the quantity it puts on the market will have some bearing on the price it receives. For example, a farmer producing potato seed of a new and spectacularly successful variety, of which he holds the only stock (i.e. he has a complete monopoly), will know that if he only lets a little on to the market he can charge a very high price. If he wishes to sell a greater quantity, the price will have to be lower. The farmer's AR, TR and MR schedules might look like those in Table 4.2.

Notice that the MR and AR (price) schedules are no longer the same, as they were for the firm in perfect competition back in Table 4.1 and Fig. 4.2. Here MR is lower than AR at each level of supply. TR does not go up in direct proportion to quantity sold because the price per tonne has to fall. At low quantities, increasing the supply to the market

increases TR, and MR is positive but falling. Beyond a certain volume the fall in price necessary to sell the potatoes more than offsets the greater volume sold, so that TR falls; MR is negative. These relationships are illustrated in Fig. 4.6.

It can also be shown that the price elasticity of demand for the product (seed potatoes) will be greater than –1 at quantities corresponding to a positive MR and a rising TR, but will decline with increasing output; it will be –1 when MR is 0 (and TR maximal) and between –1 and 0 when MR is negative (and TR declining).

The farmer in a monopolistic situation will maximize his profits by the same process as the firm in perfect competition – by equating his MC to MR. The important difference is that, because MR and price (AR) are not the same in this

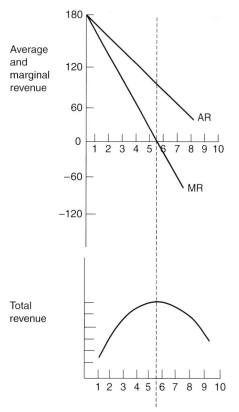

If MR is positive, then increased output will increase TR. If MR is negative, then increased output will decrease TR. The dotted vertical line marks the output corresponding to maximum TR and where MR is zero.

Fig. 4.6. Revenue of a firm in imperfect competition.

Table 4.2. Total revenue (TR), average revenue (AR) and marginal revenue (MR) of a firm in imperfect competition.

Tonnes of seed offered for sale (i)	Price needed to sell the stock (AR) (£/t) (ii)	TR (i) × (ii) (£)	MR (£)
1	180	180	180
2	160	320	140
3	140	420	100
4	120	480	60
5	100	500	20
6	80	480	−20
7	60	420	−60
8	40	320	−100

legislation exists to protect some monopolies. Copyrights on published material allow the authors of books and publishers to reap the fruits of their labours by banning cheaper pirate editions until a specified time after the author's death. Similarly, patents protect inventors of new drugs and trade marks on branded goods can be registered. From this it may be concluded that monopolies are the butt of criticism and yet may not be all bad.

Arguments against monopoly

The case against monopolies is based largely on the theory that they result in the consumer being exploited, having to pay a higher price than under perfect competition and purchasing a smaller quantity. The 'exploitation' argument follows logically if one supposes that an industry in perfect competition, producing as efficiently as possible, is taken over instantaneously by a profit-maximizing monopolist. With such a take-over the production costs of each unit (firm) would not be altered; only the individual entrepreneurs would be rolled into one, or 'concentrated'. The implications of this are shown in Fig. 4.7. On taking over, the monopolist becomes aware of the difference between AR and MR and to maximize profits reduces output until MC = MR. This difference was not apparent to the former independent producer in the perfectly competitive industry as, whatever quantity was produced, the price did not vary. (AR was the same as MR and both were constant with increasing output.) The straight replacement of many independent entrepreneurs by a monopoly is to the consumers' disadvantage, as they now have to pay a higher price and consume less as the monopolist restricts output. The monopolist does this because, although average costs rise as output is reduced, average revenue rises even more, so that profit (revenue minus costs) increases.

Under the perfect competition which existed before the take-over by the monopolist in our simple model, production took place in the most efficient way because in the competitive process high-cost inefficient firms were undercut, with the result that they earned less than normal profits and ceased production. Normal profit (that profit required to compensate entrepreneurs for the uncertainties they must face) is usually considered as a cost because, without it, entrepreneurs will stop producing, in much the same way as the wages of workers must be met or production will cease.

instance, he does not now produce that quantity where MC = price, but a smaller quantity. For this quantity the price charged will be shown by the AR curve; it is clearly above MC while, under perfect competition, price was equal to MC. The implication of the different prices and profits which occur under monopoly and perfect competition will be returned to soon. Note also that at the level of output which maximizes the monopolist's profit the price elasticity of demand of his product is greater than –1 (i.e. perhaps –2 or –3).

The monopolist has the choice of either setting a price for his potatoes and letting demand determine how much he sells, or of limiting the quantity and letting competition between buyers determine the price. He cannot fix both price and quantity sold simultaneously.

What will happen if a competitor emerges? If a rival farmer steals seed potatoes and propagates from them, or develops a closely similar strain which is a good substitute, then the pure monopoly is broken. Competition between the established and new firms will cause prices to fall and the profits made by the monopolist will be reduced. To prevent this, and to protect their position, monopolists will discourage the entry of competitors by a wide variety of means – not all of which would be open to our potato producer example. As an established producer, the monopolist will be in a good position to wage a price war against a new entrant by cutting prices temporarily to a level which the competitor cannot withstand because of his initial high costs and low sales. Another method would be for the monopolist to advertise his product heavily so that any potential competitor would need to spend a great deal on making his product known to buyers in order to take sales away from the established producer.

In the real world one frequently hears criticism of the power which monopolies are alleged to wield over defenceless consumers yet, on the other hand,

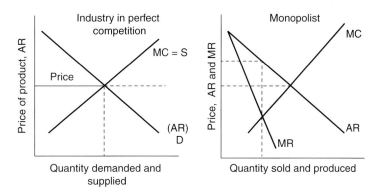

Fig. 4.7. Take-over of an industry in perfect competition by a monopolist.

Each remaining firm was producing at its minimum-cost output, so that consumers were paying the lowest prices possible with the techniques of production then known. The price received from customers (AR) just equalled average costs of production (AC including normal profit). Yet each firm was earning a profit sufficient to compensate its entrepreneur for the uncertainties of production.

If the monopolist who displaces the state of perfect competition restricts output to a level where price (AR) is above average costs of production (AC), it follows that greater rewards are being earned than would be necessary to compensate the producer for the uncertainties being faced, because this compensation element has already been taken into account when calculating AC. This profit above normal profit is termed **monopoly profit** and is shown by the labelled area in Fig. 4.8. Obviously, the monopolist will attempt to obtain this profit by restricting supply if he is profit-motivated, and consumers will be equally anxious to prevent monopoly profit being extracted from them. However, we shall now show that the case against permitting monopolies is not so cut and dried as our simple model would suggest.

Arguments for monopoly

An instantaneous process as described above is far from what normally happens in the process of monopoly creation. When a producer develops monopoly power in an industry it is usually a gradual process, based on the ability of larger firms to undercut the prices of smaller ones because of the cost advantages of large-scale production. The sources of these economies of size (technical economies, marketing economies, etc.)

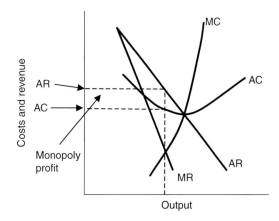

Fig. 4.8. Monopoly profit.

are discussed in Chapter 5, but as a preliminary the following are among the more obvious advantages associated with a monopolist producing on a larger scale:

1. Monopoly makes possible a more precise planning of production. The reduction in uncertainty that results from the elimination of competitors means that monopolists can utilize the best scale of production and the least-cost combination of inputs. Underused units can be shut down and output concentrated in the most efficient way, with a saving in cost. Think of the advantages that might flow from combining two competing but underused animal feedstuff factories in a county.
2. A more economical distribution of the product becomes possible. In our feedstuffs example only one team of salesmen would be needed, not two competing teams, and one fleet of lorries could be organized in such a way that wasted journeys, due

to rival lorries covering the same route, each only partly loaded, could be eliminated. With 'public utilities' (gas, electricity, telephone, etc.) the duplication of fixed equipment by competitors would most likely be undesirable and costly (several sets of telephone cables or gas pipes under each road, etc.). Under a monopoly these 'public utilities' can be distributed more cheaply.

3. A monopolist may be more able and willing to undertake research in pursuit of improved production techniques and new products because of the size of research organization that can be supported and the sure knowledge the fruits of this research can be enjoyed undisturbed. On the other hand, critics of monopolies would allege that the removal of competition dulls the spur for research.

If major economies are achievable through arranging production in larger units, then these will offset and maybe exceed the price-raising effect of a monopolist reducing output to exploit his monopoly power (see Fig. 4.9).

There are obviously both positive and negative aspects of monopolies, and instances may occur where monopoly is, as well as being to the monopolist's advantage, also in the best interests of the consumer. Other instances may exist where monopoly power is clearly exercised against the consumer, making control by society over these monopolies desirable. Some, like the monopoly held by oil-producing countries over the world supply of this valuable energy source, cannot be effectively countered except by the development of oil substitutes. Some commentators would argue that the monopoly of oil supplies by a few countries and their policy of raising prices is ultimately beneficial to the world in general, in that it slows down the depletion of this non-renewable resource, hence protecting the interests of consumers in the future.

Many countries in their histories have organized industries that have high fixed costs and that are strategically important as state-owned monopolies; for example in the UK from the 1940s to the 1980s these included the 'public utilities' (such as supplies of electricity, gas and water, and telephone services), rail transport and coal mining. Public ownership was intended to ensure economies of scale were achieved but that, at the same time, customers were not exploited. During the 1980s almost all of these (except the postal service) were transferred back to the private sector in pursuit of greater efficiency, but with restrictions of their pricing behaviour to protect the consumer (exercised by official regulators). On balance, returning public utilities to the private sector (under suitable regulation) and stimulating competition between providers appears to have resulted in an improved service and at costs that have probably been lower than would have happened under the old state monopolies.

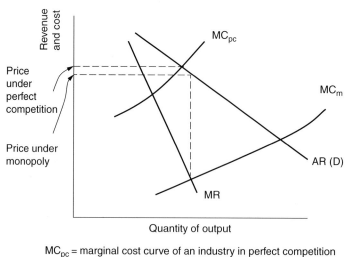

MC_{pc} = marginal cost curve of an industry in perfect competition

MC_m = marginal cost curve of an efficient monopolist

Fig. 4.9. The replacement of an industry in perfect competition by an efficient monopolist.

Where privately owned industries become dominated by a single or few firms, legislation is available to ensure that they do not exploit their monopoly power excessively (e.g. competition regulations in the EU and anti-trust legislation in the USA).

Use of Monopoly Power in Agriculture

We have already noted (Chapter 3) that market prices for agricultural commodities can be manipulated upwards by restricting the supply reaching the market. While individual farmers cannot hope to affect price by withholding their output, if they act collectively price increases are attainable. But what if they were to set up a cooperative collective marketing organization though which all farmers promised to sell their output? This could take the broad view and analyse the demand situation as any monopolist might. For example, a potato marketing body, if it were entirely under the control of potato growers, might attempt to restrict the supply of potatoes, probably by giving individual growers quotas that they must not exceed, to that level where the industry's marginal cost curve cuts the industry's marginal revenue curve. To be successful, such efforts would require the banning of imported potatoes (an effective competitor) and strict policing of growers; it would always be in the interests of an individual grower to produce more as prices rose due to the overall contraction of supply. Competition from other foods which are acceptable substitutes for potatoes (rice, bread, etc.) would also weaken a marketing cooperative's monopoly position.

When the profit-maximizing monopolist was discussed earlier, it was stated that the price elasticity of demand for the product at the 'optimum' level of output (where MR = MC) was above unity (−1). The actual figures are below −1 and for potatoes is only −0.13. A conclusion might be drawn that farmers are not acting collectively to exploit their potential monopoly power over the supply of food as much as they could. In practice, it would appear unlikely that seriously restricting supply could succeed because such attempts, as well as being undermined by dissidents in their own ranks, by imports and by substitutes, would present consumers with unpopular and politically unacceptable price rises. However, the EU's Common Agricultural Policy has given power to marketing groups in certain circumstances to reduce the volume of produce coming on to the market when prices are particularly depressed.

Discriminating monopoly

A discriminating monopoly exists where a seller can sell an identical article on two (or more) markets at different prices in order to raise its income level. Examples of this type of behaviour are exhibited by: (i) gas and electricity suppliers (different charges for domestic consumers and industry); (ii) compact disc (CD) manufacturers who re-issue old recordings at a lower price (retailers cannot buy re-issues to compete with the full-price CD until a time has elapsed and the original deleted from the catalogues); (iii) car manufacturers who try to charge different prices in different countries; and (iv) 'cheap' day returns on the railway. Also UK universities can apply the principle when they choose to charge foreign students much higher fees than those paid by home students for attending the same courses.

The object of this discriminatory activity is to increase revenue above the level attainable without discrimination. For discrimination to be effective the two markets must have different demand characteristics (otherwise the exercise is pointless) and must be capable of being isolated from each other, by legal or other actions, so that 'seepage' between markets is not allowed.

Historically, the use of a discriminating monopoly has been of particular relevance to the activities of the milk marketing cooperatives; the former Milk Marketing Board of England and Wales sold milk for liquid consumption at one price and identical milk for cheese and butter manufacture at a much lower price, with manufacturers prevented by legal restraints from reselling milk to consumers at the higher price. The demand for liquid consumption was relatively insensitive to price, giving it a steep demand curve, whereas the demand for milk for manufacturing was much more price sensitive, with a shallower slope to the curve (see Fig. 4.10).

In order to make maximum profit, the discriminating monopolist should equate MR with MC. The Milk Marketing Board realized that its MR curve was made of two components: (i) the MR from sales for liquid consumption; and (ii) the MR from sales for manufacturing. To maximize its revenue, it had to ensure that the MR in each sector of its market was the same (i.e. that the last litre sold for liquid consumption brought in the same amount of extra revenue as the last litre sold for manufacturing). This is another example of the Principle of Equimarginal Returns, first encountered in Chapter 2. In order that

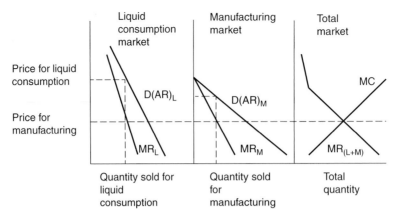

Fig. 4.10. A discriminating monopolist.

Box 4.3. Price discrimination in practice.

In practice a marketing authority may not be concerned with equating MC with MR. It is likely to face the situation of having a given quantity of product and having to dispose of it in the most profitable way. It will still achieve this by equating MRs in its several markets, although the level of this MR may not be of its own choosing.

The discriminating activities of car manufacturers, which enabled them to charge substantially different prices in different countries, have been severely eroded in recent years by the growth of the Single Market for goods and services in the EU. This has taken away many of the barriers to the international trade in cars. However, the manufacturers have not encouraged cross-frontier purchase, for example by offering different specifications and levels of guarantee in different countries and, for the UK and Irish markets, being slow to supply models with right-hand drive.

MRs in each market were equated, different prices (AR) had to be charged for liquid consumption and for manufacturing, as will be clear from Fig. 4.10.

A Monopsonist – the Single Buyer

We have considered the effect which a monopolist, either one firm or many firms having a marketing agreement with each other (often termed a 'cartel'), can have on a market. Next we must briefly mention the effect of a single large buyer – a monopsonist. Once again using a simple model we pose the question of what happens when, in a perfect market, a large number of buyers band together and buy collectively. Such an occasion would arise when a large number of independent small dairies, hitherto competing against each other for the supply of milk from farms, merge into a single buying unit. Let us call this unit the 'United Dairy'. The supply curve of milk from farms is not altered by such action – the relationship between the quantities farms are

willing to supply and the prices at which these quantities will be supplied is not affected by the nature of the buyer.

In order to call forth a larger supply of milk from farmers, 'United Dairy' will have to raise the price it offers to farmers and, as a monopsonist it becomes aware of a difference between the average cost of buying milk and its marginal cost (i.e. the increase in total costs it faces as a result of calling for more milk). This was not apparent to the former independent dairies, as each could always buy as much as it wanted at the market price. For example, if the 'United Dairy' wishes to increase its supply from farms from 10,000 l to 11,000 l, and to achieve this it has to increase the price paid to farmers from 20p/l to 25p/l, the total cost of buying milk will rise from £2000 (20p × 10,000) to £2750 (25p × 11,000). The marginal cost of the extra 1000 l is £750 so that at 75p/l it is considerably more expensive than the average 25p/l.

Armed with this knowledge the 'United Dairy' will come to a different buying decision from that which a large number of small, independent price-accepting dairies would reach. This is illustrated in Fig. 4.11. The monopsonist will buy less milk and pay a lower price than would be the case without a monopsonist. If the supply of milk were very inelastic, which is usually the case in the short run, so that the curves showing the costs of milk were steeper than shown, the quantity taken by the monopsonist would fall only a little, but the farmers would be paid a much lower price.

The demand curve (D) in this case is derived from the extra revenue the dairy can achieve by selling more milk (termed its **marginal revenue product**). The monopsonist is thus behaving in a normal profit-

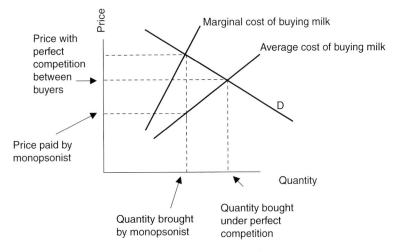

Fig. 4.11. Replacement of buyers in perfect competition by a monopsonist.

Box 4.4. Discriminating monopsonies in UK agriculture.

For much of the 20th century the UK markets for several agricultural products (including milk, sugarbeet, hops and blackcurrants) were dominated by single large buyers. That is, either complete or largely monopsonistic conditions applied. Most of these monopsonies have now been dismantled. They arose mostly through attempts by governments to regulate the markets in these products, largely for the benefit of farmers. The monopsonistic buyers also functioned as monopolists in reselling the product. Hence the former Milk Marketing Board for England and Wales was set up as a monopsonistic buyer of milk from farms that also used its power as a monopolistic seller of milk to increase the revenue which producers received by operating a discriminating monopoly. This undoubtedly also raised the prices to consumers of liquid milk. The Milk Marketing Board was eventually broken up because it contravened the EU regulations that encouraged competitive markets and thus better deals for consumers. It was replaced by a number of smaller commercial marketing companies.

As a state-sponsored, producer-orientated organization, the question of exploitation of farmers hardly arose. However, the change of orientation brought about by its replacement by private sector dairy firms has raised the question of whether they now exercise monopsony power against farmers.

If farmers could find no alternative outlet for their goods, the existence of which would undermine the monopsonist's power (such as selling milk direct to consumers) a banding together of farmers might occur to create a monopoly of supply to counter the monopsony of purchase. The result of a tussle between a monopoly and a monopsony (a situation sometimes termed **bilateral monopoly**) is not precisely predictable by economic theory – the outcome is as indeterminate as the haggling between two millionaires over a fine painting. Psychology, politics and countless intangible and unpredictable factors are heavily involved.

---

maximizing way by buying that quantity of milk at which the cost of the last litre purchased just equals the extra revenue gained from selling that litre. This principle will be further developed in Chapter 5.

Oligopoly and Oligopsony

Attention has been focused for illustrative purpose on the two extremes of the competition spectrum – a perfectly competitive industry and complete monopoly – both of which are perhaps more important to agriculture than to many other industries. However, in the real world the situation is common in which a few firms compete with each other, with single firms representing a significant proportion of the total market and thus having some degree of influence on price levels. The upstream industries that provide agriculture with inputs (fertilizer, seeds, machinery) and the downstream parts of the food chain (including food manufacturers and supermarkets) are typically dominated by a few large players. Agriculture is thus surrounded by oligopoly and oligopsony.

The economy theory of these forms of imperfect competition is complex and cannot be explored in depth here. However, some essential points are worth making:

1. The behaviour of each firm will reflect its anticipation of how other firms will respond.
2. Competition between firms by cutting prices will lead to a lower level of prices for all, as competitors try to regain their market share by also cutting prices.

3. Consequently there will be a preference for other forms of competition, such as advertising, offering additional services to customers, etc.
4. Because of the relatively small numbers of firms involved, there will be a tendency for informal arrangements to be made among them to share markets or to agree on levels of price, collusions that generally act to the disadvantage of their customers. Such arrangements ('cartels' or 'price rings'), though perhaps in contravention of laws to promote competition, may be difficult to prove because of lack of firm evidence.

Price Movements in Agriculture

It is time to return to areas of price theory which are of immediate interest to agriculture and the rural community. Often farm operators are aware of periods when prices are depressed and of others when prices are above the anticipated levels. They also have the impression that a long-run downward trend in prices for their products has been occurring. Part of the difficulty in interpreting and anticipating price movements is that the price mechanism reflects a number of changes which are different in cause but which are occurring simultaneously. This is illustrated in Fig. 4.12. Four separate types of price movement are illustrated: (i) the long-term trend; (ii) medium-period movements; (iii) seasonal movements; and (iv) daily movements. The price of a good or service at any one time will be determined by a mixture of all these. Medium-period movements can be further divided

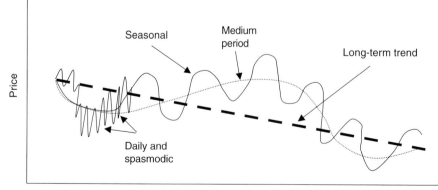

Fig. 4.12. Price movements in agriculture.

into those with prime causes lying outside the market involved and those for which the market alone is responsible.

The long-term trend

The long-term trend in agricultural prices in real terms (i.e. after having allowed for changes in the value of money) has, historically, been downwards in developed, high-income countries such as the UK and the rest of the EU. This is because the expansion in supply of agricultural products has been greater than the expansion in demand. The supply curve has shifted to the right faster than the demand curve, resulting in a fall in prices (see Fig. 4.13, which makes the point in a somewhat exaggerated way).

As time has progressed the population of the typical developed country has increased in number (increasing the market demand for food in approximately the same proportion). Households have also enjoyed higher incomes but only a small and declining proportion of this increased prosperity has been expressed in terms of extra food demand, at least as far as the basic ingredients that farms produce is concerned. (Refer to the section 'How demand for a commodity varies with income of consumers' in Chapter 3 on the low income elasticity of demand for food for a more detailed treatment.) On the other hand, rapid technological advances (new varieties, new machines, new fertilizers, etc.) have meant that increasing outputs have been possible from the existing resources of land, capital and labour in agriculture. Workers have even been able to leave farming for other industries and cultivated areas shrink, and yet farm output has increased.

The implications of this trend are that:

1. Consumers have paid a progressively lower real price for the raw material component of their food (i.e. the basic commodity excluding processing, packaging and other convenience aspects reflected in the price).
2. Lower rewards have been earned by resources in farming as product prices have declined, putting pressure on them, particularly labour, to shift to other more profitable industries (see Chapter 6 in the section 'Mobility of the factors of production and unemployment').
3. Strong political pressures have arisen from those who have an interest in agriculture (notably farmers and landowners) for the government to aid the industry by supporting prices against the long-term trend (see Chapter 10).

The downward trend in prices and the resulting pressure to transfer some of the resources in agriculture to other industries is the way the price mechanism works to reallocate the nation's productive resources among their various alternative employments in a dynamic, capitalist, free-enterprise economy.

The market for agricultural products and the market for agricultural inputs (labour, land and capital in its various forms) are linked. Rapid increases in farming's productivity, which increases supply, coupled with a fairly restricted demand for its products are bound to require an outflow of resources, signalled by falling prices both for the products and for the inputs that production uses. The price system is performing a valid economic function if it indicates the direction of change and assists in resource movements.

Medium-period booms or slumps

Medium-period booms or slumps are periods of high or low prices, the causes of which are specific and are not self-regenerating from within the producing industry.

For example, several times in the history of the UK, wars have severely restricted the supply of food from abroad. In these periods the supply curve of food to the British market, which comprises both home-grown and imported food, shifts to the left (shown in Fig. 4.14) as the foreign element is eliminated, raising prices greatly from P_1 to P_2 and causing a boom for British farmers. In the short run they are unable to increase their output much (their short

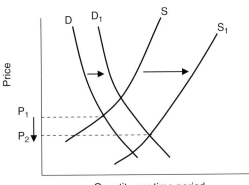

Fig. 4.13. The long-term trend of farm commodity prices in developed countries.

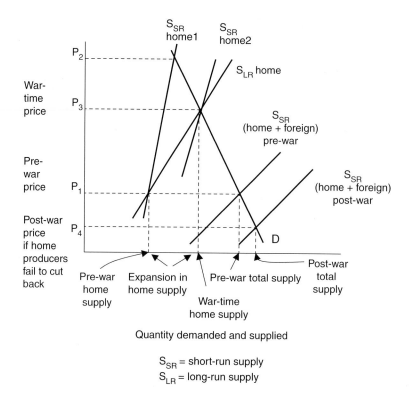

Fig. 4.14. The effect of an interruption to the foreign supply of food on prices (hypothetical).

run supply curve is steep – $S_{SR\ home1}$). But given time they have expanded their output as indicated by the less steep long-run home supply curve ($S_{LR\ home}$) and prices have settled at P_3. On cessation of hostilities and resumption of imports British farmers did not immediately contract their production to its original level (in the short run, supply is inelastic), so total supply has been greater than hitherto and prices have dropped (to P_4), causing an agricultural slump.

Slumps can also be caused by an increased availability of imports – this happened when better transport made cheap imported food available to the British market from North America and Australasia at the end of the 19th century. General depressions in the economy can also cause agricultural slumps – consumers coping with reduced incomes may only cut back their demand for food by a small amount, but if supply is inelastic the resulting price falls can be considerable.

Price cycles

The long-term price trend is part of the factor-allocation process, and some medium-period price movements can be viewed as performing shorter-term reallocation (such as attracting resources into farming to boost food production during wartime). In contrast, medium-period **cyclical price movements** which particular agricultural markets themselves foster, albeit unwittingly, do not serve an economic purpose; indeed they can lead to a more wasteful use of productive resources.

An example of a cycle is the regular changes seen in blackcurrant prices in the UK after the Second World War; prices reached a peak, fell to a trough, rose again to a peak only to fall again. Historically, each peak was separated by about 10 years. Similar but shorter cycles have been seen with pigmeat production, beef and a number of other commodities.

Cycles are most likely where:

1. A lag occurs between deciding to produce and the first goods coming on to the market. This means that in the short-run supply cannot adjust to price rises, however they are caused. With blackcurrants the delay involved might be of the order of 3 years – the time from planting to the first significant yield.

2. Entrepreneurs believe that present prices are a good indication of future prices. Present high prices encourage them to launch into preparations for later production by establishing new plantations, oblivious of the fact that similar behaviour by others will so increase the total supply that, when this comes on to the market, prices will be pushed down, causing their decision to produce more to appear much less profitable in retrospect. Once prices have fallen, producers may decide to cut their losses by ceasing production altogether and grubbing up their plantations, believing that the low prices will continue indefinitely. Their action, of course, reduces supply and hence raises prices, and the cycle starts again.

3. The demand for the good is relatively inelastic, so that variations in supply cause marked changes in price. (If demand were infinitely elastic supply changes would not cause any price variations.)

4. Random price variations caused by external influences are commonly experienced – these can initiate and reinitiate cycles. The supply of agricultural goods can be affected randomly by weather. For example, blackcurrants in the UK are particularly susceptible to late frosts, and the variation in yields that result can cause wide price fluctuations from year to year. When these spasmodic price changes are necessary to allocate an unusually light crop or to clear the market in a glut year, they serve some economic purpose. They may be counterproductive, however, in that they can initiate price cycles which serve no useful allocative function.

The stages of a typical cycle are illustrated by Fig. 4.15 and further explained in Box 4.5.

Price cycles are considered to be wasteful because of this regular migration of productive resources into and out of a line of production. Entrepreneurs never get the chance to allocate their resources in the most profitable way because of constantly changing prices. During high prices resources are attracted from other lines of production, a movement which would not occur under more stable prices, and so these other lines suffer. Nationally the most productive pattern of factor use is never reached. Taking an example from the pig cycle, farmers may erect piggeries during high pig prices rather than spend the available money on dairy equipment which could well be the better long-term investment. During low price periods capital assets may be abandoned – disused piggeries put up in times of undue optimism are not an unusual

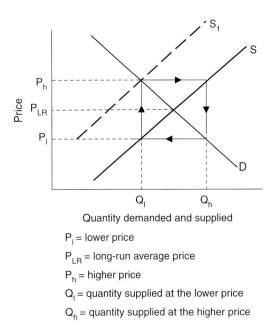

P_l = lower price

P_{LR} = long-run average price

P_h = higher price

Q_l = quantity supplied at the lower price

Q_h = quantity supplied at the higher price

Fig. 4.15. A price cycle.

Box 4.5. A price cycle (with reference to Fig. 4.15).

Price P_{LR} indicates the long-run average price. The dashed supply curve indicates an effect such as adverse weather which raises prices to P_h and initiates the price cycle. Farmers are encouraged to expand their supply up along the supply curve to Q_h, but to induce buyers to take this quantity the market price must fall to P_1. At this low price suppliers are only willing to supply Q_1 and resources are taken out of production. However, this reduced quantity results in a higher price, P_h, and farmers are again attracted to switch resources into this line of production.

The price cycle will continue if the D and S curves are of the same slope. If D is *more* elastic than S (i.e. the D curve is the less steep) successive cycles will contract and price fluctuations lessen until the long-run price is reached, although before this happens it is quite likely that the cycle will be restarted by some external influence such as adverse weather. If D is *less* elastic (steeper) than S, the cycle will continue to expand and price variations widen. The graph of such cycles looks rather like a spider's web and the theory is often termed the **Cobweb Theorem**.

Markets and Competition

sight in the UK and represent a waste of productive resources. In addition, consumers do not view gladly fluctuations in the prices of their purchases (e.g. bacon), particularly of essentials without adequate substitutes, as it causes fluctuations in their real incomes.

In that cycles cause a less-than-optimal allocation of resources, the whole economy suffers a reduction in its income. It will be in the general interest to prevent cycles becoming established. The enlightened entrepreneur will always make high profits by going against the cycle, timing his or her production so that it reaches the market when prices are approaching their peak. This may well involve investing in buildings or growing stock when product prices are low and other entrepreneurs are withdrawing from production. Wider education and information on the existence of the cycle and its causes should encourage more entrepreneurs to take counter-cyclical action, dampening out the price variations.

Another way of obviating price cycles, used particularly in the blackcurrant-growing industry, is for farmers to sell their products on a long-term price contract (in this case often for 7 years). The grower benefits, as he can plan his production with an important element in his decision-making process (product price) settled, and the buyer (principally a drinks manufacturer) is secured of supplies and knows the cost of one of the inputs – the fruit. While economic theory would suggest that the contract price might be somewhat below the long-term average non-contract price (a payment for the reduction in risk involved, see Chapter 6 'Factors of Production and their Rewards'), this can be offset by the greater efficiency which better organization based on known prices permits. In times of rapid inflation, price contracts may need an adjustment clause to take into account the changing value of money. One other feature of such contracts is that the price cycle may be aggravated for those producers who are not contracted, particularly if contracts are for specified quantities so that surpluses have to be disposed of on the open market.

A further method of dampening price cycles is for the government to set guaranteed prices for farm produce to provide an element of stability. A simple system of guarantees will not prevent peak prices, but will prevent great slumps, and the average price received will rise (see Fig. 4.16). A buffer stock system, whereby the government buys and stores goods when prices show a tendency to fall below a predetermined level, and puts them back on the market when prices show a tendency to rise above a certain level, can prevent this, although such a system demands careful judgement if permanent dumps of goods are not to accumulate.

Seasonal price variations

Seasonal variations in price particularly affect agricultural products, as farming is based on biological growth and not manufacture. Milk in winter will cost more to produce than milk in summer because of the cost of preserving and storing food

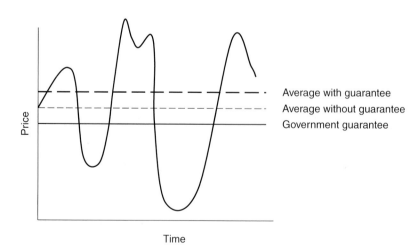

Fig. 4.16. The effect of a guaranteed price on average price over time.

for the cows. Producers require more revenue per litre to supply a given quantity – the supply curves for summer milk and winter milk are different, the winter curve lying above the summer one (see Fig. 4.17).

Another cause of seasonal variation in price is seasonal taste changes on the part of consumers. For example, there seems little reason why the supply curve of turkeys at Christmas should be any different from that in other seasons because turkey rearing is not dependent on the weather. However, the higher prices at Christmas can be explained by the extra seasonal demand shifting the demand curve to the right (see Fig. 4.17b).

Very short-term price variations

The last price variation to be considered is that relating to the very short term. Daily price variations in the free market for strawberries during the harvesting period will be necessary to clear the market daily. An increased supply, caused perhaps by a spell of fine weather which speeds ripening and makes picking easy, will tend to push prices down. Unless prices are allowed to slip downwards, goods will remain unsold, not a pleasant prospect with fruit which deteriorates rapidly. With goods which are non-perishable, daily variations in the market price to clear the market will be small or non-existent, as storage is practicable. This can be seen applying in the store animal trade where farmers buy back their own animals if they judge the price to be insufficient and try elsewhere on another day.

Exercise on Material in Chapter 4

Answer the questions in the manner indicated. Graph paper is required to complete some of the questions. Answers and explanations are given in Appendix 2.

4.1. Cross out the inappropriate alternatives:
In a perfectly competitive industry:
  (a) there are *many/few* producers
  (b) entry to the industry is *free/restricted*
  (c) each producer *can/cannot* influence market price.
In a complete monopoly:
  (a) there is *one/are several* producer(s)
  (b) entry to the industry is *free/restricted*
  (c) the producer has *considerable/no influence* on price.
To a producer in a perfectly competitive industry:
  (a) marginal revenue *is/is not* constant with varying levels of output
  (b) marginal revenue *is/is not* identical with average revenue
  (c) the demand curve which he faces is infinitely *elastic/inelastic*.
To a producer in an imperfectly competitive industry:
  (a) marginal revenue *is/is not* constant with varying levels of output
  (b) marginal revenue is normally *greater/less* than average revenue
  (c) the demand curve for his products is normally *down-sloping/up-sloping*
A perfectly competitive industry normally faces a demand curve which is *infinitely/less than infinitely* elastic.
The marginal revenue of a perfectly competitive industry normally *is constant/declines* with increasing output.

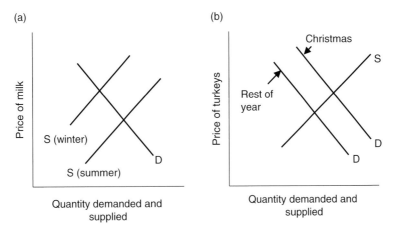

(a)   (b)

Price of milk — S (winter) — S (summer) — D — Quantity demanded and supplied

Price of turkeys — Christmas — Rest of year — S — D — D — Quantity demanded and supplied

Fig. 4.17. Seasonal variations in prices. (a) Caused by a shift of supply (S). (b) Caused by a shift of demand (D).

4.2. A village shop, operating under imperfect competition, finds that by varying the price it charges for rubber boots it can sell different quantities. It finds that it faces the following demand schedule. Derive from this the shop's total revenue and marginal revenue schedules. (NB: From a seller's standpoint, price is synonymous with average revenue.)

Price per pair (average revenue) (£)	Quantity demanded/ week	Total revenue (£)	Marginal revenue (£)
2.75	1		
2.45	2		
2.20	3		
1.90	4		
1.65	5		
1.45	6		
1.20	7		
1.00	8		
0.80	9		
0.60	10		

(a) Plot average and marginal revenue schedules (this is not essential but helpful).
(b) The shopkeeper obtains his supply of boots from the wholesaler at £1 per pair. This price does not vary with quantity. Hence the marginal cost of acquiring boots by the shopkeeper will be £1. If maximum profit is made by the shopkeeper when he equates marginal cost with marginal revenue, what price will he need to charge for boots to make maximum profits? (i.e. What is the price of boots when marginal revenue is £1?)
What quantity of boots will be sold at this price?

Using the formula given below, estimate whether the price elasticity of demand at this most profitable price is greater or less than unity.

$$E_{Dp} = 1/\text{slope} \times P/Q$$

4.3. A country has 50% of its resources employed in agricultural production, and 50% of its resources in other forms of production. Half of the national income of £100m is used in buying goods from agriculture and half in buying goods from the other forms of production. An improvement in productivity increases the national income to £110m. If the income elasticity of demand for agricultural products is 0.4 and for the products of the other industries is 1.6:

(a) What is the extra that will be spent on agricultural products?

(b) What is the extra that will be spent on other products?
(c) What will be the final income to the agricultural sector?
(d) What will be the final income to the other sector?

Use your present knowledge to cross out the inappropriate words in the following statement.

> After the rise in national income the half of the nation's resources in agriculture is earning *more/less* than the half in the other industries. Resources should be transferred *to/from* agriculture *to/from* the other industries if the returns to the nation's resources is to be equated in each sector.

4.4. Cross out the inappropriate alternatives in this statement:

> If a firm in perfect competition is attempting to maximize its profits it should produce that level of output where the addition to total costs caused by producing the last unit of output (called *average/marginal* cost of production) just equals *average/marginal* revenue. In perfect competition this will also *equal/not equal* the price of the product. A monopolist *will/will not* achieve maximum profit by pursuing the same policy, and in his case the marginal revenue at his optimum level of output will be *greater than/less than/equal to* the price which he charges for his product.

4.5. Cross out the inappropriate alternatives in this statement:

> Compared with a non-discriminating monopolist, a monopolist engaged in price discrimination will generally have a *smaller/greater/the same* revenue. This is because the discriminating monopolist equates the *average/total/marginal* revenue for his product in each of the markets between which he discriminates.

4.6. A producer of milk on an island finds that he can operate a discriminating monopoly. He can sell milk either for liquid consumption or to the island's cheese factory. By charging different prices for milk for liquid consumption and for manufacturing, the producer finds that he can increase his revenue above the level he would receive if he charged one price.

(a) What are the two chief conditions necessary before a discriminating monopoly can operate?

(b) His average revenue schedules for the two markets are shown in the table. Construct total revenue and marginal revenue schedules.

Liquid consumption				Cheese making			
Quantity sold (gal/day)	Average revenue (£)	Total revenue (£)	Marginal revenue (£)	Quantity sold (gal/day)	Average revenue (£)	Total revenue (£)	Marginal revenue (£)
1	0.45			1	0.29		
2	0.40			2	0.28		
3	0.35			3	0.27		
4	0.30			4	0.26		
5	0.25			5	0.25		
6	0.20			6	0.24		
7	0.15			7	0.23		
8	0.10			8	0.22		
9	0.5			9	0.21		
				10	0.20		
				11	0.19		
				12	0.18		
				13	0.17		
				14	0.16		
				15	0.15		
				16	0.14		
				17	0.13		
				18	0.12		
				19	0.11		
				20	0.10		

Draw graphs for each of the two markets showing average and marginal revenues. Use the same vertical price scale. (It may be helpful to place them side by side on a single large piece of graph paper – not provided.)

If the producer's marginal cost curve cuts the combined marginal revenue curve at an output of 12 gal/day, indicating that marginal revenue in each market should be £0.15 to maximize profit:

(i) What price should the producer charge for liquid milk?
(ii) How much milk will he sell for liquid consumption at this price?
(iii) What price should the producer charge the cheese factory for milk?
(iv) How much milk will he sell for cheese making at this price?
(v) What will be the total revenue?

5 Production Economics: Theory of the Firm

NOTE: Chapter 5 contains essential material for readers whose prime interest is business management. However, those with major interests in other areas may prefer to skip on first reading to the last section on 'Time and Scale of Production'. The omission of the first part of this chapter will not undermine their understanding of subsequent chapters.

Introduction – the Scope of Production Economics

Production economics studies how one sector of the economic system, firms, allocate their resources in the pursuit of given objectives. Parallels exist between this area of economic theory and the Theory of Consumer Choice already encountered in Chapter 2. When the Theory of Consumer Choice attempted to explain how individual consumers allocated their purchasing power, it was assumed that the goal of any consumer was to maximize his/her satisfaction. Similarly, in production economics the assumption is made that the goal of a firm is the maximization of profit and that they are 'in equilibrium' when this is achieved. However, the two areas of theory differ in that profit, unlike personal satisfaction (or utility), is fairly easily measured and expressed in monetary terms. In this chapter the term 'profit' is used rather loosely and in an accountancy sense to refer to the difference between revenues and whatever costs are being considered at the time.

The relative ease with which monetary profit can be measured, and hence the success or failure of a management decision assessed, has led to areas of the subject becoming highly mathematical. However, here the use of mathematics is minimal.

Production economics recognizes that in practice firms do have other objectives (or goals) besides profit. These were discussed in relation to the Theory of Supply (Chapter 3) and included the avoidance of risk, the prestige of the business and the personal preferences of the operator. For the sake of simplicity, however, production economics assumes that the sole objective of production is the maximizing of profits. Non-profit motives can be incorporated later.

For simplicity we will also assume that the firm is managed and run by a single person who also owns the business and who benefits if it makes a profit or who loses if it does not; this person is called the 'entrepreneur'. As we will see later, this simple business form is common in EU agriculture, though partnerships and other forms are also found in which the responsibilities for decisions are shared (see Chapter 6). The resources at the entrepreneur's disposal to generate profit are the firm's stock of capital assets (machinery, buildings, etc.), its raw materials, its labour force and land, and its management capacity (which in our simple model may just be the farmer). The entrepreneur must allocate and organize the other resources so that they are used in the best possible way – that is, so that the highest profit possible is generated from them.

Production economics is often called the Theory of the Firm because it attempts to explain the behaviour of profit-maximizing firms and can be adequately described by modifying the general definition of economics given in Chapter 1 of this text.

> Production economics, or the Theory of the Firm, is the study of how firms allocate their scarce resources between alternative uses in the pursuit of profit maximization.

Questions facing the entrepreneur

In the pursuit of profit an entrepreneur will have to make three major decisions: (i) what to produce; (ii) how to produce it; and (iii) how much to produce. To a large extent the resources at his or her disposal, including the entrepreneur's own management preferences and abilities, will answer these questions. For example, a farmer with land, labour, a stock of machinery and access to a certain amount of borrowing from banks, etc. will naturally not consider producing motor cars or television sets because he does not have the right types or quantities of resources required. Only products lying within the farmer's range of abilities and resources will be considered.

Given time, the farmer *could* sell his land, stock, machinery, etc. and set up as a manufacturer of, say, washing machines if he had the ability to manage such an enterprise. This illustrates the importance of time in any decisions made by firms. Generally, production economics is concerned with short or medium time periods in which the entrepreneur is unable to change radically the nature of the resources they possess and in which changes to their quantity and allocation are small (or marginal). Examples of problems studied by production economics are: (i) the best balance between crop and animal production on a farm; (ii) the cheapest combination of foodstuffs to enable a cow to give a certain yield; or (iii) the most profitable yield of cereals (it will not necessarily be the highest possible yield) at given levels of grain prices and fertilizer costs.

Major relationships studied in production economics

The choice of the best combination of enterprises (or products) on a farm with a given level of resources is termed 'optimizing the product–product relationship'. Choosing the best combination of inputs (factors of production) to produce a given level of output (e.g. the cheapest ration to produce a given milk yield) is termed 'optimizing the factor–factor relationship'. Choosing the most profitable level of fertilizer (input, or factor of production) on cereals (output, or product), given the level of the other inputs, is termed 'optimizing the factor–product relationship'. An entrepreneur in practice is faced with the problem of optimizing simultaneously these three relationships. The farmer must select the best combination of wheat and grass (and other enterprises) for his farm and at the same time choose the most profitable yields of milk and wheat, etc. to aim for while choosing the lowest cost methods to achieve these yields.

Optimizing the Factor–Product Relationship

A classic example to illustrate the optimizing of this relationship is fertilizer and crop yield. Assume that a farmer has just planted a 5 ha field with barley: how much fertilizer should he use to give the most profitable yield? Fertilizer application is the only factor which he can alter to affect yield since all the other factors are already fixed in quantity – the size of the field, the quantity of seed he has already put in the ground, his regular labour force (so that he cannot send extra men in to hold up ears that become lodged), etc.

He will know (from experience and information from other sources) that by using more fertilizer he will get higher yields. At first the response of the crop is great but beyond a certain level of application the response to additional units of fertilizer is smaller. If very high levels of fertilizer are used, its further use may actually *reduce* total yield because the crop may be poisoned, or grow so succulent that it collapses in the slightest wind. A schedule of possible levels of fertilizer use (input) and yields of barley (product) is given in Table 5.1 (all data in this chapter are hypothetical but chosen to make points about real-world phenomena). The increase in the product caused by the last (marginal) unit of fertilizer at each level of fertilizer use is called the **marginal product**, and this is also shown.

> The marginal product (MP) of input x is the increase in total product (TP) produced by the last unit of that input, all other inputs remaining constant.

Diminishing returns

Beyond a certain level of fertilizer use the marginal product declines. This is found so generally in many production contexts that it has led to the formulation of the Law of Diminishing Marginal Product (sometimes called the Law of Diminishing Returns).

Box 5.1. The simple production function.

The relationship which output (barley yield) bears to the quantity of input (fertilizer) can be described mathematically, and this mathematical relationship is called a **production function**. In our example the production function in its simplest symbolic form might be written as follows, with 'f' standing for 'is a function of' or 'depends on':

Barley yield = f (fertilizer application, all other inputs held constant)

Table 5.1. The yields of barley achieved with a range of fertilizer usages on a fixed quantity of other factors of production.

Units of fertilizer per hectare	Yield of barley (units per hectare), i.e. total product	Marginal product	Average product (nearest whole number)
0	0	–	0
1	6	6	6
2	18	12	9
3	40	22	13
4	76	36	19
5	106	30	21
6	130	24	22
7	150	20	21
8	160	10	20
9	166	6	18
10	170	4	17
11	172	2	16
12	170	–2	14
13	166	–4	13
14	160	–6	11
15	152	–8	10

Law of Diminishing Returns: Beyond a certain level of output, when successive units of a variable input are added to a fixed quantity of other inputs, the addition to total product caused by each successive unit of variable input declines.

Other examples in farming of diminishing returns can be seen in the overfeeding of animals or in stretching the use of machines beyond their designed capacity such as might occur when an increased number of cows is put through a small one-man milking parlour.

This law is very similar to the Law of Diminishing Marginal Utility experienced in the Theory of Consumer Choice (Chapter 2). As with that law, diminishing returns (falling marginal products) do not necessarily start at once. Table 5.1 shows that increasing returns accompany the first few units of fertilizer, but this is soon reversed. It is obvious that, in the real world, diminishing returns do occur. If they did not, it would be possible to grow the whole world's food requirements on 1 ha of land.

Table 5.1 also shows the **average product** of the fertilizer – this is the total product (yield) at each level of fertilizer use divided by the total quantity of fertilizer needed to produce it. Notice that average product rises, then falls. When 6 units of fertilizer are used, average product at 22 units of barley per unit of fertilizer is highest. From a *technical* viewpoint, this is the most efficient level of fertilizer usage to aim for (output per unit is greatest), but it may well not be the level at which most profit is made. *Economic* efficiency is concerned with this *most profitable* level, so economic and technical efficiency mean different things.

Total product (TP), marginal product (MP) and average product (AP) at different levels of variable input (fertilizer) are shown in graphical form in Fig. 5.1. Note that:

1. The point where MP starts declining (called the point of maximum marginal product, or point of diminishing returns) corresponds to the point of inflection of the S-shaped TP curve (i.e. where with increasing units of fertilizer input the TP curve stops bending left and starts bending right). This is easily understood once it is pointed out that the MP is the same as the *slope* of the TP curve, and at the point of inflection the slope of the TP curve is at its greatest.
2. MP falls to zero when TP is at its highest because the slope of the TP curve at this point is zero. Further units of fertilizer *reduce* total barley yields; the TP curve takes on a negative slope and MP becomes negative.
3. AP at each level of input use is numerically the same as the slope of a line drawn from the origin to the TP curve. The greatest slope this line can have is where it is tangential to the TP curve, and hence AP will be at its highest at this level.
4. The falling MP curve cuts the AP curve when average product is at its peak, so that at this level of fertilizer usage MP and AP are equal. Obviously the line drawn to the TP curve from the origin (which gives AP) when it is tangential to the TP curve has the same slope as the TP curve (which gives MP).

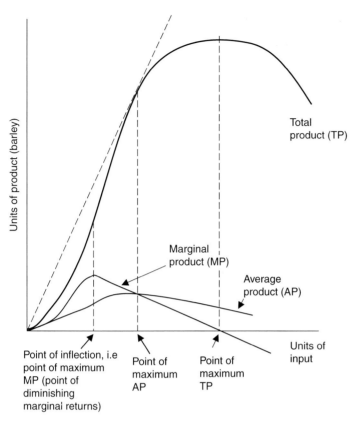

Point of inflection, i.e
point of maximum
MP (point of
diminishing
marginal returns)

Point of
maximum
AP

Point of
maximum
TP

Fig. 5.1. Barley yields and fertilizer applications, all other inputs held constant.

With higher levels of fertilizer than are shown on the graph, TP and AP can be expected to decline further until they again meet at zero yield, somewhere beyond the right-hand edge of the figure.

Figure 5.1 is a generalized case of the relationship between quantities of one variable input and the quantity of product; it contains both increasing and diminishing returns and negative marginal products. In reality, factor–product relationships could be found from which increasing returns and/or negative marginal products were absent, implying differently shaped TP curves, but for illustrative purposes it is convenient to use a model containing all these features.

What is the most profitable level of fertilizer to use?

The most profitable level of fertilizer use (in other words, optimizing the factor–product relationship) can be worked out from a diagram similar to Fig. 5.1. However, to do this it is necessary to know both the price the producer can get for a unit of his product

(in our case, barley) and the cost per unit of the variable factor of production, fertilizer. We also initially assume that the producer operates under perfect competition, not an unreasonable assumption in the case of farming, so that however much barley is produced, the price per unit obtained will always be the same and, however much fertilizer is bought, the cost per unit will not change. It is possible to incorporate complications concerning falling product prices with greater outputs and discounts for bulk purchasing, etc., but we will avoid them at present.

Product and value product

If the price of the product, barley, is known, then instead of expressing the barley produced by different levels of fertilizer in physical units, we can, by multiplying by the market price, express the product in terms of its *value*.

Product in physical units × Price obtained per unit = Value product

If, for convenience, we assume that the price barley fetches is £1 per unit, then the figures appearing in Table 5.1 for TP, MP and AP at different levels of input (fertilizer) use can be simply re-labelled **total value product** (TVP), **marginal value product** (MVP) and **average value product** (AVP). These are reproduced in Table 5.2.

From Table 5.2 it can be seen that, for example, when 9 units of fertilizer are used, £166 worth of barley is produced (TVP); the last (9th) unit of fertilizer produces £6-worth of barley (MVP) and the average value of barley produced per unit of fertilizer (AVP) is £18 at this level of fertilizer input.

Figure 5.2 shows the TVP, MVP and AVP in graphical form. The curves are identical to Fig. 5.1 except that the vertical axis is labelled in money terms (value product) and TP is changed to TVP, etc. The vertical axis is also labelled 'costs' and Fig. 5.2 contains two lines, TFC and MFC, which were not in Fig. 5.1. **Total factor cost** (TFC)[1] is the total cost of the variable factor, fertilizer, which we are considering at every level of fertilizer use. It is calculated at each level of fertilizer by the following:

Number of units of fertilizer × Cost per unit of fertilizer = TFC

Because cost per unit of fertilizer is constant, when TFC is plotted against the quantity of fertilizer used the result is a straight line. **Marginal factor cost** (MFC) is the cost to the producer of his last unit of fertilizer or, put more precisely, the increase in the cost of fertilizer resulting from using one additional unit. Because the cost of the fourth unit of fertilizer is the same as the cost of the tenth or any other unit of fertilizer (i.e. MFC is constant for all levels of fertilizer use) the MFC curve is a straight line parallel to the horizontal axis.

The most profitable level of input

The 'total' approach

The biggest profit will be made when the total value of what is produced (TVP) exceeds by the largest difference the total cost of the factor required to produce it (TFC). In Fig. 5.2 there are two ranges of fertilizer input, labelled 'Loss', where TFC (the cost of fertilizer) *exceeds* TVP (the value of barley produced). Obviously no rational farmer would use such loss-making levels of fertilizer. There is, however, a range of fertilizer usage between these two Loss areas where TVP is greater than TFC, i.e. a profit will be made. The farmer has to decide where in this range is the *most* profitable level of fertilizer to use.

> **Box 5.2. Alternative terms.**
>
> Some texts prefer to use the term 'revenue' rather than 'value' in this context. Total value product becomes total revenue product and so on.

Table 5.2. Value of barley produced from a range of fertilizer uses.

Units of fertilizer per hectare	Total value product	Marginal value product	Average value product
0	0	–	0
1	6	6	6
2	18	12	9
3	40	22	13
4	76	36	19
5	106	30	21
6	130	24	22
7	150	20	21
8	160	10	20
9	166	6	18
10	170	4	17
11	172	2	16
12	170	-2	14
13	166	-4	13
14	160	-6	11
15	152	-8	10

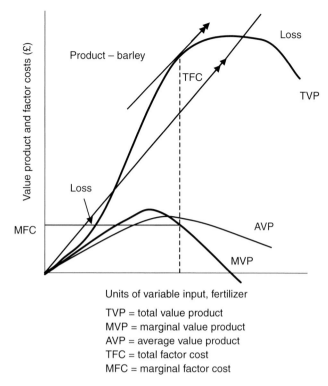

TVP = total value product
MVP = marginal value product
AVP = average value product
TFC = total factor cost
MFC = marginal factor cost

Fig. 5.2. Cost of a variable factor and value of the product at different levels of input of the variable factor.

As profit is the difference between TVP and TFC, the greatest profit will be made where their difference is greatest. On the graph in Fig. 5.2 this means where the *vertical* distance between the two curves is largest. (This point can be found by drawing a line parallel to TFC and tangential to TVP. The slopes of the TVP and TFC curves will be the same.) This corresponds to the optimum, or most profitable, level of fertilizer to use. Note that the most profitable level of production is not the level at which TVP is greatest, i.e. where the value of barley produced is highest and where yields are greatest. This is because, to achieve these high yields, disproportionately large (and hence costly) quantities of fertilizer are needed because of diminishing returns.

The 'marginal' approach

The most profitable level of input to use can also be calculated using the cost of additional units of fertilizer (MFC) and the increase in the total value of the product (MVP) that is produced by this additional input. Where the value of the extra yield of grain (MVP) is greater than the cost of the last unit of fertilizer (MFC) used to produce it, the farmer will make a profit on that grain. If he is aiming at maximum profit he will consider using more and more fertilizer until the point is reached where the value of the extra yield of grain (MVP) has dropped because of diminishing returns to a level where it only just offsets the cost of the last unit of fertilizer used to produce it (MFC). At this point MFC = MVP. Maximum profit is made. Note that at this optimum use of fertilizer MVP is falling and is below AVP. If more fertilizer were to be used the value of the extra grain (MVP) would be less than the cost of the unit of fertilizer used to produce it, and the farmer's total profit would start to fall.

From Fig. 5.2 it can be seen that the level of fertilizer input where MFC = MVP coincides with the level which was decided as the most profitable (optimum) when TFC and TVP were considered. The two approaches give the same result. In real-world situations the marginal approach is often the more applicable – farmers are concerned with the returns they can expect from using *extra* fertilizer, giving *extra* food to cows, etc. in pursuit of higher profits.

Changes in costs and prices

The optimum quantity of a variable factor of production (input) will depend on: (i) the price of the factor; (ii) the price of the product; and (iii) the nature of the production function (TP curve). What is an optimal quantity of input under one set of conditions may well not be optimal if prices change, or the production function shifts.

If the price of the variable factor (fertilizer in our example) fell, the slope of the TFC line in Fig. 5.2 would become less steep and the MFC line would fall. This would result in the MFC and MVP curves intersecting at a point corresponding to higher levels of fertilizer input, and the maximum vertical distance between TFC and TVP would also be shifted further to the right. This is shown in Fig. 5.3 where it can be seen that, to achieve maximum profit, when the price of the factor drops, more is used (an increase from OF_1 to OF_2).

> ### Box 5.3. The MVP = MFC rule.
>
> MVP is the same as the slope of the TVP curve. Similarly, MFC is the slope of the TFC. Hence if the greatest level of profit indicated by the maximum vertical difference between MVP and TFC is when their slopes are the same, then MVP must equal MFC at this maximum profit position.

The effect of a rise in the price of the product is to alter the level of all the value product curves in Fig. 5.3. It is as if the curves were all stretched upwards. (This is shown in Fig. 5.4.) A similar effect is caused when a new higher yielding variety of barley is introduced which yields greater weights of grain than the old at the same levels of fertilizer use. The value product curves can be read as physical output curves simply by dividing by the price of the product. The greatest profit level now occurs at a higher level of fertilizer usage.

How many optima are there?

When the price of the factor or of the product changes, the optimum usage of fertilizer changes too. For every cost/price situation there will be an optimum quantity of variable factor to use. Furthermore, in agriculture the weather has the effect of lowering the production function in a poor year and raising it in a good year and this will also shift the optimum quantity of fertilizer. Before deciding on the quantity of an input to use, a farmer – or entrepreneur – has to weigh in his mind not only the present cost of the input but also what price he will get for his product and the likelihood of shifts in his production function that are beyond his control. It is small wonder that, in view of the many unknowns, guesses and rules of thumb are used by farmers in place of precise estimates of what will be optimal.

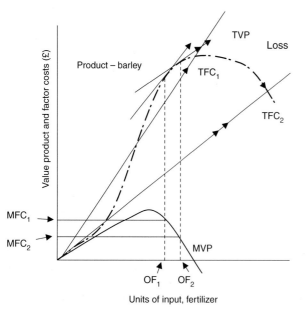

Fig. 5.3. Changes in cost of inputs.

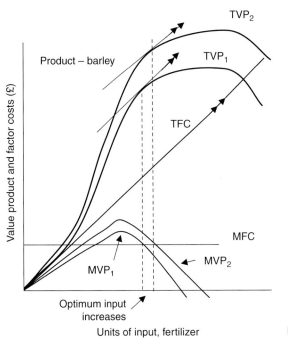

Fig. 5.4. Change in the price of the product.

An alternative approach to optimizing the factor–product relationship

So far we have been answering the question 'How much of one variable factor (input) should I use to generate most profit, all other factors being fixed?' An alternative approach is to ask 'What level of *output* of a product should I aim for to make most profit?' Situations in which such questions might be asked would be in deciding the best yields to aim for with animals or crops.

For an example we will take the problem of deciding on the most profitable level of output of eggs from poultry from the range of performance which can be arrived at by altering the feeding level. Note that all other factors of production (labour, housing, etc.) are assumed to be held constant – only the quantity of food is varied to give the different levels of output per bird. Consequently, food cost is the only variable cost and all other costs are assumed to be fixed, not varying according to the quantity of eggs produced. For the time being we will ignore these fixed costs and simply attempt to answer the question: 'Given that we already have a fixed quantity of buildings and equipment, fully stocked with birds, etc., with food being the only factor which we can vary, and given that we know the price of

eggs and the cost of food, what is the most profitable level of performance to aim for?'

The costs of the food required to produce a range of egg-laying performances are shown in Table 5.3. In addition to the total food cost column (which should also be labelled **total variable cost**, or TVC, because food is here the only variable input) there are columns for the average food cost per egg (which similarly could be labelled **average variable cost**, AVC) and the **marginal cost** of production. While the derivation of the average food cost column is obvious, being the total food cost divided by the number of eggs produced, the marginal cost figures require an explanation.

> Marginal cost (MC) of production is the addition to total costs caused by producing the last (or marginal) unit of output.

Note that the MC falls first with increasing output and then increases. This is closely linked with increasing and diminishing returns to the input, food, discussed earlier. MC eventually rises because perhaps the birds are not genetically capable of assimilating larger quantities of food efficiently and wastage occurs. Like MC, AVC also falls and then rises although the point of lowest AVC is not at the same output as the lowest marginal cost. This is easily seen when the figures are plotted

Table 5.3. Costs of food in the production of eggs at different levels of output per hen (all other inputs are assumed to be constant).

Output (eggs/hen/year)	Total food cost (TVC)[a]	Average food cost (AVC)[b]	Marginal cost
49	80	–	–
50	81	1.6	1
99	109.5	–	–
100	110	1.1	0.5
149	149	–	–
150	150	1.0	1
199	250	–	–
200	253	1.25	3
249	420	–	–
250	427	1.7	7
299	700	–	–
300	713	2.3	13

[a]TVC, Total variable cost.
[b]AVC, Average variable cost.

(Fig. 5.5). Note that, while we have hitherto placed units of *input* on the horizontal axis of graphs relating input to output, we now reverse the labels, and units of *output* appear on the horizontal axis.

The optimum level of output

To calculate the most profitable level of egg production two further lines must be drawn relating to total revenue (TR) and marginal revenue (MR). On the TVC curve the total revenue line must be drawn; this shows the revenue which is brought in by the various levels of egg production and is calculated as follows:

Total revenue = Number of eggs produced × Price received for eggs

It is shown in Fig. 5.5(a) as a straight line because we assume that eggs are sold on a perfectly competitive market. The most profitable level of production is where TR exceeds TVC (total food cost in this example) by the greatest vertical distance. (This point can be found by drawing a tangent to the TVC curve parallel with the TVR line.)

Marginal revenue (MR) is the increase in total revenue caused when output is increased by 1 unit.

In other words, MR is the extra revenue brought in by the last egg produced. Under conditions of perfect competition the producer will get the same price for his eggs irrespective of how many he sells, so the MR of the fourth egg will be the same as the MR of the 400th egg. MR is constant with different levels of output.

MR has been superimposed on the MC curve in Fig. 5.5(b). Is obviously in the interest of a farmer's profit for him to keep increasing his output of eggs by feeding his birds more as long as the extra revenue produced is greater than the extra cost incurred, i.e. while MR is greater than MC. Profit will be *reduced* if production is expanded beyond where MR = MC because the additional costs exceed the extra revenue. Maximum profit is hence made at the output where MC = MR. This corresponds to the same optimum output shown in Fig. 5.5(a).

Price and cost changes

As has been shown earlier, the optimum solution of the factor–product problem is only optimal for a particular set of price levels. A fall in product price will shift the TR and MR lines. The MR line will fall, intersecting the MC line at a lower level of output. If the product price rose, MR would rise and intersect MC at a higher level of output. Hence, for a firm aiming to maximize its profits in perfect competition, its MC curve is the same as its supply curve. Changes in the price of the variable input, food in our example, will shift the TVC, AVC and MC curves and alter the optimal level of output accordingly.

Reconciling the methods of optimizing the factor–product relationship

Two ways of optimizing the factor–product relationship have been examined. The first attempted to find the optimum level of one variable *input*.

(a) Total variable cost (TVC)

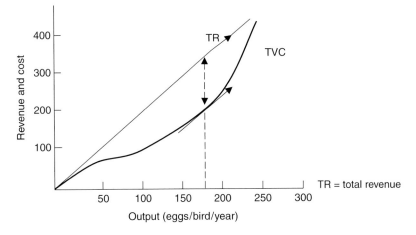

(b) Average variable cost (AVC) and marginal cost (MC)

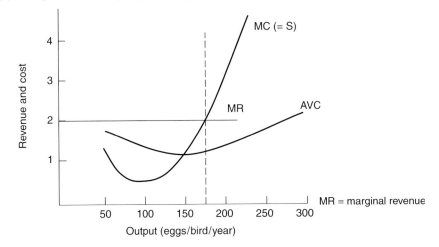

Fig. 5.5. Costs of production of eggs and optimum level of output (food is assumed to be the only variable input).

Box 5.4. Supply and marginal costs.

For a firm in perfect competition the supply curve is the same as the MC curve because it also shows the quantities the firm is willing to produce (and hence supply to the market) at different levels of product price. This point was also made in Chapter 4 when referring to Fig. 4.3. However the MC curve is only the supply curve when price is greater than average variable cost (see later) because no production will occur unless these costs are covered. In this context price must be greater than the lowest point on the average cost curve of feed before any production will occur.

The second attempted to find the optimum level of *output*. If the same one product/one variable input example had been used throughout, it would have been found that the results were perfectly compatible. By using the optimum quantity of fertilizer on barley found in the first approach, we should have produced that quantity of barley which we would have shown to be the optimum quantity to produce by the second approach. In producing the optimum quantity of eggs, our second example, we should have used the optimum quantity of food. This is because both approaches to each problem are based on the same production function and the same prices of inputs and outputs.

Optimizing the Factor–Factor Relationship

Up to this point we have been looking at the most profitable level of one variable input (or factor of production), with all other factors of production held constant, or 'fixed'. Optimizing the factor–factor relationship involves choosing the most profitable combination of inputs *when more than one input is variable*. For example, if a cow can be fed on different combinations of silage and barley to yield, say 5000 l/year, optimizing the factor–factor relationship involves devising the cheapest (or least cost) combination of the two feeds.

Table 5.4 shows the quantities of milk that can be produced from a cow by different combinations of quantities of silage and barley, all other factors of production being held constant. The production function describing the relationship between milk

yield and silage and barley usage might be written in its simplest, symbolic form as follows:

Milk yield = f(silage usage, barley usage,
all other inputs held constant)

Notice that in this example each factor is subject to diminishing returns: for instance, choose a level of silage usage (say 4 t) and go up the column of milk yields produced by 1, 2, 3, 4 t of barley. The increase in yield produced by each extra tonne of barley progressively diminishes. The same diminishing returns can be seen if a level of barley usage is selected and the increase in yield with increasing silage usage is examined.

The figures in this table are presented in graphical form in Fig. 5.6 in which a line has been drawn connecting all those combinations of silage and barley which produce 5000 l of milk/year. It is called the 5000 l **isoquant**. Other lines could have

Table 5.4. Yields of milk (l/year) produced from combinations of two variable inputs, silage and barley.

Barley (t/year)	Silage (t/year)					
	2	4	6	8	10	12
6	4500	**5000**	5375	5500	5550	5575
5	4125	4875	5250	5450	5530	5560
4	3500	4500	**5000**	5375	5500	5550
3	2500	3750	4500	**5000**	5375	5500
2	500	2000	2500	3750	4500	**5000**
1	–	–	–	2000	2500	3750

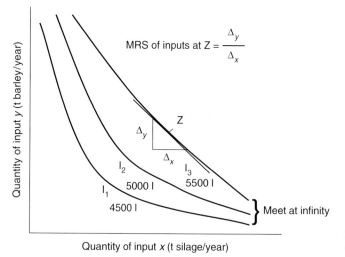

Fig. 5.6. Marginal rate of substitution (MRS) of inputs.

been drawn corresponding with yields of 4500 l, 5500 l, etc.

> An isoquant is a line connecting the various combinations of two inputs which produce the same level of output.

Isoquants have a number of properties:

1. They are normally convex to the origin. This means that isoquants are normally steeper near the vertical axis and less steep further away. In this example, situations where all barley and no silage is used or vice versa are not technically possible, so the isoquant never touches the axis (see Fig. 5.6).
2. As long as each input and the product are easily divisible into small units, a new isoquant can be drawn between any two existing isoquants. Isoquants further from the origin imply higher levels of output.
3. The number of units of barley which must be given up to keep yield the same if silage usage is stepped up by 1 unit is the marginal rate of substitution (MRS) of inputs. (It is of course the same as the number of *extra* units of barley which must be fed to maintain yield if one less unit of silage is fed. The MRS between two inputs is also sometimes called the 'rate of technical substitution'.) The MRS is a technical relationship which, in the example chosen, is determined by the biological processes of the dairy animals (see equation at bottom of page).

The MRS changes as we move along the horizontal axis away from the origin. When little silage but a lot of barley is fed, using one extra unit of silage will result in a large saving in barley usage for a given level of milk yield, but if the balance is already heavily weighted towards silage, using one further unit will result in little less barley usage. The MRS at any point is numerically the same as the slope of the isoquant (see Fig. 5.6); it should be accompanied by a negative sign since the change in quantity of one of the inputs is negative.

Notice that these characteristics of isoquants are very similar to those of the indifference curves encountered in the Theory of Consumer Choice, Chapter 2; for an individual, a curve corresponded to a particular level of satisfaction given by combinations of two goods, whereas the isoquant corresponds to a particular level of output for a firm.

Figure 5.7 shows the straight-line isoquant of two inputs which are perfect substitutes for each other. Apparently oil and alcohol are perfect substitutes in the production of synthetic rubber and all of one, or all of the other, or combinations of the two can be used. Furthermore, at any given level of output, 1 unit of one input can always be replaced by the same number of units of the other, irrespective of the combination in use. This means that the MRS is constant along the entire isoquant; as the MRS is numerically the same as the slope of the isoquant, this also does not alter, producing the straight line.

Figure 5.8 shows the opposite case to Fig. 5.7. Only one combination of the two inputs can be used to produce a given level of output, and so the isoquant takes the form of a single point. An example might be the production of a specific alloy from two elemental metals.

Cost and the optimum combination of inputs

Figure 5.9 is the same as Fig. 5.6 but with two additional lines – **iso-cost** lines labelled '£50' and '£60'. The £50 iso-cost line joins all the combinations of silage and barley, the two inputs, which can be bought for the same £50 outlay by the producer. (It is equivalent to the budget line encountered in the Theory of Consumer Choice.) Similarly the £60 iso-cost line joins all those combinations which cost the producer £60.

> An iso-cost line joins the various combinations of quantities of two variable inputs which can be purchased for the same total cost.

Note that there are two combinations of silage and barley which can be bought for £60 and produce 5000 l (where the £60 iso-cost line and 5000 l isoquant cut), but only one combination that will allow 5000 l to be produced for £50. This is where the £50 iso-cost line and the 5000 l isoquant are tangential. This is the optimum combination of silage and barley for the producer to use to produce 5000 l, since it is the least-cost way of achieving that output.

At the point of tangency the slopes of the isoquant and the iso-cost lines are the same. It has already been established that the slope of the isoquant at any point equals the marginal rate of substitution of inputs (MRS) at that point (see Fig. 5.6 to refresh understanding of the MRS).

$$MRS = \frac{\text{Change in number of units of input on vertical axis required to keep output constant}}{\text{Change in 1 unit of input on horizontal axis}}$$

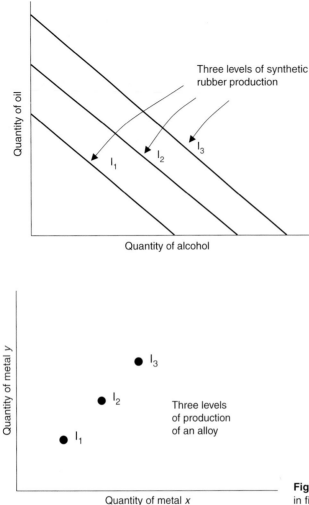

Fig. 5.7. Isoquants where inputs are perfect substitutes.

Three levels of synthetic rubber production

Quantity of oil

Quantity of alcohol

Quantity of metal y

I_3

I_2

I_1

Three levels of production of an alloy

Quantity of metal x

Fig. 5.8. Isoquants where two inputs can only be used in fixed proportions.

At the optimum combination, because the slopes of the isoquant and iso-cost lines are the same, the MRS will numerically equal the *inverse* of the ratio of the prices of the two inputs. That is, if the price of barley is double that of silage per tonne, then at the optimum the MRS will be 0.5.

$$\text{Slope of the isoquant} = \text{MRS} = \frac{\text{Change in input } y}{\text{Change in input } x}$$

$$\text{Slope of the iso-cost line} = \frac{(-)\text{Quantity of } y \text{ bought for £50}}{\text{Quantity of } x \text{ bought for £50}}$$

$$= \frac{(-)\text{Price of } x}{\text{Price of } y}$$

Therefore, at the combination of inputs

$$\text{MRS} = \frac{(-)\text{Change in input } y}{\text{Change in input } x} = \frac{(-)\text{Price of } x}{\text{Price of } y}$$

Furthermore, it can be shown that, when implying infinitely small changes in x and y: at the optimum combination of two variable inputs the saving in cost resulting from reducing the quantity of one input by an infinitely small amount will be exactly offset by the extra cost of using more of the other factor necessary to maintain the level of production. This statement can be extended to any number of variable factors. When implying more than infinitely small changes in x and y, the fall in costs resulting from using less of one input will be more than offset by the extra cost of the other input, i.e. the combined costs of both inputs to produce a given level or output will rise.

Changes in the price of inputs: price substitution effect

Returning to the silage and barley example, if the price of one input, say barley, fell, a greater quantity could be purchased for £50. Thus the slope of the iso-cost line would shift, as shown in Fig. 5.10. For £50 the producer can now produce 5500 l. If the farmer still wished to produce only 5000 l, the new least-cost diet is found by shifting the iso-cost line,

keeping its slope constant, until it is tangential to the 5000 l isoquant. Note that in this new least-cost diet, barley will have been substituted for silage, with less silage but more barley in the optimum mix than under the previous set of prices.

Figure 5.11 shows the effect that a change in relative prices can have on the choice of inputs where straight line isoquants are involved. In the first situation (iso-cost line 'a') all of input y is used and none of x. When relative prices change (producing iso-cost line 'b') no y and all x is used – a small change in relative price will cause a large change in the optimum choice of inputs. Only when the iso-cost line and isoquant are of identical slopes will a combination

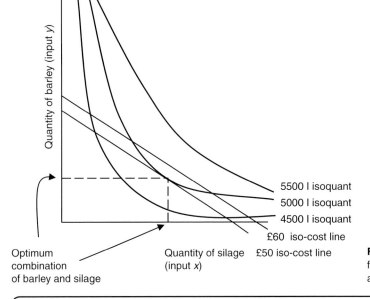

5500 l isoquant
5000 l isoquant
4500 l isoquant
£60 iso-cost line
Quantity of silage £50 iso-cost line
(input x)

Optimum combination of barley and silage

Fig. 5.9. Isoquants and an iso-cost line for milk production with silage and barley as variable inputs.

Box 5.5. Marginal value products at the optimum.

It can also be shown that at the optimum combination of two variable inputs, the ratio of their prices will equal the ratio of their marginal value products, i.e. the addition to total output resulting from the last unit of each input.

$$\frac{\text{Marginal Product of } x}{\text{Marginal Product of } y} = \frac{\text{Price of } x}{\text{Price of } y}$$

In rearranged form this becomes

$$\frac{\text{MP } x}{\text{Price of } x} = \frac{\text{MP } y}{\text{Price of } y}$$

This can be extended to greater numbers of inputs. At the least-cost combination the addition to total output resulting from the marginal unit of each variable input will be proportional to each input's price. For example, if one input has a price per unit double that of another, at the least-cost optimum the value of extra output generated by the marginal unit will be double.

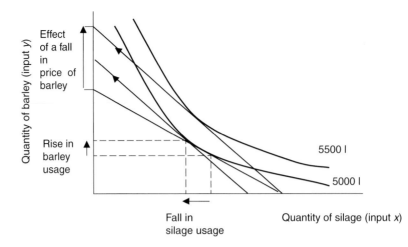

Fig. 5.10. A change in the relative price of inputs.

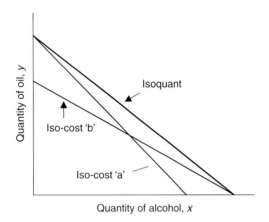

Fig. 5.11. The effect of a change in input price where two inputs are perfect substitutes.

of x and y inputs be 'optimal', and under that situation *any* combination on the isoquant is optimal because all combinations cost the same. An example is the use of alcohol or oil as the bases for some chemical processes. Apparently manufacturers shift easily between the two dependent on relative costs.

Figure 5.12 shows the case of a single-point isoquant (the hypothetical case of an alloy quoted earlier). Here changes in the relative prices of inputs have no effect on the least-cost combination of inputs to produce a given level of output.

A less extreme form of insensitivity to changes in the relative prices of inputs is encountered with 'lumpy' inputs (i.e. those which can only be had in relatively large indivisible units, such as tractors). Figure 5.13

shows the case of the quantity of labour (man-days) and tractors required to produce a given quantity of hedge cutting. The isoquant is a series of discrete points corresponding to whole numbers of tractors. Because of the 'corners' on the isoquant formed by this series of points, the cost of labour can alter considerably, changing the slope of the iso-cost line, before the least-cost combination is affected and a different number of tractors is used. (This is admittedly a superficial treatment of the 'lumpy input' problem and ignores such questions as the possibility of only partly using the capacity of the machine to replace labour. However, no mention at all would be a greater error.)

The effect of expanding the scale of operation

Figure 5.14 shows isoquants for a range of milk outputs and iso-cost lines enabling the least-cost combination for each output to be found. A line connecting these optima for various levels of output is called the **expansion path**. Note that the inputs are not necessarily in the same relative proportions at different levels of output. Obviously the prudent entrepreneur wishes to operate at the best scale of production, and how far he or she proceeds along the expansion path is determined by the revenue brought in by each level of output (i.e. the revenue corresponding to each isoquant) and the level of cost, represented by the iso-cost line, necessary to achieve it. The profit-maximizing entrepreneur must aim for that output which gives the biggest difference between revenue and costs. Knowing

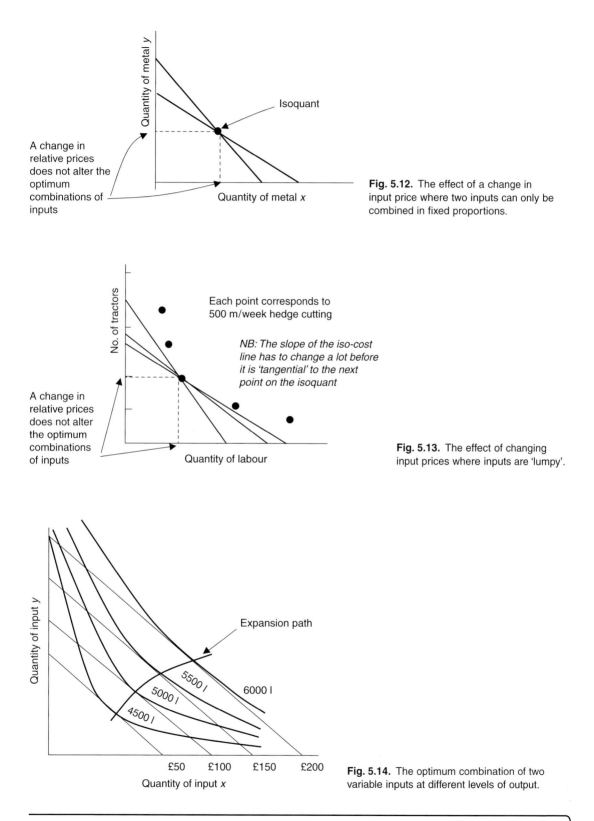

A change in relative prices does not alter the optimum combinations of inputs

Fig. 5.12. The effect of a change in input price where two inputs can only be combined in fixed proportions.

Quantity of metal *y*

Isoquant

Quantity of metal *x*

No. of tractors

Each point corresponds to 500 m/week hedge cutting

NB: The slope of the iso-cost line has to change a lot before it is 'tangential' to the next point on the isoquant

A change in relative prices does not alter the optimum combinations of inputs

Quantity of labour

Fig. 5.13. The effect of changing input prices where inputs are 'lumpy'.

Quantity of input *y*

Expansion path

5500 l

5000 l

6000 l

4500 l

£50   £100   £150   £200

Quantity of input *x*

Fig. 5.14. The optimum combination of two variable inputs at different levels of output.

the optimum combination of variable inputs for each level of output, and hence the lowest *combined* cost for each level, he or she can decide the optimum level of output using the marginal cost and marginal revenue approach described in the section 'Optimizing the Factor–Product Relationship'.

Optimizing the Product–Product Relationship

'Optimizing the product–product relationship' involves the entrepreneur allocating the limited resources at his or her disposal in order that they may bring in the greatest revenue. For example, a farmer with a farm of a given size, a fixed labour force, stock of machinery and buildings and a fixed amount of working capital has to choose to allocate them between, say, potato production and sugarbeet production; he can produce all of either commodity or some of each. Figure 5.15 shows the various combinations of potatoes and sugarbeet that can be produced in a given period using the same bundle of resources – those which exist on the farm in question. This curve is therefore called an

iso-resource curve because it relates to the same lot of resources (land, labour, capital, entrepreneurship). It is also called the **production possibility boundary** as, while any combination of two products in question lying inside or on this curve can be produced with the existing resources and technical knowledge, combinations lying further out from the origin are not possible. We need not concern ourselves with combinations lying inside the curve, only with those right on the boundary, since we can assume that the profit-maximizing farmer will aim to keep his productive resources fully utilized.

Potatoes and sugarbeet are **competitive products** because if more of one is produced *less* of the other can be produced. The slope of the iso-resource curve is negative and it is numerically the same as the marginal rate of substitution of products; the MRS between competitive products has a negative coefficient since a positive change in the amount of one product produced must be accompanied by a negative change in production of the other. This MRS between products is also called the **marginal rate of transformation** between products, or the **rate of technical transformation** (see equation at bottom of page).

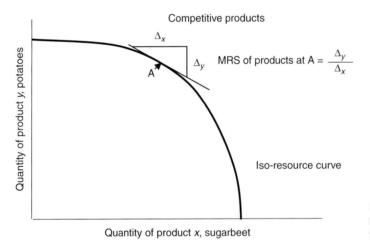

Fig. 5.15. The combinations of two products that can be obtained from a given bundle of resources.

Marginal rate of subsitution of products	= Number of units of product y that have to be forgone for an increase in production of 1 unit of product x
	$= \dfrac{\text{Change in } y}{\text{Change in } x}$
	$= \dfrac{\Delta_y}{\Delta_x}$

Figure 5.16 shows that the slope of the iso-resource curve is not always negative. If a wholly cereal farm devotes part of its acreage to growing grass (and producing hay) its total cereal production may *increase* because of higher yields per hectare. This is caused by the grass 'break' enabling cereal diseases to be held in check, soil structure to be improved, etc. Over the range AB in Fig. 5.16 hay and wheat are **complementary products** since producing *more* hay enables *more* wheat to be produced. (MRS is positive over this range.) However,

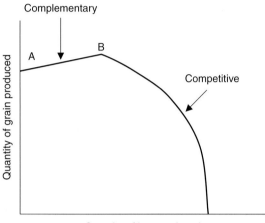

Fig. 5.16. Iso-resource curve of complementary products.

beyond B the extra cereal yield per hectare is insufficient to offset the loss of production caused by switching hectares from cereals to grass, and hay and cereals become competitive products, the MRS between them taking a negative sign.

Revenue and the optimum combination of products

Not all combinations of products will bring in the same amount of money to the entrepreneur. The optimum combination of products (or enterprises) produced from a given bundle of resources is that which brings in the greatest revenue. Figure 5.17 is the same as Fig. 5.15, except for the introduction of **iso-revenue** lines labelled £900, £1000 and £1100.

> An iso-revenue line connects the combinations of quantities of two products which bring in the same revenue.

An iso-revenue line can be constructed by joining the quantity of the good on one axis which could be sold for, say, £1000 with the quantity of the other good on the other axis which could be sold for £1000. Figure 5.17 shows that two combinations of potatoes and sugarbeet could be produced to bring in £900, but only one combination to bring in £1000. This is the highest revenue attainable and hence this combination of products is the best (or is optimal).

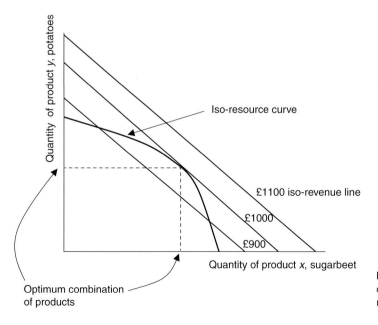

Fig. 5.17. The optimum combination of two products from a given bundle of resources.

At the optimal point the equations at the bottom of the page apply.

At the optimum combination of products x and y:

$$\text{MRS} = \frac{\Delta \text{ Product } y}{\Delta \text{ Product } x} = \frac{(-)\text{Price of } x}{\text{Price of } y}$$

It can be shown that, when considering infinitely small changes in x and y:

At the optimum combination of two products from a given bundle of resources any rise in revenue caused by producing an infinitely small amount more of one product will be exactly offset by the loss resulting from the reduction in production of the other product.

When considering greater-than-infinitely-small changes in x and y, the rise in revenue caused by producing more of one product will be *more than offset* by the loss resulting from the reduction in production of the other product. This can be extended to more than two products, and this optimum allocation of a given quantity of productive factors between a range of possible uses is yet another manifestation of the Principle of Equimarginal Returns encountered already in Chapters 2 and 4.

Changes in relative prices of products

A change in the relative prices of products will cause the slope of the iso-revenue line to change (Fig. 5.18) producing a new optimum combination of product quantities. If the price of one product rises, it will be favoured in the new combination.

Straight line iso-resource curves

Where there are no diminishing returns the iso-resource curve will be a straight line. For example, if a factory could easily switch from producing plastic dustbins to producing plastic buckets with no diminishing returns, small changes in relative price would cause a complete swing from one good to the other (Fig. 5.19).

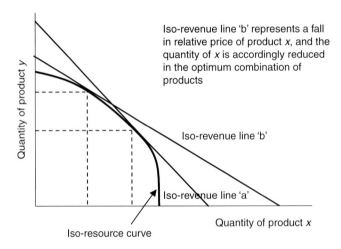

Iso-revenue line 'b' represents a fall in relative price of product x, and the quantity of x is accordingly reduced in the optimum combination of products

Iso-revenue line 'b'

Iso-revenue line 'a'

Quantity of product y

Quantity of product x

Iso-resource curve

Fig. 5.18. The effect on the optimum combination of products of a change in relative product prices (non-constant MRS of products).

Slope of iso-resource curve = Slope of iso-revenue line (Note : both are negative)

$$\text{Slope of iso-resource curve} = \text{MRS} = \frac{\Delta \text{ Product } y}{\Delta \text{ Product } x} \text{ (with the same bundle of resources)}$$

$$\text{Slope of iso-revenue line} = \frac{(-)\text{ Quantity of } y \text{ which will bring in } \pounds n}{\text{Quantity of } x \text{ which will bring in } \pounds n}$$

Complementary products

Because of the 'bump' on the iso-resource curves for these products, large changes in relative prices may be required before major changes in the optimum combination of products will occur (Fig. 5.20). With the complementary products example cited above (grass for making hay, and cereals), a fairly heavy loss on the hay crop viewed alone would be required before no grass would be grown because of the boost it gives cereal production.

Time and Scale of Production

The three important relationships (factor–product, factor–factor, product–product) have been examined

in some detail. It is now necessary to return to a comment that was made at the start of this section on production economics – the importance of *time*.

Fixed costs and variable costs

The fact that some costs can be considered fixed while others are variable has already been mentioned in a variety of contexts. A **variable cost** of production is one that varies with the level of output. A **fixed cost** is one which does not vary with output.

Which costs are fixed and which are variable depends on the time period being considered. If a farmer has rented a farm and hired a labour force, he is committed to pay the rent and wages whether

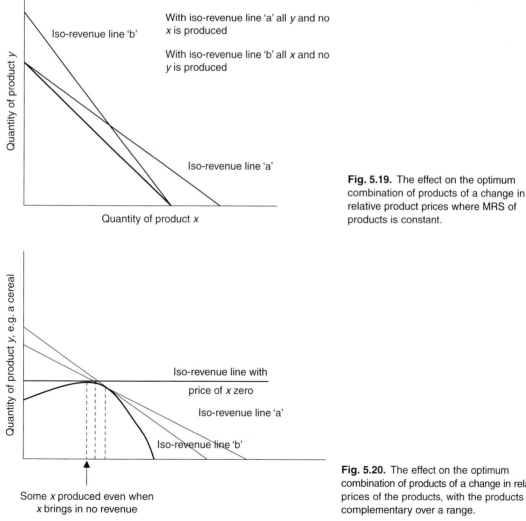

Fig. 5.19. The effect on the optimum combination of products of a change in relative product prices where MRS of products is constant.

Fig. 5.20. The effect on the optimum combination of products of a change in relative prices of the products, with the products complementary over a range.

he decides to produce 4 t of cereals/ha or 8 t/ha. Conversely, seed costs are variable when viewed before planting – they will depend on the quantity of grain the farmer decides to produce. In the long run all costs are variable – the farmer could increase or reduce the size of his labour force or the area of his farm. On the other hand, the shorter the time period the more costs become 'fixed'. An hour before harvesting cereals, the grower cannot change anything to increase yield except the harvesting process – he is committed to all the costs of production except those of harvesting. Given more time he could have varied fertilizer use, herbicide use, seed rates, etc., but in the short term these are 'fixed'.

Table 5.5 shows the fixed and variable costs faced by the operator of a brewery at different levels of output. The fixed costs, which do not vary with the number of barrels of beer produced per month, consist of the lease of the factory, the monthly wage bill, depreciation of equipment, etc. The variable costs consist of barley costs, malt costs, hop costs, colouring and water costs.

Costs from Table 5.5 are plotted in Fig. 5.21, which shows how total costs vary with output, and Fig. 5.22, which shows average costs and marginal cost.

Note from the table and graphs that:

1. Because increases in output spread fixed costs over more barrels of beer, average fixed costs (AF_xC) per barrel fall consistently.
2. Total variable costs (TVC) rise with output but not in a fixed proportion, so that average variable cost (AVC) per barrel first falls, then rises. AVC initially declines because inputs are being combined more efficiently, but it eventually rises because of inefficiencies from such factors as the overwork of the equipment, overstretching of the labour force, etc.

Table 5.5. Fixed, variable, total and marginal costs of production for beer manufacture.

Output (barrels/ month)	Fixed costs		Variable costs		Total costs		Marginal cost of producing the last barrel[a]
	Total	Average per barrel	Total	Average per barrel	Total	Average per barrel	
100	240	2.40	250	2.50	490	4.90	4.90
200	240	1.20	450	2.25	690	3.45	2.00
300	240	0.80	615	2.05	855	2.85	1.65
400	240	0.60	800	2.00	1040	2.60	1.85
500	240	0.48	1010	2.02	1250	2.50	1.50
600	240	0.40	1290	2.10	1530	2.53	2.70
700	240	0.34	1580	2.26	1820	2.60	3.20
800	240	0.30	2000	2.50	2240	2.80	4.20

[a]For an explanation of how marginal costs were estimated, see Box 5.6.

Box 5.6. Calculating marginal cost (MC).

The MC of the *n*th barrel of beer is calculated by subtracting the total cost of *n–1* barrels from the total cost of *n* barrels, to give the addition to total costs attributable to producing the *n*th barrel. Because fixed costs are constant with output, MC is also the same as the increase in variable costs, i.e. VC of *n* barrels minus VC of *n–1* barrels. The outputs shown (100, 200, etc.) in Table 5.5 do not contain the information necessary to calculate the MC of (say) the 300th barrel, but the following figures serve as an example of how the MC of the 300th barrel is derived.

Output	Fixed costs	Variable costs	Total costs	Marginal costs
299	240.00	613.35	853.35	–
300	240.00	615.00	855.00	1.65

3. Average total cost (ATC) per barrel is a combination of fixed plus variable costs. At first the fall in average variable cost with output reinforces the fall in average fixed cost, enabling average total cost to fall. However, the eventual rise in average variable cost more than offsets any fall in average fixed cost, so that average total cost rises.

4. The marginal cost (MC), i.e. the addition to total cost attributable to the last barrel produced, first falls and then rises. Note that in the graph in Fig. 5.22 the rising MC passes through the lowest point of the AVC and ATC curves.

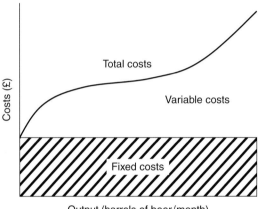

Fig. 5.21. Fixed, variable and total cost of beer production.

What is the lowest price at which beer will be produced?

The beer producer will equate MC with MR to maximize profits. Remember that in perfect competition MR = AR (see Chapter 4). To continue in production long term, the entrepreneur must receive a price (average revenue, AR) for his beer so that both his variable and fixed costs are covered. These are lowest at the lowest point of the ATC curve – he must get at least this price in the long run. However, in the short term he will continue to produce if his variable costs are covered, so the minimum short-run price he can accept is the lowest point on the AVC curve. In other words, although the brewer has to have a price of £2.50 per barrel to permanently stay in business, he will not stop production if prices slump temporarily unless they fall below £2.00 per barrel, because any price above £2.00 covers his variable costs and leaves a small margin to help pay the fixed costs which he has to bear whether he produces much or little beer (see Fig. 5.22).

Economies and diseconomies of scale

Figure 5.23 shows ATC curves for three different sizes of dairy enterprise – assume that these, which are hypothetical rather than actual, cover both fixed and variable costs. These relate to three different quantities of fixed factors of production: (i) the smallest size corresponds to an enterprise based on a small

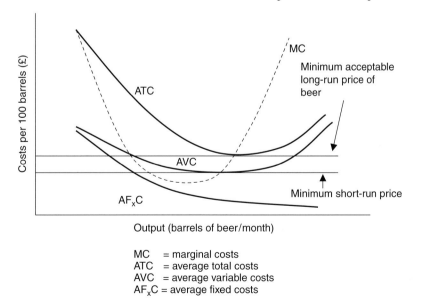

MC   = marginal costs
ATC   = average total costs
AVC   = average variable costs
AF$_x$C = average fixed costs

Fig. 5.22. Average and marginal costs of beer production.

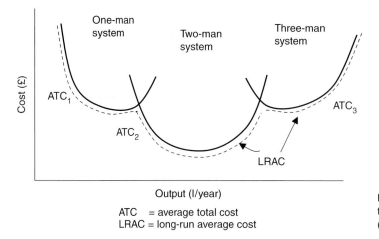

ATC = average total cost
LRAC = long-run average cost

Fig. 5.23. Costs of production at three scales of milk production (hypothetical).

> **Box 5.7. Economies of size and of scale.**
>
> Strictly, the term 'economies of size' should be used in this context rather than 'economies of scale', as the latter is reserved for situations where all inputs are increased in the same proportion. This is clearly not the case with this dairy herd example, as it is implied that one farmer can choose between the three different sizes of herd and so the input he, himself, represents is not increased. This is the case in almost all practical situations.

milking parlour designed to be operated by one man; (ii) the medium-size enterprise to a larger parlour operated by two men; and (iii) the largest to a three-man system. Once a farmer has built a one-man system he is stuck with it – the costs of running it are 'fixed' and he cannot readily scrap it and set up a two-man system. These three sets of equipment plus staff represent three possible scales of production where some costs are fixed (depreciation and probably labour) whereas others (e.g. feed) are still variable. Although there is some flexibility in the number of cows each system can handle, both the underutilization of the fixed equipment and overstretching its designed capacity will result in higher average costs of milk production. Only when a period of, say, more than 5 years is reviewed can the farmer think of changing the scale of his enterprise.

In Fig. 5.23 the line of dashes shows the minimum average cost at which it is possible to produce each level of output, given the time necessary to alter the scale of production, i.e. when all costs can be considered variable. It is termed the **long-run average cost** (LRAC) line.

In Fig. 5.23 the level of the ATC curve is lower for the two-man dairy enterprise than for the others. This means that such a system could produce at a lower cost per litre than either larger or smaller scales of production. Being larger than a one-man operation has both advantages and disadvantages (which will be discussed below), but on balance the two-man enterprise benefits in terms of lower costs; going larger than a two-man system incurs disadvantages which more than offset any advantages, so that costs are again higher.

With greater numbers of possible scales of operation the LRAC line (showing the lowest possible average costs of production at each level of output when scale is not 'fixed') becomes much smoother than in Fig. 5.23. This is shown in Fig. 5.24. When the LRAC line is falling **economies of scale** are said to operate, and when rising **diseconomies of scale** operate. If over a range the LRAC is horizontal and flat, neither can be said to operate. With the LRAC curve is associated a long-run MC (LRMC) curve.

The situation in agriculture appears, in reality, to be somewhat 'L' shaped rather than the flat 'U' shown in Fig. 5.24. That is, there are economies of size up to a certain size (empirical evidence for the UK suggests at the size of farm that can be operated by two or three people), beyond which there are no clear advantages or disadvantages from being larger. This gives an LRAC curve as in Fig. 5.25.

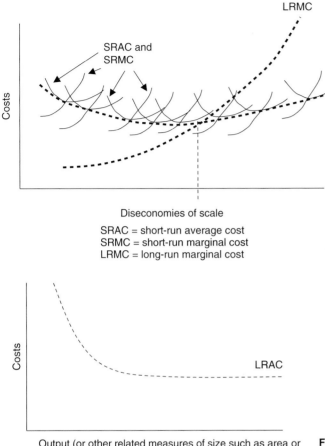

LRMC
SRAC and
SRMC
Costs
Diseconomies of scale
SRAC = short-run average cost
SRMC = short-run marginal cost
LRMC = long-run marginal cost

Fig. 5.24. Economies and diseconomies of scale.

Costs
LRAC
Output (or other related measures of size such as area or labour input)

Fig. 5.25. The general pattern of economies of scale in farming.

Sources of economies of scale

Economies of scale arise from a range of sources, some of which are more relevant to farming than others. They are:

1. Economies in the use of *land*. Except with some farm crops, it is rare that doubling the output of an enterprise requires double the land area, so that costs per unit of output will be lower than with the larger scale of production.
2. Greater efficiency in the use of *labour*.
   (i) Labour can specialize, develop skills and hence produce more. Specialization (sometimes called the 'division of labour') only becomes possible with the growth of firms. For example, on a one-man farm the farmer may have to milk cows, drive the tractor and manage the business. On a larger farm specialists in each type of job can be employed who should be better at their individual jobs than the all-rounder and the farmer can concentrate his efforts on managing the business. Considerable savings in time are also made because one person has no longer to keep switching between jobs which may require different clothing or levels of hygiene, or to be in different places.
   (ii) With higher output, more machinery and more powerful machines can be used. For example, one man can drive a small tractor or a much higher powered model, and the cost of his labour can be spread over a greater volume of output with the larger machine.
3. Economies in the use of *capital*. Larger machines and buildings are usually cheaper in relation to what they can do than smaller ones. Some examples will suffice:
   (i) A 40 m length of fencing will enclose a square plot of 100 m^2. However, an 80 m length of fence will enclose a square plot of 400 m^2, so that the

cost of fencing per square metre enclosed falls with increasing size of plot. Similarly, doubling all the dimensions of a square building (including its height) will increase its internal volume eight times. If the same materials can be used, the larger building will have a much lower cost per unit of volume. For similar reasons, the price of tractors per horsepower generally falls with increasing size of tractor (at least until the very largest tractors where their small-scale production increases their price).

(ii) Some units of capital are expensive and indivisible and are only justified by large-scale enterprises. For example, a large farm may be able to fully utilize a machine for laying drains in fields whereas a small farm may have to hire a contractor and machine, at a greater cost per hectare. Another example is a large business may use a computerized accounting system and the cost per invoice handled will be lower than a manual system used by a smaller firm.

4. *Economies of administration.* Increasing the scale of production does not necessarily need a proportional increase in administrative staff. A secretary can as easily write an order for 50 t of raw material as for 5 t. Managers can often as easily manage a big business as a smaller one. Also firms which are large can often afford higher salaries for their management and, for a little extra salary, can attract the best managerial talent, which more than earns its extra cost.

5. *Economies of material.* The waste product of a large firm may often be turned profitably into a saleable by-product. A trite example is that manure from a one-horse stable may be difficult to dispose of, but a large racing stable can sell its manure profitably for mushroom compost.

6. *Economies of marketing.* Large-scale purchasing of inputs often achieves discounts. In selling, large-scale firms are often able to secure an outlet using a contract, thereby enabling them to specialize and thus to lower their average cost. For example, a chain store may be willing to take lettuces from a grower on contract at a predetermined price if the grower is large enough to be able to guarantee the quality and quantity he can supply. With commodities where a travelling sales force or advertising is important, larger sales mean that these selling costs are spread over more units.

7. *Economies of finance.* Large firms can often borrow more easily and at preferential rates. Most farms are small private businesses, unable to finance investment by issuing new shares as could a large public company (see Chapter 4).

8. *Economies of research and development (R & D).* Large firms may be able to afford R & D departments which can give their products a competitive edge (more relevant to the industries supplying agriculture than to farming itself).

9. *Other economies.* A large firm may be able to provide social and other welfare facilities and so retain a band of loyal, happy and productive workers. There are also the advantages to the firm (if not to

---

Box 5.8. Economies from matching equipment.

The advantages coming from the ability to fully utilize indivisible resources are further illustrated by the better matching of units of equipment. For example, a production process using four machines in succession and an output of 10 units/h may seriously underutilize some machines while others are used to full capacity. In the following figures the *indivisibility* of machines A, B and D (it is not possible to have a third of machine A, etc.) leads to a situation where larger-scale production results in economies. The bottleneck in production is machine C. The lowest output where all machines could be fully utilized, implying the lowest average cost per unit, is 120 units/h (the lowest common multiple of 10, 20, 30, 40) when four type A machines, six type Bs, 12 type Cs and three type D machines would be employed. Thinking of machines not individually but as they fit into the whole process of production is called a **systems** approach.

	Machine			
	A	B	C	D
Actual usage (units/h)	10	10	10	10
Potential capacity	30	20	10	40

---

society) which may accompany the establishment of monopolies (refer to Chapter 4). Monopoly profits are often closely associated with large-scale production and market domination.

In agriculture the very small farm frequently suffers primarily from its inability to utilize fully the labour and essential equipment found on it. For example, the minimum labour force of one man (the farmer) may not be fully utilized yet, because of his *indivisibility*, it is not possible to have less labour on the farm. Growth in business size which takes up this spare capacity can result in considerable economies.

Disadvantages of large-scale production

Against these advantages can be ranged an array of possible disadvantages from large-scale production, again not all of which are relevant to farming.

1. With increasing size a business often has to introduce some formal organization. Rules and regulations become necessary. The close contact between manager and workers is lost and the cost of 'unproductive' administrators may cancel out economies of scale.
2. The separation of management and workers lengthens the chain of consultation so that changes by the management cannot be implemented quickly and views, ideas and observations of the operators are not rapidly communicated to the managers. In contrast, on the vast majority of farms a worker can easily contact his employer if he notices something wrong with a crop or animal.
3. Because large businesses are usually run by salaried managers and not the owner, the incentive for efficiency and hard work is diluted. This can be obviated by profit-sharing agreements or piece-work payment (e.g. some dairymen/cowmen are paid according to how much milk a herd produces).
4. Large complex firms demand a high standard of management ability. Lack of such talent is usually quoted as the most important limiting factor to the size of businesses. Overstretching mediocre talent can result in costly mistakes. An agricultural example might arise if a cowman could easily manage a 100-cow herd but made expensive oversights when the herd was raised to 150 cows.
5. The division of labour (specialization) often associated with large-scale production (e.g. in the motor industry) is alleged to reduce workers to the level of mere machine-minders with no interest in their product's quality and encourages industrial unrest. Specialization also puts great potential to disrupt into the hands of a few key workers.

In addition to the above, the following factors contribute to the persistence of small-scale production:

- Craftsmen-entrepreneurs take pride in the quality of their work and may not wish to expand. For example, the reward a craftsman violin maker using traditional methods derives from his activities comes in part from a satisfaction in creating beauty with his own hands, and he may eschew larger-scale factory methods even though they are technically feasible (and are used by some firms) for violin manufacture.
- Where personal service is essential (e.g. water diviners) or where hand craftsmanship is the prime characteristic of the product (e.g. roof thatching).
- Where the market for the product is limited. For example, a dry-cleaning shop in a small village will tend to remain a small business. Similarly, artificial legs are not demanded in vast quantities so large-scale production is not justified.
- Most large firms had their beginnings as small firms and some present small firms *may* be embryonic large ones.

If expanding the scale of operations to gain certain advantages, say in marketing a product, brings with it diseconomies in the form of inefficiencies in the production process, firms can be expected to take steps to avoid these diseconomies. Hence, in the complex motor industry we find that some of the technical processes of production are subcontracted out to other – usually smaller – firms. Also, where diseconomies are suspected, manufacturers often set up duplicate production plants around the country so that the best scale of production is not exceeded, but the advantages of large-scale selling techniques can be used to dispose of all the identical products from the various factories. In agriculture, farmers form groups to gain the advantages of bulk buying or selling and cooperate in machinery-sharing syndicates; these forms of horizontal integration allow cost or price advantages to be enjoyed while still allowing the farms to remain largely independent businesses.

Is there an optimum scale of operation?

The question is often asked, 'Is there an optimum scale of production?' or, more specifically, 'Is there an optimum size of farm?'. To attempt to answer such a question, we would need to know the shape of the LRAC curve for the type of production under consideration. If this were a flat U shape, with economies of scale merging directly into diseconomies and with no region where neither operated (as back in Fig. 5.24), under a situation of perfect competition, firms would compete in price with each other until the product was produced at the lowest possible cost corresponding to the lowest point of the LRAC curve. All firms would be of the same scale – that scale whose lowest point on the short-run average cost curve corresponded to the lowest point of the LRAC curve.

However, in the real world, competition is not perfect and technical innovation and changes in the relative prices of factors of production go on which constantly shift what could be thought of as the 'ideal' size of the firms. In the industry which is probably close to perfect competition, agriculture, economies of scale are not vividly apparent beyond the size of farm requiring two or three men. For most types of farming in the UK (and probably also in the agricultures of many developed countries) the LRAC curve appears to be L shaped as shown in Fig. 5.25. While very small farms can generally be shown to suffer from serious cost disadvantage on average, so that, when moving to larger sizes, considerable economies of scale are apparent, beyond the two to three man size neither economies nor diseconomies seem to operate. This could be one explanation why large, medium and relatively small farms can apparently compete successfully with each other and are all viable at the prevailing level of prices for agricultural products.

A feature of agriculture in most developed countries over the last few decades has been the general fall in the number of small farms and rise in the number of large ones. Farm enlargement in order to reap economies of scale no doubt forms part of the explanation for this movement. However, upward adjustments in the size of agricultural firms are often difficult to make because of the characteristics of the factors of production used by farming. Most notably, farmers often find it impossible to acquire additional land, and agriculture requires proportionally more land than most industries. The nature of factors of production, with special attention to their use by agriculture, is discussed in Chapter 6.

Change in the Supply of Farm Products – Technological Advance

So far in this chapter it has been assumed that the technical relationships between the inputs to production and the outputs remain the same, that is, the state of technology remained constant. However, instances have been given where this is not the case – for example, the development of new higher-yielding cereal varieties. Technological advances are important in that they result in an increase in the quantities that producers are willing to supply at each level of price – a rightward move of the supply curve (Fig. 5.26). As was seen in Chapter 4, this leads to the long-term downward movement of prices for agricultural commodities, with a squeeze on revenues and lower profits. As a result, agricultural producers are faced with a need to adapt to the new situation or leave the industry, as many do, particularly those on smaller farms. Despite falling prices, productivity continues to rise in developed countries, for example in the UK it has grown over the last half-century by about 2–3%/year on average but with periods above or below the trend (it increased by 18% between 1997 and 2005 but has changed little since).

Technological advance

The term 'technology' interpreted literally means the sum of knowledge of the means and methods of producing goods and services; 'technological advance' implies a growth in this knowledge. However, it is not simply the advance in knowledge which is of significance, but the fact that some of

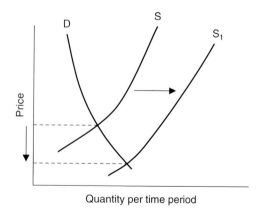

Fig. 5.26. The effect of technological advance on the industry supply curve.

the new 'means and methods' are put into actual production, and this carries implications both for farming and for the country's wider economic and social fabric. Some advances in knowledge do not prove practical, at least in the short term: the automatic, driverless tractor is a case in point, although developments in geolocation may be changing this. Others are seized on and applied rapidly, such as new heavier-yielding varieties of cereals. The factors influencing the rate at which 'advances' are taken up by farmers will be considered later.

The essential nature of an adopted technological advance is that it improves the relationship between inputs and outputs so that, at the farm level, the margin between costs and revenues is greater than under the old system. This can be achieved in several ways, described below. Because farm firms are in business for a run of years, and because agriculture by its nature is subject to fluctuations, innovations which merely reduce these fluctuations and the risks which go with them may also be attractive to farmers. However, the net result at the industry level is that agriculture's aggregate supply curve is pushed permanently to the right, manifest by greater output and a downward pressure on farm product prices. New techniques, or 'innovations', are therefore watched closely by policy makers because in large part they dictate the future structure of the agricultural industry and the shape of governmental agricultural policy.

Impacts of advances in technology on the three central relationships

The effect of technological advances can be viewed from each of the three relationships examined above (factor–product, product–product and factor–factor). These are illustrated in the thumbnail diagrams of Fig. 5.27 (to refresh understanding refer back to the relevant sections earlier in this chapter). In the factor–product relationship, a new heavier-yielding cereal variety will increase the amount of output from each level of fertilizer input, raising the total value product (TVP) and other product curves. The implication for optimizing the relationship is that the optimum level of fertilizer to use will increase with the new variety (Fig. 5.27a).

Turning this relationship around to find the optimum level of output to aim for, we find that the technological advance lowers the cost curves of producing cereals (Fig. 5.27b), which again will increase the optimum at which the producer should aim.

In the factor–factor relationship, technological advance moves the isoquants nearer the origin, as fewer inputs are required to produce the same amount of output. There is no guarantee that the isoquant will move inwards in a parallel way, so the balance between the two inputs is likely to change. Also, the producer is likely to expand up his scale line, increasing output to the point at which MC = MR (Fig. 5.27c).

For the product–product relationship, technological advance pushes out the iso-resource curve (production possibility boundary), as now more can be produced from a given bundle of inputs. Again there is no particular reason why it should move out in a parallel way, as improvements in the two enterprises are unlikely to be the same. This has an implication for the new balance between the enterprises in the optimum mix (Fig. 5.27d).

A particular form of technological advance common in agriculture is where new machines are invented (large tractors, combine harvesters) that are expensive but enable costs to be lowered if advantage is taken of their potential larger capacity. This is explained in Box 5.9.

Types of innovation seen from the farm level

A useful classification of innovations is according to their impact on inputs and output at the farm level. Each will carry different implications for structure and policy. Table 5.6 shows that advances can either increase output or, at least initially, leave it unchanged, while they can also be input saving, input requiring, or leave inputs unaltered.

The simplest types of advance to understand are those which primarily either affect output *or* inputs but not both, at least in the initial stages. For example, a heavier-yielding variety of cereal seed will increase output but not the amount of other resources used (land, labour, machinery, fertilizer). A better way of organizing office procedures (paying bills, contacting merchants to sell produce, etc.) will not have a direct impact on the farm's physical output but might mean that the farm secretary need only be employed for fewer hours, saving on inputs and costs. In practice, of course, there may be indirect impacts: the new variety of seed may be able to respond profitably to higher levels of fertilizer thereby using more inputs as well, and the wages not paid to the farm secretary can be used to finance higher stocking rates on the farm, so that the examples given in Table 5.6 can only be approximations.

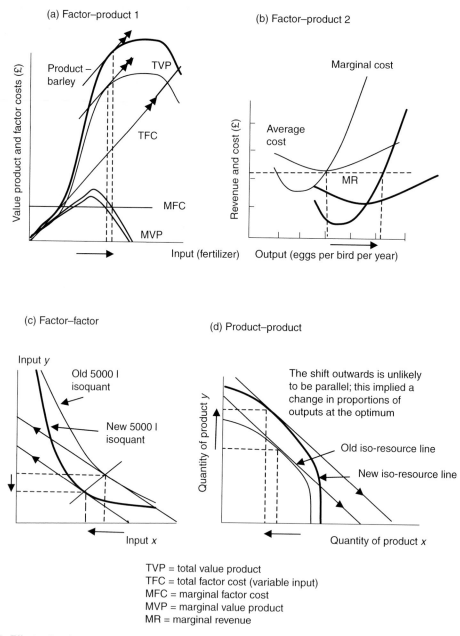

(a) Factor–product 1

Value product and factor costs (£)

Product – barley
TVP
TFC
MFC
MVP
Input (fertilizer)

(b) Factor–product 2

Revenue and cost (£)

Marginal cost
Average cost
MR
Output (eggs per bird per year)

(c) Factor–factor

Input y

Old 5000 l isoquant
New 5000 l isoquant

Input x

(d) Product–product

Quantity of product y

The shift outwards is unlikely to be parallel; this implied a change in proportions of outputs at the optimum

Old iso-resource line
New iso-resource line

Quantity of product x

TVP = total value product
TFC = total factor cost (variable input)
MFC = marginal factor cost
MVP = marginal value product
MR = marginal revenue

Fig. 5.27. Effects of technological advance on production relationships.

A reduction in one form of input may be partly compensated by more of another. This may happen, for example, in the hoeing of field-scale root or vegetable crops if a small army of labourers working with hand tools is replaced by a tractor-mounted multiple hoe; the amount of capital involved may rise but a great deal of labour is displaced so that the total cost of hoeing will be less. A similar situation has occurred with the mechanized harvesting of blackcurrants in England; with mechanization it is quite conceivable that output might fall (by less effective picking or through bush damage lowering next year's crop) but the technology would be used if costs were

Box 5.9. How a new technology with higher fixed costs can reduce average costs of production.

The diagram represents the cost curves of two technologies – for example, an older smaller combine harvester and a newer larger one. TC lines represent total costs of the two technologies at different levels of output. These are comprised of two components: (i) fixed costs (F_xC) related to initial acquisition, such as depreciation, and other costs that do not vary with output; and (ii) variable costs (VC), such as fuel costs, that vary with output.

Average total costs (ATC) at each level of output are shown by the slope of a line drawn from the origin to the TC curve at that output. The new technology has a higher fixed cost than the old one. The diagram shows that the lowest ATC possible using the new technology is lower (less steep slope) then with the old technology. Thus ATC will be lower if the greater output potential of the new technology is utilized.

However, there is no advantage from using the new technology to generate the same output as achieved the lowest costs using the old technology. A line drawn to the new TC curve at the old output will be steeper (i.e. ATC will be greater). Thus the potential of the new technology needs to be exploited before average costs fall.

The diagram can also be used to show average fixed costs (AF_xC) and average variable costs (AVC) at each level of output. The slopes of lines drawn from the origin to the (total) F_xC lines will show AF_xC – this will always fall with higher levels of output for each of the technologies. Average variable cost (AVC) is shown by the slope of a line drawn from where VCs commence to the TCs curve; the slope of this line first falls and then rises with increasing output. Marginal costs (MCs) are shown by the slope of the TC curve at each output.

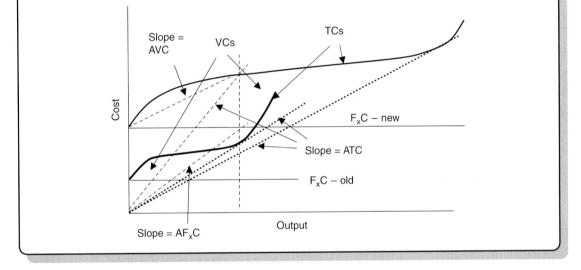

Table 5.6. Initial impact of technical change.

	Output		
Input	More	Unchanged	Less
More	Large-capacity tractor/combine harvester	Short-life buildings	
Unchanged	Improved seed varieties	(No technological advance implied)	
Less	Improved fertilizer; better labour training; more efficient seed drills	Improved office organization; artificial insemination of cows; replacement of hand hoeing by tractor hoeing	Mechanical harvesting of crops

sufficiently reduced. If risk reduction is also considered, a farmer might prefer to dispense with the uncertainties of hired gangs of labour to harvest crops of vegetables or fruit if machine harvesters become available, even if this means using more inputs and/or getting lower yields.

Other types of innovation from the start imply changes both in inputs and in outputs. Without an increase in both, the innovation is not attractive. Examples are the rotary parlour and the large-scale combine harvester. The initial costs of these items is greater than the types they replace, but their attractiveness lies in the large capacities they can handle – potential number of cows milked per day or tonnes harvested. If this capacity is utilized the average milking or harvesting costs are below alternative methods (see Box 5.9 for the explanation why a machine with a high initial cost can lower average total costs). For those farmers already operating at a sufficiently large scale, the new techniques are rapidly adopted; the lower levels of costs enjoyed are generally responded to in the form of further output. For those farmers operating on a smaller scale, the pressure will be to expand up to that size necessary to enjoy the new technique, implying more of the other inputs (cows, fertilizers) and a greater level of output. Eventually the increased supply may cause prices to drop and for some farmers still using the old techniques to cease production, but at least initially the effect of the innovation is output-expanding. It is important to note that most of the recent advances in agriculture appear to have been of this type.

A brief reference is necessary to the type of technical advance whose only effect is to use more factors of production. At first sight this might seem an unlikely candidate for incorporation by farmers into their farming systems, except perhaps where risk reduction was involved. More inputs but no greater output might imply higher costs and lower profits, hence preventing its uptake. However, the paradox may be explained by realizing that some inputs such as buildings are used up over a number of years. For example, low-cost housing for dairy cattle made from lightweight timber and plastic sheeting may pose an attractive alternative to higher-cost, traditional concrete-and-steel structures, but the cheaper alternative is less durable; for illustration, let us assume a life of 20 years as opposed to 60 years for the conventional building. If, say, the new-technology housing costs only half that of the traditional structure and the farmer's planning horizon is 20 years, corresponding to the life of the new-style housing, the new technology may be chosen as being the preferable investment. However, over 60 years this type of housing will need replacing three times whereas the old-technology building would last this long without renewal. Hence, looking back and assuming that costs indicate the quantity of resources, the technological advance has, over 60 years, resulted in *more* resources being used to provide the same amount of housing.

This example begs the question of the incorporation of further advances in technology at each replacement phase over the period and, indeed, the whole process of valuing resources. Fortunately in agriculture such examples are rare, although in food packaging it is by no means an irrelevant issue, with new materials of the throw-away variety replacing reusable glass bottles or metal containers.

While some sorts of innovation do not in themselves need an increased output to be realized, in practice this almost always happens in some form. A better method of manufacturing fertilizers will lower their price, more will be used and cereal output will expand. Even improved farm office organization can give the farmer more time for attention to field operations, with an increase in production at the whole-farm level.

Implications for the agricultural industry

Technological advances which at the farm level may require more inputs and generate more output do not necessarily carry the same implications at the industry level. For example, many advances require large-scale farming to reap their benefits, but the total national area of farmland as an input is not usually expansible. Consequently these innovations have been taken up by the already large farmer, by farmers specializing so that land is released from other enterprises, and by some farmers taking over larger acreages from others leaving the industry or reducing their scale of operations. In industrialized countries with free markets in land, it is usually found that most of the land coming on to the market is bought by local farmers expanding their areas, farmers who most frequently already occupy medium or large farms. With the input capital, however, no such natural limitation on the amount available applies and it is clear that increasing quantities are used.

On the output side, again the expansion by some farmers will be in part offset by others quitting the enterprise. With those products facing very inelastic demand, and with very little demand growth arising from population changes, the result of a tendency to increase supply coming from technological advance will be a severe drop in price and the shedding of productive resources as they search for more profitable employment. The EU's Common Agricultural Policy (CAP) before the reforms of 1992 applied vigorous and expensive market interventions to keep up prices by buying and storing farm produce – thereby creating cereal, beef and butter mountains and wine lakes – and retaining resources in agricultural production to a greater extent than if prices (and hence profitability) were allowed to decline in response to expanding supply. Despite such market support it is evident that, in real terms, agricultural product prices declined and continue to do so, though this can be masked by what is happening on international markets.

The upshot is that *some* expansion of output and *some* shedding of resources normally accompanies technological advances, with the balance varying between commodities. In the UK part of the rising supply has displaced imports, the degree of self-sufficiency in all major products having risen in the post-war period.

Implications for the national economy

Taking a view even broader than agriculture, at the national level all technological improvements can be considered as output-increasing since, even if they do not result in a rise in the industry to which they most directly relate, they release resources for greater output elsewhere. National income rises as a result.

This effect on national output assumes, of course, that the released resources do not remain idle. Labour shed by agriculture has proved an important resource for other industries, and in many countries this applies to capital too. These are only released, however, when improvements in farming mean that food production has expanded through technological improvements to the point that it becomes possible to withdraw the resources without endangering the national food supply.

In developing countries the take up of technological advances in farming may be even more vital, since they have to contend not only with a desire for a growth of industries other than farming, requiring resources, but also with feeding a rapidly expanding population and increased demand resulting from higher incomes (see Chapter 3). Fortunately some of the necessary technological advances are already known about in richer countries, so that the poorer ones do not need to reinvent the knowledge about irrigation techniques, pest control, storage environments, etc. However, a danger exists that they will copy the technology of industrialized countries where labour is relatively

scarce and capital relatively plentiful, the opposite of what obtains in many low-income countries. Increasing care has been shown to ensuring that the technologies which such countries try to take up is appropriate to their relative factor endowments (especially their plentiful labour supply) and the skills of their populations.

Industry, regional and national consequences of technological advance

If a higher-yielding variety of cereal becomes available, or a waste-preventing trough for feeding pigs or a labour-saving device for cleaning milking equipment, it is not too difficult, knowing the innovation's characteristics, to predict the outcome for the enterprise in question. However, the consequences for the whole farm are less clear cut: labour saved from dairying may leave milk output unchanged but might be used to expand other enterprises or may simply be wasted if no other uses for it can be found. At more aggregate levels the outcome is even harder to predict because of the multiplicity of factors to be considered. Too narrow an approach can be wildly misleading. For example, when in the early 19th century the Luddites in England destroyed new textile machinery that threatened their jobs, they failed to foresee that in practice the machinery would result in more, not less, employment, though admittedly of a different sort.

However, in hazarding a prediction of the industry-wide implication of a technological change, the following characteristics should be in the forefront of consideration:

- the effect on output of the product, and the impact on price this has;
- the effect on total costs of production;
- the nature of the short-term supply for individual factors of production; and
- the effect on supply and demand of other products.

The income of farmers will correspond with the margin between the revenue coming from selling output and the sums paid out to other sectors of the economy for fertilizer and fuel (and where hired labour and tenanted land is involved, wages and rent). We know that most agricultural products face an inelastic demand at national level (see Chapter 3), so that an increase in output of $x\%$ causes the market price to fall by *more than* $x\%$, with the result that the returns coming back to farmers actually fall with an increase in output. Innovations which increase output volume without raising total costs must, on this basis, *lower* the income of the industry as a whole although, for the individual who first spots the innovation and applies it, his single income will benefit in the short term. However, if the market is international, a country may be able to expand its output without depressing price by much, if at all, and revenue can increase. Thus the UK Government is keen for its farmers to be innovative and improve their productivity and competitiveness within the EU Single Market for farm commodities, the implication being that it may be possible to increase the UK's share. Of course, if other Member States also increase their productivity and output, prices are likely to fall.

Innovations which reduce costs without any effect on output raise income at the industry level. With those which lower total costs but raise output the income of the industry could go either way, depending on the relative magnitudes of cost saving and revenue contraction due to the expanded output encountering inelastic demand.

Technological change can spread its impact through the price mechanism far beyond those farms which take up and apply the new technology. For example, improved transport and storage of tomatoes from Spain might seriously affect British growers. A new strain of wheat, suitable for growing in East Anglia because of an improved toleration of dry conditions, by expanding cereal output could depress prices and hence have income-reducing consequences for cereal growers in the wetter West of England where there is generally no call for such plant varieties. The pleas of such growers and farmers for import bans or other product price support mechanisms can be easily understood – they feel that they should not be penalized for the innovative activities of others. But the market system operates in a way which ensures, in an industry such as agriculture which has many independent operators, that the benefits of lower-cost production methods get taken up, with ultimate benefit to society although in the process it is inevitable that farmers wedded to old technology come under severe economic pressure.

The 'treadmill' of technological advance

The notion that innovations spread remorselessly and inevitably through the agricultural industry has been encapsulated by the idea of a 'treadmill'.

> The average farmer is on a treadmill with respect to technological advance. In the quest for increased returns, or the minimization of losses … he runs faster and faster on the treadmill. But by running faster he does not reach the goal of increased returns; the treadmill simply turns over faster.
>
> (Cochrane, 1958)

The more progressive farmer will always be on the look-out for means of improving his profit, and this generally results in higher output. The average farmer will thus find himself faced with product prices that tend to decline in real terms, so that he has to adopt the new technology in order to guard his income position and ensure his survival. It is the *average* farmer who is on the treadmill; it is the innovator who is responsible for its direction of turning. The unaware or those unwilling to adapt find themselves squeezed out of the industry as their old-style technology proves unviable at the lower level of prices and their incomes too low to continue in farming (this argument is explained in greater detail in Box. 5.10). There will be a residue of these hanging on to farming because they have few or no opportunities for employment elsewhere and an unwillingness to invest their capital in activities other than farming. In many cases they will be the elderly, the small farmer and those in the more remote regions. It is towards this disadvantaged group that agricultural policy is frequently directed, with the intention of supporting their incomes or aiding their outflow from farming.

While technological change can result in a lower income to the industry as a whole, it would be wrong to conclude that this *inevitably* leads to lower farm incomes at the individual level. If innovations require farms to be larger, then a reduction in aggregate income could result in a maintained or even increased average income per farm if a sufficiently large number of people leave the farming industry. However, average figures tell us nothing about the distribution of incomes. In the process of adjustment it seems inevitable that there will always be some farms which are unviable, with incomes squeezed to the level that their occupiers wish to leave farming.

Rates of diffusion of new technology

Some innovations spread through farming faster than others, and some experience a period of dormancy – while they are known about they

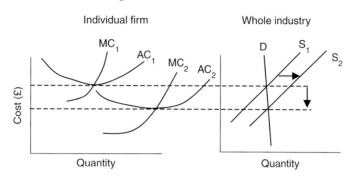
have to await some event or change in prices to trigger their adoption. Some farmers are habitual innovators, always the most progressive and the first to buy a new type of machine, while others cling on to old methods so that their farms come to resemble museums. This prompts the question of what factors influence the rate of take-up of innovations and which farmers first adopt them. This is of obvious importance if, for example, it is government policy to promote an efficient modern agriculture and thereby aim to aid the incomes of the farming community.

Information

The first step necessary for the take-up of any new technology is the dissemination of information about it. Commercial advertising literature and the farming newspapers and broadcasts play an important part in creating awareness, and these avenues can be augmented by publicly funded information services (such as the Farm Advisory Service that exists in all EU Member States, with its extension workers and advisers on environmental practices) and private sector consultants. As will be seen below, different groups of farmers tend to use different channels of communication, with the more progressive tending to be most likely to use source material (scientific reports) whereas the least progressive will tend to rely on the farming press and talk from farming friends and relations.

If the government wishes to affect, say, the small, less progressive farmer, it will need to choose the appropriate means of communication or its efforts will be wasted. This should force the designers of agricultural policy into clarifying who they are aiming at – the large, progressive farmer who dominates total agricultural output or the backwoodsman who may have an unacceptably low income. The latter may be particularly hard to reach by either public or commercial advisory services.

Uncertainty

Any innovation by its very nature will involve some element of uncertainty. Even when extensive field trials of some new crop variety or machine have been undertaken, there may be surprises when it is put into the hands of commercial farmers. For this reason alone one might expect innovations to be taken up first by the bigger, diversified, more prosperous farmer since usually they are best able to take the risk. As more and more farmers take it up more becomes known, the uncertainties are reduced and the more risk-averse farmers (i.e. those less willing and able to take risks) grow willing to try the new method.

The economic climate can be expected to affect the rate of innovation; in times when incomes are subject to instability and uncertainty because of some influence such as changing government policy, farmers might be expected to steer clear of adding to their insecurity by adopting innovations of an unproven nature. While simple cost-saving exercises might still be found attractive, major changes and those requiring heavy capital expenditure would be avoided until more certain times returned.

Capital aspects

It is commonly found that innovations such as new seed varieties which simply replace one form of working capital with another are taken up far faster than changes which require new capital assets and the abandonment of the old. One useful classification of technological changes is as follows:

1. Those that result in capital saving, such as the development of the artificial insemination scheme for cattle in the UK, which meant that most smaller dairy farmers needed no longer buy and keep a bull.
2. Those that imply little or no capital change, such as an improved type of fertilizer or new seed variety.
3. Those that require much additional capital, such as a new type of milking parlour.

While the first two can be rapidly taken up, the third spreads much more slowly. Not only must

an expensive new building be financed, perhaps competing with other potential investments on the farm, but existing outmoded equipment may still be viable. For example, once a milking parlour has been installed its cost becomes of little more than historic interest; its scrap value is probably zero and if a new type of parlour comes on the market the farmer decides whether to continue using his existing equipment on the basis of its current and future operating costs (labour, feed, etc.) in relation to the price of milk. An investment appraisal could well favour the continued use of the outmoded equipment, although with its gradual wearing out and, most probably, a fall in the price of milk brought about by expansion of those farmers who have just acquired the new technology, there may come a time when abandonment in favour of investing in the new equipment becomes the preferable option. This could mean scrapping the old before it reaches the end of its physical life.

Management requirements

Management ability, described as the power to predict and foresee outcomes and plan accordingly, must be present before any innovation can be put into effective operation, but some new techniques or products clearly demand greater or different management skills than others. Some innovations require little additional expertise – a heavier-yielding variety of oats would probably be within the capabilities of any current oat grower. A whole new crop, such as grain maize, lupins or midnight primrose would require rather different skills and an intensive, environmentally regulated piggery yet others. Where the skills required are close to those already being exercised, the innovation is the more likely to be taken up rapidly by the industry, but those where new skills have to be acquired by training, or perhaps by hiring them in, will be adopted less readily, and perhaps not at all by most farmers.

It is interesting to note that farmers are increasingly called upon to use management skills which, to the layman, appear to be primarily non-agricultural. For example, tourism finds an increasing place in the income-generating activities of farms in the West of England and in the uplands. Farm computers, electronic animal supervision, processing and packaging, and financial control all require talents in general business management rather than specifically agricultural skills, suggesting that increasing levels of farmer education are required to cope with the advance of technology. Consonant with this is that in the USA it seems that innovation proceeds most rapidly in areas: (i) where formal education is most available; and (ii) where farm families have more time and money for travel, study and meeting fellow managers of high ability. In the UK it has been noted that formal higher education (university degree courses, etc.) is associated with more rapid adoption of scientific advances and a greater willingness to undergo specialist technical training associated with innovations; it is the experience of being exposed to higher education which seems to be important rather than the subject that is studied. Furthermore, higher education is more commonly encountered among the occupiers of large farms, though the line of causality is not self-evident. Probably the larger farmers, who are well placed to be innovators, are also in a favourable position to ensure that their children receive a formal education, which reinforces the rate of innovation on large farms.

Factor and product pricing

The point was made earlier that an advance in knowledge does not necessarily imply that changes will necessarily take place in the production techniques actually used by farmers. There are countless inventions which never prove to be practical – such as man-powered flying machines and Brunel's atmospheric railway – and agriculture has its fair share of these. Even if they can be made to perform satisfactorily in the technical sense, they may often lie dormant until a change in economic conditions causes them to be attractive to farmer–businessmen for adoption. This is well illustrated by automated field machinery: the technical knowledge has long existed to enable fields to be sprayed or cultivated by unmanned tractors but these devices have not been much used (save in experimental circumstances) because they were not commercially viable. However, were the cost of labour to rise and/or the cost of automation to fall (i.e. if a change in the relative price of inputs were to occur) then the automated system could prove attractive. This is illustrated in Box 5.11. There are signs that recent technological advances in geolocation equipment has so lowered the price that precision application of plant chemicals of machines guided by satellite navigation may become the norm, at least on large arable farms.

Of course the reverse could also happen. If labour were to become very cheap, and some politicians advocate a fall in real wages in order to reduce unemployment, some of the old techniques which have been abandoned because of high labour costs could well be reinstated. In the extreme, one might see again ploughing done by men and horses. This degree of retrenchment is unlikely, but there is a move to protect the quantity of employment on farms, primarily for social reasons, by removing the distortion to input prices that has often been caused by government grants and excessive tax allowances on capital items; many of these have now been abolished in the UK.

Model of the process of innovation by farmers

The process by which technological advance is taken up is thought to consist of four basic stages: (i) awareness of an innovation; (ii) the formation of interest and the gathering of information; (iii) the 'mental acceptance' of the innovation (that it is something which should be taken up); and (iv) the actual adoption of the innovation.

These stages sweep in waves through the farming community, but not uniformly to all farmers. Some will acquire knowledge earlier than others and some will accept the desirability of a new technique faster than their neighbours. Some farmers will be in a position in which they can incorporate the innovation into practical farming more rapidly than others. The spread of knowledge is likely to be faster than the rate of physical uptake, and this will depend on the type of innovation – where it originates, whether it is commercially promoted, the sorts of management required, capital aspect and so on. The progression of each step through the farming community in generalized form is shown in Fig. 5.28. It is commonly found that the awareness and adoption curves are both approximately symmetrical in the form of an S.

We have already explained that the rate with which innovations are taken up by farmers will be affected by factors such as the amounts of capital they require and the necessary management skills. Figure 5.29 illustrates this, with the first 50% of farmers taking different times to adopt the various innovations. The 'fast' curve might be illustrated by a new wheat variety and the 'slow' curve by rotary milking parlours.

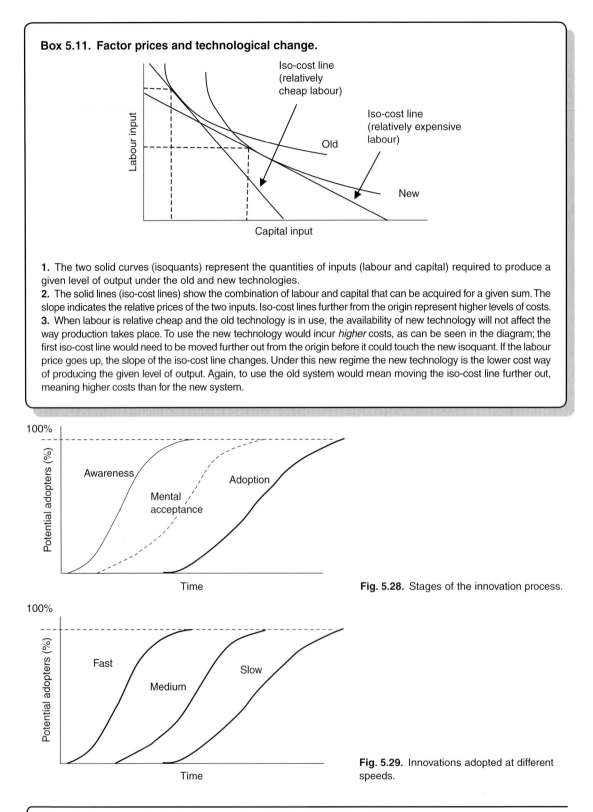

Box 5.11. Factor prices and technological change.

1. The two solid curves (isoquants) represent the quantities of inputs (labour and capital) required to produce a given level of output under the old and new technologies.
2. The solid lines (iso-cost lines) show the combination of labour and capital that can be acquired for a given sum. The slope indicates the relative prices of the two inputs. Iso-cost lines further from the origin represent higher levels of costs.
3. When labour is relative cheap and the old technology is in use, the availability of new technology will not affect the way production takes place. To use the new technology would incur *higher* costs, as can be seen in the diagram; the first iso-cost line would need to be moved further out from the origin before it could touch the new isoquant. If the labour price goes up, the slope of the iso-cost line changes. Under this new regime the new technology is the lower cost way of producing the given level of output. Again, to use the old system would mean moving the iso-cost line further out, meaning higher costs than for the new system.

Fig. 5.28. Stages of the innovation process.

Fig. 5.29. Innovations adopted at different speeds.

Farmers as fast or slow adopters

Our earlier review of what factors could affect the rate of spread of an innovation led to the suggestion that some farmers would be more ready to adopt changes than others. This is reflected in the shape of the adoption curve, with some farmers being innovators while others are very slow to adopt, described technically (and vividly) as 'laggards'. If, instead of the cumulative frequencies used to generate the curves in Fig. 5.29, the number (or percentage) of farmers adopting a particular innovation *per time period* is considered, the pattern of early or late adoption is even more clearly portrayed (Fig. 5.30).

First to come are a small number of innovators, followed by larger numbers of early adopters and then the majority. The last to adopt are the 'laggards'. Overall the distribution takes on an approximately bell-shaped 'normal' distribution. It is generally found that farmers who are laggardly or innovative with respect to one innovation tend to have this attitude towards many other new technologies. This notion of innovative or laggardly behaviour can also be extended to communities, and those areas of the country or types of farmer which are fast or slow adopters tend to hold their relative positions over the years.

There are some recurrent personal characteristics of individual farmers who are persistent innovators, laggards or between the two extremes. Empirical evidence suggests that size of farm, income, level of education and off-farm experience are positively linked with innovative behaviour. In contrast, laggards are usually found on small farms, have the lowest incomes and tend to be elderly.

Perhaps less obvious is the marked association with the channels through which communications are received. As noted above, innovators tend to have the greatest contact with scientific information sources, to use impersonal channels of information, and to interact with other innovators. Early adopters tend to have greatest contact with local change agents (including consultants, extension workers and commercial technical advisers) and to be competent users of the mass media. However, laggards tend to depend for their information on neighbours, friends and relatives and to be suspicious of advisers. This is of great relevance to government policies encouraging the uptake of productivity-improving innovations (see Box 5.12). If the intention is to target this policy on the small elderly farmer there is little point in publishing lots of scientific literature, as the target farmers do not read it. Rather it will be necessary to approach such farmers through demonstrations, neighbours, friends, relatives and the farming press, a more difficult task than posed by the big farmer who is already aware of information sources, able to accept the risks associated with innovation and is encouraged by the taxation systems (and sometimes publicly financed grant aid) to invest in them.

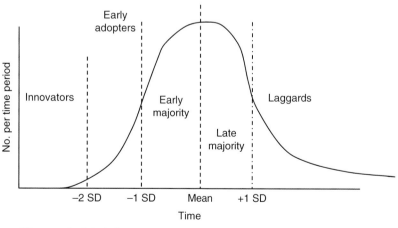

SD = standard deviation

Fig. 5.30. Types of adopter.

Box 5.12. Case study – Public provision of advice to farmers in Wales.

The Welsh Government funds a system called Farming Connect (FC) to provide advice to farmers as a way of promoting the uptake of innovations and good practice in Welsh agriculture. It covers technical matters (such as nutrition and health issues in sheep), skills training (safe use of machinery and use of the Internet), business management, and awareness of strategic issues (such as the likely impact of global warming on farming). This knowledge transfer programme includes a mix of tools: (i) mass media (contributions in the national magazine directed at farmers, technical advice available on the Internet, printed fact sheets, etc.); (ii) demonstration farms; (iii) farm walks; (iv) discussion and networking groups (including ones for females only); (v) surgeries on specific issues such as planning consent and succession; (vi) assessment of skills needs and guidance to training courses; (vii) development of farm plans (which often lead to the use of other grants); and (viii) advice on how to maintain the good agricultural and environmental conditions on farmland that are a condition for receiving the main form of direct income subsidy under the CAP. Most of these services, organized by FC staff, are free to Welsh farmers but, for some one-to-one forms of advice that are delivered by consultants, they are required to make a financial contribution (typically 20% of the consultant's charge).

The wide range of activities is designed to enable all types of farmer to benefit, though it is recognized that there are still some who are hard to reach, typically those in the more remote parts of the country, the elderly and those on small farms. Word of mouth by relatives or demonstrations by trusted neighbours are probably the most effective ways that some of these can be involved with ways of improving their own farm's performance.

FC underwent an independent evaluation of its effectiveness in 2013. This raised several issues:

- Which is the target group of this spending of public funds, those farmers in greatest need of advice (often the small farmers) or those who control most of the land and who thus could make the biggest improvement to national performance?
- Which knowledge transfer activities give best value for money? (This will depend on who the advice is meant to reach.)
- How does the performance of farms that have received FC advice differ from those that did not?
- Has the system generated longer-term benefits, such as empowering farmers to find solutions to their own problems by increasing their ability to search out information, increasing their networking skills and willingness to share knowledge with others, and raising their ability to draw on public resources (collectively viewed as increasing 'social capital')?

By considering such issues it should be possible to enhance the performance of FC.

Under the CAP, all Member States are required to have advisory systems. A main part of the rationale is that these are necessary so that farmers can deliver environmental and other public goods, but countries can add other services (as happens in Wales).

The future direction of technological change

It will have become apparent that most of the technological advances mentioned in this chapter have not originated from within the agricultural industry itself but in other sectors – the chemical and fertilizer industry, farm machinery manufacturers, computer programmers, plant scientists, animal geneticists, etc. In the 19th century there was a tradition of farmers coming together to develop a scientific approach to farming and, more recently, farmers have shown themselves adept at making adjustments and improvements to machines and in breeding their own high-performing animals. But, fundamentally, the main flow of new technology adopted by farmers has come from elsewhere. Some of this will emanate from commercial companies trying to make a profit by inventing and developing new technology (everything from genetically modified (GM) crops to the latest chemicals to treat sheep footrot). From the mid-20th century much has come from research institutions and universities financed by government. A case could be made that the innovations generated much benefit to the economy as a whole and this justified the use of public money in this way. Furthermore, in the UK (and many other EU Member States) there has been a tradition of publicly funded extension services to transfer knowledge of innovations to farmers. Changes in perceptions of who should bear the cost and who gets the benefit led to a major cutback in the public finance of research and extension systems, but there are concerns in the second decade of the 21st century that this has resulted in a slow-down in the rates of innovation and productivity increase. It is likely that some of the withdrawn public funding will be restored.

Finally we must raise a question about the direction of technological advance. It has been stated earlier that most advances in agriculture over the past half-century have been of the output-increasing and input-using type, and especially of the sort which used more capital but less labour. To some extent, this form of innovation has been encouraged by government policies to support the prices that farmers receive. The development of new varieties, improved machines, etc. that results from the extra profits that plant breeders and engineering firms can make because of agricultural policy is termed **induced innovation**.

Part of the reason may have been the subsidies to investment that existed in the EU for much of the last half-century, creating a favourable climate for labour-saving capital-using innovations by lowering the price of capital relative to labour costs. Also, because agricultural scientists have tended to regard more output (heavier crop yield per hectare or more milk per cow) as an indicator of their success, the direction of technological advance has been encouraged perhaps towards this sort of development. Since the mid-1990s financial incentives for production-increasing investments have been greatly reduced in the EU. Those that remain are largely to do with enabling farmers to diversify into non-agricultural activities, on or off the farm, and for environmental purposes. Where public funding for research is restored, it is likely to focus on how innovation in agriculture can make an enhanced contribution to: (i) conserving biodiversity; (ii) protecting vulnerable landscapes; (iii) reducing climate change; (iv) reducing energy use from burning hydrocarbons; (v) improving animal welfare; and (vi) contributing to human health and well-being. This is a much more sophisticated research agenda than simply producing more food.

Note

[1] TFC is often used as an abbreviated form not only of total factor cost, as here, but also of total fixed cost. To distinguish between them, later in this chapter TF_xC is used for the latter.

Exercise on Material in Chapter 5

NOTE: Students who have omitted the first part of Chapter 5 prior to 'Time and Scale of Production' should commence these exercises at 5.5.

Answer the questions in the manner indicated. Graph paper is required to complete some of the questions. Answers and explanations are given in Appendix 2.

5.1. The production function
   (a) Draw the graph of a typical one product-one variable input production function, exhibiting both increasing and diminishing returns, with average and marginal product curves.
   (b) Using it, answer the following:
      (i) At the point of inflection of the production function .... product is at its maximum.
      (ii) At the point where a line from the origin is tangential to the production function .... product and .... product are equal.
      (iii) When total product is at its maximal .... product is zero.
      (iv) When total product is declining with increasing quantities of input .... product is negative and .... product is positive and *declining/increasing*.
      (v) The point where marginal product is greatest is also called the point of ....
      (vi) Where, apart from when only 1 unit of input is used, do total product and average product coincide?

5.2. You are given the following information about the production of a firm (see table overleaf). Complete all the columns.
   (a) Does the firm appear to be producing under perfect or imperfect competition?

   (b) What are its fixed costs?

   (c) Plot marginal revenue, average total cost, average variable cost on the same graph. Also plot marginal cost, but to get over the difficulty of 'lumpiness' of the input, plot

Output (units)	Total revenue	Marginal revenue	Total costs	Fixed cost	Variable cost	Marginal cost	Average variable cost	Average total cost
0			110					
1	50		140					
2	100		162					
3	150		175					
4	200		180					
5	250		185					
6	300		194					
7	350		219					
8	400		269					
9	450		349					

the marginal cost of each unit at its mid-point (e.g. plot the marginal cost of the second unit of output half-way between 1 and 2 units on the horizontal axis).

(d) A firm is in equilibrium when *average/marginal/total* cost equals *average/marginal/total* revenue. (Cross out the inappropriate words.)

(e) At what output is the firm at equilibrium? (Take the nearest whole-number of units of output.)

(f) Estimate the profit (net revenue) earned by the firm at equilibrium in two ways:
   (i) Total revenue (........) minus total costs (........) = ...
   (ii) Average revenue (........) minus average total costs (........) multiplied by the number of units of output (........) = ...

(g) What is the minimum price of the product at which the firm will produce, and what will be the quantity produced:
   (i) in the long run?
     Price
     Quantity
   (ii) in the short run?
     Price
     Quantity

5.3. A chicken farmer can alter the composition of the food he uses for his flock by substituting barley for oats. If he increases both, the yield of eggs increases. The isoquants he faces for two levels of production are given below (tonnes of oats and barley):

(a) Plot these two isoquants.
(b) If the price of oats and of barley are the same at £30/t, what is the least-cost combination

230 eggs/bird/year		250 eggs/bird/year	
Oats (t)	Barley (t)	Oats (t)	Barley (t)
10	0.5	10	0.8
8	0.7	8	1.4
6	1.2	6	2.2
4	2.2	4	3.8
3	3.4	3	5.6
2	5.2	2	10.0
1	11.0		

of barley and oats the farmer should use in the food to produce 230 eggs/bird? (Hint: construct an iso-cost line and slide it until it is tangential to the 230 egg/bird isoquant.)
Tonnes of oats .... Tonnes of barley ....

(c) What are the relative proportions of the two cereals in the food? Oats/barley

(d) What is the combined cost of oats and barley in the food if they both cost £30/t?

(e) What is the least-cost combination to produce 250 eggs/bird if the price of oats equals that of barley?
Tonnes of oats .... Tonnes of barley ....
Are these in the same proportions as (c) above? *Yes/No*

(f) What is the least-cost combination to produce 230 eggs/birds if the price of oats drops to half that of barley?
Tonnes of oats .... Tonnes of barley ....

(g) Does the fall in the price of oats result in more or less oats being used in the mix of cereals in the poultry food?

5.4. A farmer, with his resources on a 100 acre farm, can produce either all wheat or all hay for sale, or various combinations of the two. The possible combinations are:

Hay (t)	Wheat (t)
0	200
50	250
100	260
150	200
200	140
250	50
260	0

(a) On graph paper plot his production possibility curve, with wheat on the vertical axis.

(b) From the table estimate his marginal rate of substitution of products:
  (i) between producing no hay and producing 50 t.

  (ii) between producing 150 t of hay and 200 t of hay.

(c) What are the most profitable levels of wheat and hay to produce under the following sets of relative prices of wheat and hay? (Hint: construct iso-revenue curves to find these quantities by connecting quantities of hay and wheat which bring in the same revenue.)
  (i) The price of wheat and hay are the same per tonne.
     Wheat (t) .... Hay (t) ....
  (ii) The price of wheat is double that of hay.
     Wheat (t) .... Hay (t) ....
  (iii) The price of wheat is treble that of hay.
     Wheat (t) .... Hay (t) ....

(d) If the farmer knows that he will be unable to sell his hay at any price, will it still pay him to produce some? (Assume that it can be destroyed at no cost.) *Yes/No*

Students who have omitted the first part of Chapter 5 should start here.

5.5. What of the following best describes a fixed cost in production economics terminology?
  (a) The cost of any input whose price per unit has been fixed by long-term contract.
  (b) The cost of any input which rises in a fixed proportion with output.

(c) The costs of production which would have to be borne even if the firm temporarily ceased production.

5.6. (a) Cross out the inappropriate alternatives from the following statement:

If marginal cost is rising with increasing output, average total cost *will be falling/will be rising/may be rising or falling.* Average fixed cost will be *rising/falling.* The minimum price which the entrepreneur will be able to accept in the short run will correspond to the lowest point on the *average variable cost/average total cost* curve, but in the long run price must be at least the lowest point on the *average variable cost/average total cost* curve. The rising marginal cost curve cuts both the *average fixed cost/average variable cost* curve and the *average total cost/total cost* curve at their lowest points. To make maximum profits an entrepreneur will select that level of output at which *average revenue/price/marginal revenue* equates with *average total cost/average variable cost/marginal cost.*

(b) Give three sources of economies of scale.

(c) Give three sources of diseconomies of scale.

5.7. Technological change:
  (a) Describe what is meant by a technological advance.

  (b) What is the effect of a technological advance on the following:
    (i) the factor–product relationship (in terms of shifting production functions and cost curves)?

    (ii) the factor–factor relationship (in terms of shifting the iso-quant)?

    (iii) the product–product relationship (in terms of shifting the production possibility boundary)?

  (c) Name three factors that affect the speed at which new technology is taken up by the operators of firms.

(d) Name three characteristics of entrepreneurs that tend to make them faster adopters of new technology.

(e) Which are the correct components of this description of the 'treadmill' of technological advance, as expounded by Cochrane? (Cross out the inappropriate alternatives.)

The *small/average/large* farmer is on a treadmill with respect to technological advance. In the quest for *reduced costs/increased output/increased returns*, or the minimization of losses … he runs *slower and slower/faster and faster* on the treadmill. But by running *slower/faster* he does not reach the goal of *reduced costs/increased output/increased returns*; the treadmill simply turns over *slower/faster*.

(f) Give three likely implications of the treadmill of technological advance for the structure of the agricultural industry.

6 Factors of Production and their Rewards: Theory of Distribution

Introduction

It is easy to lose sight of the reasons why economic activities take place when becoming engrossed in the technicalities of production economics, or the implications of elasticities of demand. This is an appropriate point to restate that economics is concerned with the study of how the wants that people have – for food, clothes, cars, hi-fis, entertainment and a whole host of other goods and services – are satisfied within the limits set by the resources which are available. This section examines these available resources in more detail.

The nature of production

The term **production** conjures up the image of factories making cars, freezers, clothing or similar goods and in the process employing workers, buying steel, paint, rubber, textiles and a host of other necessities. This picture is rather too narrow and urban, but contains the essential nature of production – that consumers require the resources available for production to be changed into a form in which they can satisfy wants. In car manufacture, metal, glass, plastic and other materials are assembled by a labour force working in a factory into a product demanded by the consumer. This is one type of production process and each of the items needed in it can be described as a **factor of production**. We will see later that factors can be grouped into four main categories.

Assembly is only one of the forms which production can take. Iron production from iron ore using fuel and blast furnaces changes the *form* of the metal; it is then passed perhaps to a second production process where its *shape* is changed by casting or pressing into components which may be the starting point for perhaps another process.

Nor must production be limited to tangible goods: the provision of entertainment is just as much a production process. Although the music pounded out by a public-bar pianist cannot be seen, it is the output from the combination of his physical effort, his mental ability and his instrument (machine). Similarly, transport is itself a production process, as are marketing and advertising. The provision of services (including tourism and entertainment, food processing and packing, retailing, and financial services such as insurance and banking) is becoming an increasingly important activity in the economy, including in rural areas.

Agriculture is fundamentally different from most other industries because its production processes are dependent on biological growth, not manufacture.

Specialization and exchange

A man stranded on a desert island would need to satisfy his wants by producing his own food, shelter and so on; he would be forced by circumstances to become self-sufficient. In our modern society such independence is rare, although farmers are perhaps in an unusual position in that they can be more self-sufficient than most – they can feed their own family, produce their own timber for building repair or for fuel, etc.

Most people, however, participate in production by specializing in one task or occupation, such as being a secretary or teacher or factory worker for which they receive payment, and with this money buying all the other things which they require. This is sometimes described as **dependent activity**, in contrast to self-employed people who are said to undertake **independent activity**. As has been pointed out earlier, specialization and exchange permit a higher level of production and consumption than could be achieved by self-sufficiency; the formal explanation for this will be presented when the benefits from international specialization and trade are described (Chapter 8).

Classification of factors of production

All production processes, whether of the self-sufficient sort or the product-exchange sort, involve taking a collection of inputs (for milk production these would include animal feed, water, labour for milking and stock care, land to grow fodder, buildings and equipment, management skills) and combining them in some way, resulting in an output, or product. In our example we have several products – as well as milk, which is the primary product, calves and dung are by-products. It is generally agreed that the inputs to production processes can be classified into four broad groups of factors of production: (i) land; (ii) capital; (iii) labour; and (iv) entrepreneurship.

Each main group of factors of production is considered separately in this chapter. Almost all production processes involve all four. Knocking a crab apple off a wild tree with a stick is a very simple process, but all four types of factor are present. The earth's resources (land) have produced the apple, the stick is a piece of capital, labour was involved in using the stick, and the decision to pick the apple was an entrepreneurial one.

Land as a Factor of Production

Land as a factor of production has a particular importance for agriculture because, compared with other industries, farming requires a lot of it. In the UK, agriculture (excluding forestry) takes up about four-fifths of the total land area but only engages about 1.3% of the working population. Just as we describe a firm which needs a high proportion of labour in its mixture of inputs (such as thatching, or any craft work where mechanization is impossible) as labour-intensive, so we should label agriculture as land-intensive when comparing it to most other industries.

In economics the term **Land** not only includes the soil surface, but embraces minerals under the surface, water, climate, topography – indeed everything that is a 'gift of nature' and not the result of man's past activity. The early economists paid much attention to land as a factor of production, notably Thomas Malthus (1766–1834) and David Ricardo (1772–1823). Land was considered special in that it was strictly fixed in quantity, and was thought the only factor subject to the Law of Diminishing Returns. We now know that neither assumption is true – productive land can be reclaimed from the sea, or lost through coastal erosion, and diminishing returns apply to labour and capital too.

Land still has, however, several distinguishing characteristics in modern economics justifying its separation from the other factors.

1. Land means *space* in which production processes take place. Almost all such processes demand some space – factory space for car manufacture, office space for a firm of accountants – and agriculture requires more space than most. Crops require space to grow, cows require space to graze, although the space requirement for egg production and pig rearing is much less than most other farming enterprises and this has earned them the title 'factory farming'. When the areas required to grow food for the housed animals are also taken into account, 'factory farming' may not be such an appropriate label.

2. Land means *location*. Pieces of machinery can be moved from place to place, and even buildings can be taken down and rebuilt elsewhere (e.g. some UK buildings have been transported and rebuilt in the USA), but a piece of land is actually the location it occupies. Much of the value of a building site on farmland adjacent to an expanding town arises from the uniqueness of that land's location. J.H. von Thünen in *Der Isolierte Staat* (The Isolated State) of 1826 pointed out that, all other things being equal, farmland nearest to markets would fetch highest prices because transport of produce to the market was shorter, and hence easier and cheaper. The further from the market one progressed, the lower would be the value of the land. Ease of transport and land values would mean that crops of high value per unit area (e.g. horticultural produce) would be grown nearest the town, with less intensive production further out. This would produce concentric rings around the town of horticulture, then perhaps dairying, then perhaps corn and beef production (see Fig. 6.1).

Such Location Theory breaks down in a world of easier transport (von Thünen wrote in the time of horse transport) so that the location of each crop tends now to be more heavily influenced by where the most suitable soils and climates are. Nevertheless, location is still influential in, for example, the growing of peas in areas accessible to freezing factories, or the organization of large mixed farms where the use of a field for grazing rather than potatoes may be as much determined by the fact that it is within a cow's walking distance from the milking parlour as by its soil type. Rural businesses are often hampered by the lack of a large number of potential buyers of their goods and services on

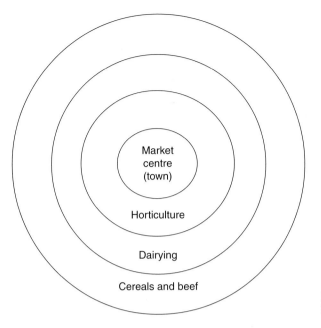

Fig. 6.1. Land use based on von Thünen's Location Theory.

In the figure:
- Market centre (town)
- Horticulture
- Dairying
- Cereals and beef

their doorsteps, so their ability to grow is constrained. They may also suffer from their remoteness through having to pay more for their inputs (delivery charges may be higher) and transport costs may make them less competitive than firms based in large towns.

3. Land is a *repository of natural forces*. This third characteristic of land is probably one which springs most readily to the minds of agriculturalists. Agriculture (using the term in its broadest sense) is unique in being so fundamentally dependent on natural forces – physical, chemical and biological – contained in the soil because it is concerned with biological growth rather than manufacture. Farmers are thus preoccupied with **soil fertility** which to them is the practical manifestation of the sources of growth. Indeed, to them the term often acquires semi-religious overtones. A **sustainable** farming system is one that takes into account the longer-term impact of present production, and tries to ensure that future generations are able to use the world's natural resources (see Box 6.1).

If land is 'a repository of natural forces', then the term can easily be interpreted as including coal, oil and other minerals. Little land, however, remains untouched by human hand. Certainly in developed countries the countryside is a result of the 'gifts of Nature', having been worked on by man to clear forests, drain marshes, plant hedges, improve soil structure and correct acidity and so forth. This past effort has resulted in improvement in levels of production, and is really a form of the next factor – capital.

Capital as a Factor of Production

The term **capital** means different things to different people. If a farmer were asked to list the items making up his capital, he would probably include the value of all the assets he owns and uses on the farm – his animals, machinery, buildings and land. He would probably deduct the size of any borrowings he may have made from banks or relations. A shareholder in a large public company might interpret his 'capital' as the value of his shares; shares indicate a part-ownership of the assets of the business to which they relate.

To an economist the term 'capital' implies something rather more precise. Goods such as shirts, shoes, tea, bread, etc. are wanted by consumers, and are hence called **consumer goods**. Other goods, such as tractors, machine tools, factory buildings and milking parlours are not wanted for their own sakes, but are wanted by producers because they help in the eventual production of consumer goods. Such articles are thus called **producer goods** or more commonly, capital goods, pieces of capital or, simply, capital.

A definition of 'capital' might be 'anything which has been produced and is used to increase the effectiveness of current productive activity'. Another is 'wealth employed in the production of further wealth'.

The essential nature of capital can be explained by a simple model. Imagine a factory producing cars (the nature of the product is not important), and imagine that this factory occupies in total of 6000 m² of floor space (i.e. the factor 'land' is present), employs 140 workers ('labour' is present) and uses 100 machine tools of a variety of types ('capital' is present). Part of the factory (1000 m² and 10 workers) is set aside to make machine tools to replace the existing ones as they wear out, and the effort of these people is just sufficient to maintain the existing stock of machines. The factory produces 500 cars/year. The total output of the factory can be represented by the following equation:

Total production = Production of cars (consumer goods) + Production of machine tools (capital goods)

Suppose the management (i.e. the factor 'entrepreneurship' is present as well) decides to increase the number of machine tools used in the factory, and that it does this by increasing the number of staff and the floor area devoted to making machine tools. It will have to divert workers and area from actually making cars to do this, but the rate of production of machine tools will rise so that they are being produced faster than they are wearing out. This means that the number of machine tools used on the shop floor will rise to a higher level, at which the increased production of tools just keeps pace with the wearing out of tools in use. With the greater number of tools, the output of cars eventually rises to 700/year. Table 6.1 summarizes the positions.

Several points can be illustrated with this model:

1. In the 'after' situation, the same total labour force and floor space is used with a *greater quantity of capital*, with a resulting higher level of output. There is no *guarantee* that building up the stock of capital in use will increase output, but in this case the factory's investment in additional capital goods had had that effect, perhaps because the labour force can do things faster (say, with power screwdrivers), or do things with machines that were previously not possible, such as lifting.

2. In the factory as a whole in the 'after' situation there is a greater quantity of capital used relative to the quantities of the other factors, which in total have not altered; the production has become more *capital-intensive*. In the car-making part of the factory, fewer people are used but more machinery; not only has this process become more capital-intensive, but there has also been a substitution of capital for labour.

Table 6.1. A simple model of increasing the use of capital.

	Before	After
Total labour force	140 workers	140 workers
Total factory space	6000 m^2	6000 m^2
Making cars	130 workers, 5000 m^2 100 machine tools	120 workers, 4000 m^2 110 machine tools
Making machine tools	10 workers, 1000 m^2	20 workers, 2000 m^2
Output	500 cars/year	700 cars/year

3. A third feature of this factory is the *division of labour* (or labour specialization). Some workers specialize in producing machine tools while others specialize in car production. Although the model does not spell it out, the probability is that, within the two divisions of the factory, some specialize in, say, electrical work while others use lathes. The division of labour has several obvious advantages. Staff members acquire skills if they specialize and the employment of people with specialized training or abilities becomes possible (e.g. designers or draughtsmen). In certain circumstances there may be substantial time savings, for example, if a blacksmith can keep his forge constantly in use, work will be done quicker than if individual workmen have to light up a forge each time they need one. Also there may be less fatigue if, for example, one man fits the bonnets to cars while another, working at a lower level, fits sumps; the alternative would be for one man to clamber up and down. Perhaps the most significant aspect of the division of labour is that breaking down the production process into small simple operations allows more mechanization and maybe even automation, where the limitations of human abilities are replaced by untiring machines.

While the division of labour can be advantageous, it also has negative features. The monotony of repetitive acts can lower efficiency and encourage industrial unrest. Workers can become alienated – they lose the craftsman's interest in his product and become mere machine minders, and a strike by a few key specialists can halt the output of complete factories. (See also the association of specialization with the economies of scale cited in Chapter 5.)

4. The fourth feature of the factory model is that of the period of *waiting* which elapses between the time when the decision to build up the stock of machine tools is made and the time when the output of cars rises to its new higher level. This is illustrated in Fig. 6.2. Note that when some men and floor space are switched from car production to tool production the output of cars falls and only recovers after the new tools have been constructed and are in use. With the enlarged number of tools the *eventual* level of car output is higher than at first, but the factory's operators have had to put up with a fall in output in the short-run to make possible a high final output level.

5. The fifth feature of the model is that of *maintaining capital intact*. This has already been touched on, but is of sufficient importance to stand as a separate feature. Capital goods wear out with time and usage – this is called **capital consumption**. While this is obvious with machines and vehicles, it also applies to roads, buildings and so on. Capital goods are only wanted by producers because such goods aid production. Clearly, a new machine with a full working life in front of it is capable of aiding production for a lot longer than a worn-out machine, so the value of the new machine to the producer is much greater than the old one. In our model, if steps were not taken to compensate for the wearing out of machine tools, after a time the factory would have found itself equipped with worn-out and hence valueless capital.

Some of the factory's resources were, however, devoted to producing replacement tools, so that the overall value of machines in use did not fall (or, in other words, the value of the total capital in use was maintained intact). Note that this meant in the short run the output of cars suffered because men who could have produced cars were making tools, although this higher level of output could not have been maintained indefinitely because the machines would have been getting older. An alternative policy for the factory (although this implies a different economic model) might have been for it to use all its men and floor space for making cars, and to set aside out of its income a sum to enable it to buy new machine tools from other factories specializing in tool making. (This implies a greater total quantity of factors of production used in car manufacture by the extent of those used by the tool-making firms.) This sum, called a **depreciation allowance**, could be calculated by knowing how fast machines wear out, and hence when replacements would be needed.

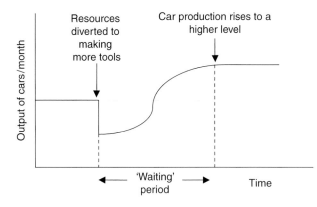

Fig. 6.2. The 'waiting' period associated with a decision to invest.

Graph labels: Output of cars/month (y-axis); Time (x-axis); "Resources diverted to making more tools"; "Car production rises to a higher level"; "'Waiting' period".

Capital in the real world

Before passing to the next factor of production, it is worth broadening the discussion of capital to the model's implication for the real world. The developing countries are keen to expand their levels of production and thereby raise the living standards of their populations. To do this they require more capital (factories, roads, etc.) yet they are caught in a vicious circle of poverty which prevents them building up their stocks of capital. Investment requires some resources to be diverted to the production of capital goods, yet if all a country's resources are already fully stretched in providing necessities for life, then little or none can be spared for capital build-up unless extreme sacrifices in living standards are made. In contrast a developed country would find it much easier because the general level of production is higher to start with. This is illustrated in Fig. 6.3.

Even though in the rich country more production per head may be considered necessary to satisfy 'essentials', this still leaves a greater margin which could be used to build up capital (or to produce 'luxuries').

The vicious circle of poverty can, however, be broken – at least in principle – by a loan from the rich to the poor for the purchase of capital equipment. This enables output to grow, resulting in a more-than-proportional increase in the poor country's margin for investment. Technical advice, which permits a greater output from given resources, also increases the margin. So would, in theory, aid in the form of consumer goods (e.g. milk powder) which should allow a country's own productive capacity to be devoted to capital production. The danger in this last method is that it tends to depress the local consumer goods industries

without a corresponding increase in capital production. The poverty cycle is illustrated, together with ways of boosting the circle, in Fig. 6.4.

A parallel situation exists in the small farm in the UK and other developed nations. Low outputs from small farms means that there is little left for investment once the basic living expenses of the farmer have been deducted from the farm's income. Low investment in livestock, machinery, buildings and so on will keep the farm's output low, and so the cycle continues. In the UK's history attempts have been made to break this cycle in various ways: (i) by the injection of capital through government grants to cover a proportion of the cost of new buildings and equipment; (ii) by making technical advice available paid for by public funds; and (iii) by supporting farmer incomes through guaranteed prices for farm products or subsidizing costs of production.

It would be wrong to leave our discussion of capital without a reference to the banking system and to the **capital market**, the latter being the market for new or existing stocks and shares. In our factory model we envisaged capital as being built up from within the production unit. Some countries have built up their capital in this way – notably the former USSR which enforced great hardship on its population during the 1920s and 1930s in terms of low living standards so that its heavy industries could be built up. At the level of the individual, many farmers pride themselves on having built up their businesses by saving hard.

However, there is often no need to depend exclusively on internally financed capital. Loans can be arranged from individuals or financial institutions; lending between members of farming families has been a common feature of agricultural finance in

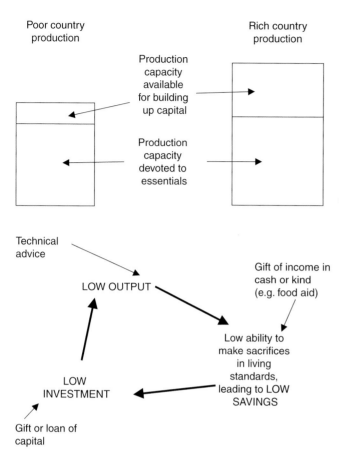

Poor country production

Rich country production

Production capacity available for building up capital

Production capacity devoted to essentials

Fig. 6.3. Resources available for building up capital in poor and rich countries.

Technical advice

LOW OUTPUT

Gift of income in cash or kind (e.g. food aid)

LOW INVESTMENT

Low ability to make sacrifices in living standards, leading to LOW SAVINGS

Gift or loan of capital

Fig. 6.4. The 'vicious circle' of poverty.

the UK. Public companies can raise funds by issuing *new* shares which members of the public or investing institutions can buy in the 'capital market', for example through the London Stock Exchange. The system of loans enables entrepreneurs who have opportunities to use extra capital profitably to acquire the necessary machines, etc. by using purchasing power borrowed for a time from others who have the ability to buy goods, but do not wish to do so. From the perspective of the borrowing entrepreneur, the interest rate on loans will represent the price that has to be paid for access to investment funds now. This cost has to be set against the extra income that can be generated in a budgeting exercise that, if it is to go ahead, has to show a surplus to the entrepreneur. But not all entrepreneurs are the same in terms of the risk they represent to the people making the loans, and higher interest rates are likely to be charged if the borrower cannot provide collateral security (such as a claim on land owned) which could be used to

pay off the loan if the investment turned out to be unsuccessful, or if the entrepreneur lacks a record of good investment decisions and repaying loans on time. (Risk is discussed in detail later in this chapter.)

Banks can create credit by extending loans in excess of the deposits they hold, as they will know from experience that all they need to have in their reserves is enough to cover the relatively few demands from investors who call for their money back. Banks will hold reserves in various forms with different degrees of liquidity (i.e. able to be turned into cash). Under stable economic conditions most depositors will be happy to leave their deposits with the banks, earning interest. Only in exceptional circumstances, such as a loss in confidence in the bank's ability to meet its commitments, will large numbers of people owning deposits want to take them away (i.e. will a 'run on the bank' occur). Clearly a banking system needs to be of good repute in order to work successfully, which is why governments take close interest in bank behaviour and intervene when necessary, such as by: (i) applying legal regulation to the amount of reserves they hold; (ii) making special loans to rescue banks that have got into difficulties; and (iii) even taking them into public ownership. The banking crisis that started in the UK in 2007 (with the failure of the Northern Rock bank, the first example in this country for 150 years) is partly attributed to the regulatory regime having been relaxed too much in the 1980s. In addition, for the system of capital financed by borrowing to work successfully, economic conditions need to be relatively stable (e.g. inflation has to be under control). (Chapter 8 considers interest rates, the money supply and inflation from an economy-wide perspective.)

Labour as a Factor of Production

A parallel can be drawn between capital and labour in that a country's productive ability will be related to the quantity and quality of both of these factors. Labour in production is possibly best thought of as **labour service**, since it is what men and women do which is their contribution to production and for which they receive payment. The *quantity* of labour service available in a country will depend on the size of the population, and the proportion of it which is available for work at any time. A rise in the population, caused by a birth rate that exceeds the death rate, or by immigration, will build up the stock of labour in a way similar to building the capital stock. The proportion available for work will in turn be affected by the country's attitude to such things as: (i) whether women with children should be expected to work; (ii) the ages of school-leaving and of retirement; (iii) the age and sex structure of the population; and (iv) the commonly accepted length of working week and holidays. (Further interesting parallels with capital might be drawn: like machines, workers age, but do they wear out? In some manual activities physical output definitely declines as people get older, but this is not necessarily true for mental tasks where age and experience may enhance some aspects of performance. Unlike machines, people learn.)

The factors influencing the supply of labour service are altered as the country's standard of living rises and the general trend is to reduce the proportion of the population in the workforce with greater affluence (school-leaving age rises, retirement age falls; the proportion of the population above retirement age rises – even the general attitude to work probably changes). However, there are also movements in the other direction: many more UK wives and mothers now go out to work than was customary in the 1950s. It might be noted that in the UK the length of the working week for agricultural workers is a few hours longer than for workers in other industries and their earnings are lower (about three-quarters of the average earnings in other industries).

The *quality* of labour service determines how productive the labour force is. A highly efficient labour force usually means a high level of output per worker. This is affected by the physical well-being of the workforce, and a certain minimal standard of health, clothing and food is required below which efficiency will fall dramatically. Working conditions are influential, as can be witnessed by the output achieved when riddling potatoes in a rain-and-windswept clamp as opposed to the output in a protected barn. A well-trained and educated workforce is conducive to high output; hence all developed countries have state-provided education and attendance is compulsory up to a given age. All EU Member States also have publicly funded schemes to encourage skills training, including ones in agriculture funded under the Common Agricultural Policy (CAP). Part of education and training can be considered an investment just as purchasing a machine tool might be an investment. The term **human capital** is often used in this context. As with other forms of capital, to educate people, resources have to be used, and after a period of 'waiting' these resources may well reap a reward in terms of the increased productivity of the educated and trained people.

The degree of capital intensity is important – one man can be more productive (in terms of output per man) on a powerful tractor than on a less powerful one, as is also the quality of the other factors with which labour is combined. A well-educated, trained worker hindered by poor equipment on poor land and managed by an inept farmer cannot hope to produce a great deal.

Many of the factors which can affect efficiency are two-edged. Education to read and write, or a technical training, can improve output per man, but the results of a more advanced education can be counterproductive if it encourages doubt about whether increasing production should be the overriding goal of society. Whether more advanced education is considered to be productive or counterproductive depends on what the objectives of education are. Similarly, social benefit payments covering unemployment and sickness can promote efficiency by providing a minimum standard of health and living during misfortune, but some people would argue that such benefits, by removing anxiety associated with dismissal and unemployment, also remove the spur to efficiency and to search for employment. However, the removal of anxiety may in itself be a goal of society, and any deleterious effect on the output of goods and services may be considered a small price to pay for this increase in well-being. The quality of life and the happiness (or well-being) of the population, in contrast to their monetary income and material consumption patterns, is something that has risen to prominence in UK political debates in the 21st century, though it is by no means a new issue and can be seen in the writings of Marx from a century and a half ago (see Chapter 1).

An interesting hypothesis on the relationship between well-being and productivity has been put forward by the Chinese. Communist China was criticized for the wasteful use of its manpower in the 1950s to 1970s, because people there were expected to switch between jobs – manual and mental, urban and rural, in leadership and on the shop floor. The Chinese justified this partly on the grounds that individuals cannot be considered fully developed unless they have the opportunity to express all facets of their ability through their work, which implies that workers should have experience of a range of jobs, and partly by the claim that a population of well-developed, fulfilled and happy members will in the long run be *more* productive in terms of output per man. This view retreated in the 1980s. Specialization and financial incentives have become major ways of achieving greater production in China, especially in agriculture where private farming in pursuit of profit is now encouraged. In the early years of the 21st century China has invested heavily in having young people educated in Western universities in subjects such as business management studies, an aim being to enable its industries to perform well and to enable the setting up of similar courses in China.

Entrepreneurship as a Factor of Production

Land, labour and capital do not produce anything unless they are organized together. Even knocking fruit from a wild tree requires the bringing together of the tree (land), stick (capital) and an arm to wield the stick (labour). Furthermore, the decision to gather the fruit must be made, and the risk taken that the attempt may be unsuccessful and hence the effort wasted. The person who takes the decision, organizes the factors of production and bears the risk is the entrepreneur. In a farm firm he will be responsible for deciding: (i) what is produced (barley or oats, etc.); (ii) how production is implemented (e.g. whether to use labour-intensive or capital-intensive methods); (iii) the choice of the scale of production (say, two 50-cow dairy herds or one 100-cow herd); and (iv) for marketing the product (when, where and how to sell). All this requires an ability to make and implement decisions and the larger the business, generally the higher the qualities that are demanded. People differ in their organizational ability and their ability to assess risks, so the quality of entrepreneurship is highly variable. While some economists would maintain that the organizational role of the entrepreneur is just a rather specialist form of labour, the risk-taking element is a unique characteristic of the entrepreneur in a capitalist economy.

When discussing entrepreneurship, we most often take as a typical example the operator of a smallish firm who is both its owner and its manager. If he or she launches an enterprise after assessing the risks involved and it proves successful, he or she benefits through reaping profits. If their endeavours are ill-founded and flounder, they will make a loss. The individual can be held legally responsible and personal assets seized if debts are not honoured. The vast majority of farms in Western countries are arranged like this (95% in the UK in 2007, 93% in Germany, 99% in Italy, 71% in France), with a sole proprietor or perhaps two partners bearing the organizational and risk-taking functions and the legal responsibility, so the convention is acceptable for a study of economics associated with agriculture.

However, it must be borne in mind that agriculture is exceptional among industries in capitalist economies. Most other industries consist of large firms, arranged as companies (corporations). A characteristic of a company is that it has a separate legal status, and the company itself can take out loans, make contracts, be sued in a court of law, and do things that an individual entrepreneur can. In a company the people responsible for managing the firm and taking the decisions (the directors) are not necessarily the people who suffer if the decisions prove to be wrong. The chief losers will be the shareholders of the firm.

In being mainly concerned in this text with agriculture in Western capitalist countries with market economies, we can count ourselves fortunate that in these the farming entrepreneur is relatively easily identifiable. However, in a study of agricultural decision making in a formerly socialist country which still has large-scale farms now arranged as companies or cooperatives, the identification of the body taking the entrepreneurial decisions might be much more difficult.

In the case of a nationalized undertaking (such as the Post Office in many countries) the individual or group who make decisions is probably difficult to identify, and the ultimate source of important decisions may be the elected government. When decisions prove wrong, then presumably everyone suffers to varying extents, though the government can be held responsible and its chance of being re-elected may be dented.

The origins of risk and uncertainty

Risk and uncertainty arise because production is not an instantaneous process. Time elapses between when an entrepreneur decides to go into a line of production and when the product finally reaches the market. The entrepreneur has to assess costs and prices and physical performances before he embarks on his enterprise, and these may change with time. Taking the case of barley production, once a farmer has planted his seed he is fairly heavily committed to the crop for the year. The cost of his inputs (e.g. fertilizer, insecticides, fuel, etc.) not yet purchased but needed for production may rise unexpectedly, the price he finally gets for his barley may be affected by things totally beyond his control (such as a change in the attitude of Russia towards buying grain from the USA, upsetting the expected prices on the international barley market), and his physical production may be affected by drought or excessive rain or uncontrollable disease. Agriculture is unusual (although not unique) in having its levels of output heavily influenced by weather conditions. Not so unusual, but of major importance, is the guiding hand of government policy which can change more rapidly than the industry can adjust. As has been pointed out in Chapter 4, the combination of long supply lags and a lot of relatively small producers acting independently are conditions conducive to cycles of low and high prices.

In addition to the factors already covered, entrepreneurs have to contend with risks that their capital equipment may be destroyed by fire, flood, etc. before its useful life is completed, that it may become obsolete by advances in technology (such as happened to the stationary baler when tractor-drawn machines were developed), and that they may themselves suffer illness or accident, jeopardizing the viability of the enterprise.

Difference between risk and uncertainty

A conventional distinction, although by no means rigorously adhered to, is that unwanted happenings can be divided into two categories: (i) those for which insurance with an insurance company can be arranged are termed **risks** (e.g. fire, personal accident); and (ii) those for which no formal insurance is available are termed **uncertainties**. Formal insurance is based on a statistical analysis by insurance companies of, say, the known number of farm fires each year, and premiums worked out on the basis of the total cost of claims, with a suitable margin to give the company a profit. The farmer benefits because, for a small, known, certain cost (the premium) a large, unknown, possible loss can be avoided.

The classification into risks and uncertainties is by no means absolute. While farmers widely use some fire or personal insurance (i.e. fires and accidents are usually regarded as risks) it *is* possible to insure against rain (as is done by some village fetes) and some diseases, although the premiums are generally beyond what many farmers considered justified – they prefer to treat such occurrences as uncertainties and take alternative avoiding measures.

Another classification of risks and uncertainties is into: (i) those that are part of normal commercial operation (such as fire risks) that farmers would be expected to insure against (as would any firm); (ii) the market risks that are the responsibility of the farmer to manage by how the business is arranged (see below); and (iii) those that cannot be foreseen and countered by the individual farmer, such as natural disasters affecting wide areas and many farmers, or catastrophic market falls caused by international developments. While governments will sometimes take responsibility for the last by putting together aids to recovery and partial compensations, the other two are seen as the responsibility of the farmer as entrepreneur to manage, using a range of tools.

Avoidance of risk and uncertainty

The entrepreneur can take steps to minimize the effect of risks and uncertainties on his business. Risks can be insured against, while a wide variety of steps can be taken against uncertainties.

The *choice of enterprises* is important, as some are prone to uncertainty. Beef production is so prone because of the long production cycle and the possibility that the returns, coming in in a single lump sum when the animal is sold, may coincide with a time when prices are depressed. Others, such as milk production are traditionally safer because a cheque from milk sales comes in at regular monthly intervals from the purchasing dairy with whom the farmer normally has a contract, and prices are known within a relatively small margin of error well in advance (though in the UK such stability was eroded when the former Milk Marketing Board was disbanded and replaced with a number of competing private-sector dairy companies).

Another method of combating uncertainty is *diversification* (i.e. the combination of several enterprises on the same farm, e.g. cereals, beef and grass-seed production). This reduces the overall uncertainty because the likelihood of all the enterprises suffering from, say, price falls at the same time is smaller. On the other hand too much diversification introduces the possibility that each enterprise will be less well managed. Enterprises which contrast in some way are good – perhaps being harvested at different periods of the year to minimize the effect of adverse weather. Such enterprises may have the additional benefit of spreading out the farm's labour requirement, avoiding seasonal peaks. But diversification need not be restricted to agriculture. Farmers control resources that can often be used in activities on the farm that fall outside what is conventionally thought of as agriculture (such as the provision of accommodation for tourists). They or their spouses can also spread their resources across a portfolio of other types of business or profession or sell some of their labour as employees, as many do (see Box 6.2).

Box 6.2. Diversification within and outside the farm business.

We think of farmers as producers of crops, animals and animal products that are unambiguously agricultural in nature but they also often use their farm's resources to generate income in other ways. On-farm diversification includes: (i) the provision of tourist services (such as farmhouse bed-and-breakfasts); (ii) small-scale food retailing and processing; (iii) handicrafts; (iv) energy generation; (v) small-scale forestry; (vi) letting out premises; and (vii) contracting for other farmers. Over one-quarter of UK farms are engaged in such activities (27% in 2007). Part of the CAP's approach at the start of the 21st century to the problems faced by European farmers resulting from falling product prices has been to encourage them to start enterprises which would not normally be considered as part of agriculture.

In UK agriculture, many farms (49% in 2007) have income coming from other (off-farm) jobs taken by the farmer or spouse. They can be described as **pluriactive** and farming is part of their overall diversification. The off-farm income can be from wages and salaries as employees, but for many it comes from other businesses or professions run in parallel with farming. Some will be connected with agriculture (cattle transporting, food shops) but many are not (lawyers, builders, manufacturing, etc.). Operating several businesses, of which farming is one, is termed **portfolio diversification**. Taking both self-employment and employment together, there seems to be almost no occupation that cannot be found combined with farming (ranging from blue-collar jobs, professionals, managers at several levels, to entertainment industry stars).

Pluriactivity can be found on all sizes of farm. It arises in many ways. While some pluriactive farmers were once full-time in agriculture and have branched out, many have moved in the opposite direction, buying themselves into agriculture from some other occupation base. Their motives are mixed. Risk reduction is one, but so is the desire to improve incomes, to invest in assets that are likely to rise in real value, to live in a rural area, to control one's own living and working environment.

Diversified activity involving farming is by no means new. It enabled many of the UK farms to survive the pressures of the agricultural depression in the 1880s. It is pervasive in Europe, and at least one-third of EU farm households (probably far more) have one or more members engaged in some non-agricultural economic activity.

A third method is *production flexibility*, which can take several forms:

1. *Cost flexibility* implies adopting production methods which have high variable costs and relatively low fixed costs. A suitable example is calf rearing, which can occur either in purpose-built, environmentally controlled rearing houses, with efficient use of food (i.e. high fixed costs in the form of interest on capital in the building and depreciation but low variable costs), or in cheap straw-and-corrugated-iron structures with relatively inefficient use of food (low fixed costs, high variable costs). Total costs in each case might be the same but, in the event of the price of reared calves failing, the operator with the cheap structure could easily terminate the enterprise by salvaging the corrugated iron for some other use and setting fire to the straw, whereas the operator with the expensive building would be forced to continue production or even expand to recoup as much of his original outlay as possible from his reduced profit per animal (see also the discussion on the 'reverse' supply curve in Chapter 3 and the decision to continue production in the short run and long run in Chapter 5). Cost flexibility is also given when substitution between variable inputs is possible: a glasshouse heating system that can readily be switched from oil fuel to gas and back again will be less subject to uncertainty caused by price changes and supply interruptions than a system solely dependent on one fuel.

2. *Product flexibility* is achieved when a firm's resources can readily switch products according to changes in product prices. A good example is the Holstein or Friesian dairy herd which can either produce dairy replacements or beef animals, depending on the type of bull used, according to whether the price of young dairy stock or beef stores is the more attractive. A herd of Hereford beef-type cows, in contrast, cannot produce animals suitable for dairying, only for beef production. Another example of product flexibility is afforded by the wide span, general purpose, floored and walled farm building which can be used for on-floor grain storage, housing dairy or beef animals, machinery storage or (if the farm is turned over to leisure use) as a covered tennis court. Such a building would fare much better in times of uncertainty than, say, a specialized poultry house which can only be used for one product.

3. *Time flexibility* is achieved by choosing those enterprises which have a short production cycle. All other things being equal, enterprises with long lags between initiation and fruition (e.g. the establishment of a plum orchard) will be more uncertain than those with short lags (e.g. cereals, or even radishes) because the longer time period gives more opportunity for changes to occur. Another form of time flexibility is exhibited by products which store easily; a farmer is given a wider choice of where he markets his output than with a highly perishable product.

A fourth general method of countering uncertainty is by the use of *informal insurance*. This may take the form of keeping an extra tractor on the farm beyond what would be considered normally necessary to act as a safeguard against unexpected breakages at critical times, or having a larger than normally justified combine harvester as an informal insurance against bad weather necessitating rapid harvesting. Pesticides are sometimes applied without the offending insects being apparent, just in case they are there and develop later. Farmers often err on the generous side when applying fertilizer because, while too little may reduce yields noticeably, a little too much will not do harm. A common practice by livestock farmers is to carry over a reserve of fodder from one year to the next in case the crop in the following year is unexpectedly light, and a type of informal insurance which is probably more legendary than fact is the mattress stuffed with money which farmers are reputed to keep as a precaution against crop failure or disease. The more modern manifestation of this legend is for farmers to maintain their bank accounts unnecessarily in credit or to arrange an overdraft limit and then stay well within it.

A fifth method of avoiding the deleterious effects of uncertainty is by arranging *contracts* for inputs or outputs and other forms of financial instruments (see Box 6.3). If a farmer can arrange with his supplier of animal feedstuff a contract to buy a certain quantity over a year at a given price he will be insulated from subsequent rises in feedstuff price, at least to the end of the year. Similarly, a calf producer may contract to sell all his calves to a rearer at a fixed price for a certain time period, and is thereby protected against falls in the market price of calves. If the cost of inputs and/or the price of products are known in advance, planning the business to operate at its most profitable level becomes much easier. Many of the precautions listed above which can be taken to minimize uncertainty become no longer necessary (such as diversification, informal insurance, cost flexibility), and the firm's resources can be utilized more efficiently. Economic theory suggests that the reduction in risk must be paid for in some way. For example, the contract price which a blackcurrant grower might negotiate for a 7-year period with a manufacturer of blackcurrant cordial could be expected to be a little below the average of prices on the open market for the same period.

Vertical integration is said to occur when two or more stages in a production chain fuse together. Contracting is a form of vertical integration, but the term is more often used when a change of ownership is involved. An example would occur if a farmer or group of farmers bought up a retail butchery business and sold their own meat through it. While the motives for such a take-over are complex, one of them could well be the desire to have a secure outlet for their product, and to be no longer reliant on prices offered by other butchers. Another example would be where a supermarket bought a horticultural holding so that it could be sure of a reliable supply of produce of controlled quality for its shelves, thereby reducing its risks. Examples of a fully integrated supply chain exist: in the UK the Co-operative Society movement has operated farms with dairy herds, large collecting dairies, a distribution system and retail milk rounds. (Note that **horizontal integration**, where firms at the same stage of production band together, is mainly to form powerful bargaining groups rather than to reduce uncertainty, though this may occur too. For example, farmers forming a buying group can buy in bulk and obtain considerable discounts.)

In addition to the ways of reducing uncertainty already discussed, entrepreneurs have an increasing amount of *market intelligence* upon which to base their decisions. While in other industries this may partly come from privately commissioned research, in agriculture it comes very largely from the government and its agencies which publish a wide range of data available to farmers. The advent of the Internet has made much more information rapidly available, though this only benefits those farmers with the equipment and skills to access it.

Mobility of the Factors of Production and Unemployment

Factors of production can be considered mobile or immobile in two senses – occupational and geographical. This classification according to mobility runs across that of factors into land, labour, capital and entrepreneurship, so we can find occupationally immobile capital, geographically mobile labour and so on. As with most systems of classification, some factors are found which do not happily fit into any single category, or which fit into several according to circumstance, but this method of describing factors is useful because it helps to explain why payments to the owners of factors vary. This will be developed later.

Occupational mobility

Occupational mobility refers to the ability with which a factor can be switched to alternative uses. Factors which are occupationally immobile (i.e. where they cannot be easily switched to an alternative use) are often called **specific**, while occupationally mobile factors are also termed **non-specific**. A milking parlour is a specific (or occupationally immobile) type of capital because it cannot be used for anything except milking cows, and if the farmer decides to produce milk no longer, the parlour will simply go out of use. In contrast, a covered yard, intended for dairy cattle, can be used easily for a wide range of other farming activities – beef housing, machinery or hay storage, even perhaps as an indoor tennis court; such a building would be considered non-specific, or occupationally mobile.

Similarly some types of land are restricted in the crops which they can grow: heavy clay soils in areas of high rainfall are probably limited to permanent grass production. Such land is 'specific'. On the other hand, a medium loam in a more kindly climate can be switched between a variety of crops, so is less specific.

Some farmers feel particularly at home with, say, sheep rearing and perhaps have little knowledge or ability with other types of farming. They would be considered specific (or occupationally immobile) entrepreneurs, whereas a farmer willing and able to vary his enterprises or even embark on a totally different type of business would be relatively non-specific.

The classification of labour into specific (occupationally immobile) and non-specific (occupationally mobile) is not quite so straightforward. Generally, specific labour implies that for which a long period of training or a special ability is required. People with special skills or training are generally unwilling to move to other occupations – a veterinary surgeon or a concert pianist is usually unwilling to take a job which does not use her skills, even though she could easily do the non-skilled job. At the other extreme, a general labourer could find a job in the, building industry, on farms or even working in shops – his or her labour is non-specific and can switch between labouring jobs easily.

We must not try to put a veneer of perfection on our classification. Specialized labour can usually do non-skilled jobs, but its skills (and the amount of human capital they represent) are then wasted and its earnings often lower. Sometimes this switch out of a skilled occupation is made necessary by technical advance – the number of skilled hand-spinners of wool needed in the textiles industry was greatly reduced by the introduction of machinery during the British Industrial Revolution. Similarly the number of farriers required on farms was reduced by the introduction of the tractor to agriculture. On the other hand, it is very difficult to switch labour *into* a skilled job, either from the pool of non-specific labour or from other types of specific labour when demand for the skill increases. This has implications for the earnings of people in the sought-after job, and these will be discussed later. Table 6.2 combines the two types of classification which have been discussed so far.

Geographical mobility

Geographical mobility implies the ability of the factor to move from place to place. With the factor 'land', the question of geographical mobility hardly arises, except that if 'land' is used to include all natural resources, then items such as coal and water can be made to move from place to place.

Some units of *capital* (e.g. tractors) are geographically mobile, but many (most buildings) are not. However, it must be recalled that even immobile capital wears out, and the money allowances which are made to compensate for this need not be reinvested on the same site (replacement structures can be erected elsewhere) so that in the long term capital *is* geographically mobile.

The geographical mobility of *labour*, which is the willingness and ability of workers to move from place to place, is significant not only from the restricted view of shifting this productive resource (labour) from where it is plentiful to where it is in short supply, but because geographical mobility has major social implications.

As a broad generalization, workers who have undergone long periods of training in non-manual skills, and are therefore termed 'specific' in the occupational sense, are willing to move from area to area for jobs: teachers are generally willing to move for promotion, and the 'brain drain' of medical staff and scientists from Europe to North America is also illustrative of geographical mobility. Also as a broad generalization, non-specific labour is generally less geographically mobile. Factory workers in declining industry in the north of England, particularly the more elderly, may not be willing or able to move to the south where jobs are available. To assume that this immobility of the 'blue-collar' worker is the result of an over-attachment to a particular area is a gross oversimplification of the case; the desire to stay in familiar surroundings, close to family and friends and the support networks they provide may be an entirely rational pattern of behaviour and in line with economic principles once it is recognized that satisfaction (or well-being) comes from both monetary and non-monetary sources.

The *ability* to be mobile is highly relevant; high-income earners are more able to finance moves, especially if this involves buying a house. Lower

Table 6.2. Examples of factors of production classified according to factor group and occupational mobility.

Main group of factor	Occupationally immobile (specific) – factors cannot easily be switched to alternative production processes	Occupationally mobile (non-specific) – factors easily switched to alternative production processes
Land	Heavy clay in wet area suitable only for permanent pasture Very sandy soil in dry area suitable only for crops capable of withstanding dry conditions Land covered with a motorway – cannot easily be changed to farming or other uses	Medium soil, well drained, in area of equable climate – wide variety of crops Park in central London – can be used for buildings or recreation or farming
Capital	Milking parlour – only suitable for original purpose Blast furnace for making steel A navy submarine	General purpose farm building – farming uses or non-farming use Estate car – can be used in a wide variety of firms for carrying goods, or for pleasure
Entrepreneurship	Elderly sheep farmer – only capable of managing one type of farm	Young businessman willing and able to embark on a wide range of business activities
Labour	Medical labour, musicians or other artists – unable or unwilling to transfer to other occupations Older skilled or semiskilled labour unable or unwilling to retrain for another job	Workers able and willing to switch jobs

earners, even if they are house owners, may be constrained if considering a shift to an area of relatively high house prices. To ease mobility, employers may have to offer relocation incentives, such as a subsidy on mortgage payments for the first few years following a move or a generous lump sum which enables new carpets and furnishings to be afforded. Labour dependent on rented accommodation will be unwilling to move unless suitable housing to rent is available at the new location. It seems that such accommodation is often scarce in places where job vacancies exist for blue-collar workers. For national policy, two options then present themselves: (i) construct new housing to let where jobs exist; and (ii) encourage new industries to establish themselves in areas where labour (and housing) is available. Both have been tried, as was illustrated in the UK by the (then) National Coal Board building houses to encourage miners to move to expanding collieries, and by the designation of parts of the country as 'development areas', including some rural ones, to which industries are encouraged to go by financial incentives.

In agriculture the mobility of hired labour (both occupational and geographical) is closely linked to housing, the worker's age and the availability of alternative employment. Often labour is deeply committed to an area and is unwilling to move to a different farm or type of farming. However, specialists such as dairymen or managers are probably more geographically mobile. When young workers wish to marry, geographical mobility may rise because they will tend to take a job where a house is available with the job. On the other hand, once established in such a tied house, the ability to leave farm work for more attractive employment in another industry may be seriously hampered if such a move necessitates leaving the house. In a period of high house prices, alternative accommodation may be impossible to rent or buy (i.e. occupational mobility is constrained). No doubt the arguments for and against tied housing, with its political connotations, will continue for a long time.

Unemployment of the Factors of Production

Unemployment of the factors of production is closely related to their mobility, although immobility is not the sole reason for unemployment. While labour unemployment is the type that most readily springs to mind, unemployment of the other factors occurs and can be described in a similar manner. There are various main types of unemployment and they differ in their cause.

Mass (or cyclical) **unemployment** is the form associated with the slumps which tend to alternate with booms in most of the Western economies. The 1929–1935 Great Depression was a particularly severe example. From 1975 the UK experienced yet another period of unemployment of this type, related to the efforts to curb inflation. The widespread banking crisis and general downturn in the economies of many countries in the latter half of the first decade of the 21st century saw the UK unemployment rate, which had been about 5% since 2002, climb rapidly from mid-2007 to 8% by 2009, where it has more or less stayed. The situation in some other EU Member States is far worse (27% for both Spain and Greece in early 2013, with an EU-27 average of 12%). The cause of mass unemployment is, in essence, a deficiency in the general level of demand in the economy, and this is correctable by government action, although other policy goals may have to be sacrificed in its pursuit. This type of unemployment will be met again in the section on macroeconomics.

Structural unemployment is the type to which factor mobility bears a direct relevance. It occurs when the demand for labour or the other factors engaged in particular uses such as in the mining industry diminishes. If the factors are occupationally or geographically immobile they will be unable to find alternative uses, so will remain unemployed. One cause of structural unemployment can be a reduction in the *demand* for the product of firms, as occurred when UK miners were made unemployed in the 1960s and 1980s because the demand for coal fell (and hence the demand for miners). At the same time labour may be in short supply in other industries because of an increased demand for their products. In other words, this unemployment arises from the continually changing structure of demand in the country. Another cause of structural labour unemployment could be the development of new *labour-saving techniques* which require fewer workers. New machinery means that UK farms can be run with a much smaller labour force than 20 years ago, and this has caused some farmers to shed labour. In areas with few other industries, rural unemployment can be a serious problem.

Historically workers have often been afraid that new machines would threaten their jobs, and the Luddites (active from 1811 to 1818) set about destroying spinning machines to try to stop mechanization of the textile industry in the UK's Industrial Revolution. However, in practice new jobs have in the past been created which have absorbed the unemployment either in the machine-making industries or by the development of other industries which have prospered as the result of the economic growth resulting from mechanization. If factors are willing to move from place to place and to change their occupation (i.e. if they are both geographically and occupationally mobile), technical unemployment will be greatly reduced. While this is generally true, it does not apply in all circumstances and in the short term. For example, in developing countries the mechanization of agriculture and other industries can, if taken too fast, worsen unemployment if alternative employment opportunities are not created at a sufficient rate. Even in developed economies, if labour markets are inflexible (e.g. if there is legislation on dismissal that makes employers reluctant to engage people, or a high minimum wage, or strong unions) there may be an incentive to develop labour-saving technologies, creating a growing pool of unemployment which will have a serious impact on the stability of society and which can only be obviated by changes to the economy, including the reform of labour unions.

Some occupations are essentially of a seasonal nature, and this can give rise to **seasonal unemployment**. The tourist trade in rural areas is inherently seasonal, the period of high labour demand often being met by students on summer vacation. Apple picking and strawberry picking take place over relatively short periods, and some 'travellers' living in caravans take advantage of their geographical and (within a range) occupational mobility to move from farm to farm and crop to crop as their services are required, thus obviating seasonal unemployment and removing the need for farms to carry large 'regular' staff to cope with the seasonal labour peaks that are characteristic of many types of agriculture.

To complete the picture, there will always remain some unemployed people, even when the demand for labour is high. This **residual unemployment** will consist of those who are incapable of being employed for physical or mental reasons or those who do not wish to take a job. People on strike could also be included as unemployed; their unemployment is 'voluntary' but may cause others to be laid off and produce some involuntary unemployment. In addition, the process by which workers become informed of job vacancies, apply and are selected takes time. Better communications (including Internet use) can reduce this 'friction' in the labour allocative process, as could, some suggest, lowering unemployment benefit to increase the necessity of taking a job. The term **frictional unemployment** is frequently used to describe unemployment resulting from the failure of labour to flow smoothly between uses.

From the individual factors themselves we must next turn to how they are allocated among the range of production processes in order to generate the goods and services which consumers demand.

Allocation of the Factors of Production and their Rewards

The definition of economics that was cited in our introduction to the subject (Chapter 1) was that 'Economics is the study of how individuals and society choose to allocate scarce resources between alternative uses in the pursuit of given objectives'. If the objective is known, then there will be some allocation of the resources (or factors of production) which enables that objective to be best fulfilled. For example, assuming that the objective is to maximize the value of the nation's production (begging the question of what this implies precisely), then labour, land, capital and entrepreneurial ability could be allocated between industries and firms in such a way as to achieve this. In Chapter 5 on production economics we saw that this optimum allocation of scarce factors between a range of alternative uses will be achieved when each of the factors is so allocated that the addition to production caused by the last unit (the marginal unit) of each factor in each use is of the same value. This is an example of the use of the Principle of Equimarginal Returns.

Take for example the allocation of vehicle drivers between industries. Being a driver in one industry requires very much the same sort of skills and abilities as in any other, and switching between industries is easy. If the last man in manufacturing industry results in a greater value of increased output than the last man in agriculture, then there will be a net gain if drivers are transferred from agriculture to manufacturing. Diminishing returns will apply so that a point will eventually be reached where the value of the marginal product of labour

in each industry will be the same and no net gain would come from further switching of drivers. Any allocative process should be aiming to achieve this optimum pattern of resource use. At this point **allocative efficiency** will be achieved, and it will not be possible to improve things by any reallocation of productive factors.

In the model of the centrally planned socialist economy adopted by the USSR and its satellites in Central and Eastern Europe (all now swept away), the deployment of the nation's factors of production between different industries and locations was decided and directed by central authority. Labour was directed from one industry to another, and capital from place to place. This process was attempting, perhaps not always consciously, to equate marginal returns to factors in each use. Provided that the central planning authority knew what types and quantities of final products its population preferred, what resources were available to it for production, and the technical nature of the production functions in each line of production (to give it an indication of marginal products), such an allocative procedure could, in theory, result in the optimum pattern of resource use. In practice, the lack of knowledge of consumer preferences, available resources and technical coefficients hampered the system and considerable quantities of resources were used up simply in providing the planners with the information they required to form decisions. One factor leading to the collapse of such planned systems at the end of the 1980s was their inability to use resources in an efficient way. Not only did they suffer from allocative inefficiency at any one time (which might be termed **static inefficiency**), they also resulted in economic growth being slowed by the inappropriate use of resources (**dynamic inefficiency**).

In a basically capitalist market economy the allocation of resources is achieved largely through the price mechanism. We have seen in Chapter 3 that the wants of consumers are reflected in what they are willing to buy (i.e. the demand they express in the market for goods and services). If, for example, consumers develop a liking for smartphones, this will be reflected in a strong demand and waiting lists for them. Some consumers will be willing to pay more than the existing prices. In order to meet this strong demand, the producers of phones will need to expand their output; to do this they must attract labour, capital, etc. from producers of other goods. This they do by offering higher rewards than other producers, creating a reward differential that they can afford because the strong demand for their products enables them to raise the price of smartphones somewhat and still sell all the units they can make. The prices of consumer goods and prices of factors of production are thus linked, and signals of consumer demand (the prices of consumer goods) are passed on by producers to the market for factors of production in the form of factor prices, so that factors move between lines of production, attracted by differentials, to try to best satisfy consumer demand. Recall, however, from Chapter 4 that the price mechanism contains imperfections which may necessitate some modification to the allocation of factors it produces (e.g. the growing of opium may need to be banned).

Demand and supply of factors of production

In a free market economy both a supply of factors (land, labour, capital, entrepreneurship) and a demand for these factors exist. The demand for factors is said to be a **derived** demand because it comes from businesses which produce goods that consumers demand. For example, the demand for dairy cows is derived from the consumer demand for milk. When the demand for factors and the supply of factors are allowed to interact, a price results. The price of labour is the wage rate, the price of land service is rent and similarly, the price of entrepreneurship is profit.

Shifts in the demand and/or supply curves for factors will affect prices in a similar way as experienced with consumer goods, and a similar set of imperfections arises when there is a sole buyer of a factor (i.e. a monopsonist), a sole supplier (as typified by a trade union operating a closed shop), when one firm or the whole industry is considered, and so on.

The market for factors is by no means completely smoothly operating, and the failure by government to recognize its shortcomings, particularly the resistance of labour to falling wages, was a major contributor to the high unemployment during the 1920s and 1930s in the UK. This will be returned to in Chapter 8. However, some aspects of the demand and supply of factors must be considered here. (The formal derivation of the demand and supply curves for a factor, and how changes in product price and productivity relate to the quantities demanded and supplied, require a somewhat more advanced treatment than is possible here. The reader is referred to one of the standard microeconomic textbooks.)

The responsiveness of the demand for a factor to changes in its price is, of course, the **price elasticity of demand** for that factor. This elasticity will depend on a number of variables:

1. *The price elasticity of demand of the product.* If the price of a factor increases (e.g. animal feedstuffs) the price of the product (e.g. milk) will tend to rise. If this results in a severe cutback in the demand for milk (i.e. milk has an elastic demand), then the demand for animal feedstuffs will be curtailed heavily too.

2. *The availability of substitutes.* If good substitutes exist, as happens when closely similar animal feedstuffs are available from rival firms, then a rise in price of one brand will cause much less of it to be used, as farmers switch to the alternatives (i.e. demand is elastic). Contrast this with what would happen if the prices of all brands were increased simultaneously.

3. *The relative cost* of the factor to the total costs of production. If the cost of water represents only a tiny proportion of the cost of milk production, then an increase in the cost of water will not cause farmers to use much less of it. In contrast, feedstuffs form a major proportion of the total costs of egg production and any cost increases will affect total costs to a significant extent.

The responsiveness of the supply of factors to changes in factor prices is dependent both on the time period in question and on whether the supply to individual firms or the national supply is under consideration. Nationally, the stock of land is almost completely fixed, irrespective of its sale price or rent, but the amount of it being supplied to the market will be sensitive to price, and an individual farm firm can increase its stock of land by outbidding its competitors. With capital a similar situation obtains: nationally the stock of capital in the form of plant, machinery and buildings is fixed in the short run at least, but single firms are not necessarily so restricted and can acquire more, changing the distribution of the national stock. The national supply of labour *can* be increased by paying higher overtime rates to encourage longer hours, or by tempting women from being full-time mothers to join the labour force in industry, or by delaying the retirement age. Individual firms have, in addition, the opportunity to tempt workers from competing firms by offering higher rewards. (Note the reverse supply curve, cited in the section on supply in Chapter 3, which can occur when the length of working week has risen to a level at which further wage rises cause *fewer* hours to be worked as people prefer to take more leisure and the same money in income.)

Dynamic and equilibrium differentials

The movement of factors of production from firm to firm or from industry to industry in order for their owners to benefit from higher rewards tends to iron out differentials. This type of differential is a **dynamic** one because it is temporary, and is eroded by movements of labour, capital and so on. Not all differentials, however, are of this type. For example, the pay of university staff is below their equivalents in industry; part of the explanation is that university teaching has some non-monetary compensations, such as greater time flexibility and freedom of expression. This differential is an **equilibrium** differential; removing it by raising university salaries to industrial levels would cause a flood of applications from outsiders for university posts. Land in city business centres can command a much higher annual rent than agricultural land in the countryside because of the higher earning power of its location and its scarcity value. Such a differential will be permanent because it is impossible to increase the quantity of land in a city centre.

Economic rent in the reward to factors of production

Economic rent is the term used to describe the differential element in the reward which the owner of an economically scarce factor of production receives above the reward which could be earned in the factor's best-paid alternative use. Thus the term **rent** is used here in a way rather different from its commercial meaning. While economic rent was first described by reference to land, it can be applied to each of the groups of factors, although the term **quasi-rent** is then sometimes used.

Economic rent can apply to any factor which is less than completely elastic in supply, at least in the short run. Economic rent can be defined as follows:

> Economic rent is the margin (or surplus) a factor earns above that payment necessary to attract it into or keep it in its present employment.

A factor of production will be kept 'in its present employment' as long as it cannot earn more elsewhere; its earnings in its best-paid alternative use form its **transfer earnings**, so we can redefine economic rent as 'the margin (or differential) a factor receives above its transfer earnings'.

To illustrate the concept of economic rent we will first take an example using the factor labour. If the horticultural industry faces a supply of labour which is less than infinitely elastic, as illustrated in Fig. 6.5, and it wishes to expand the number of persons working in horticulture from q_1 to q_2, then it will need to pay higher wages to attract workers from other forms of employment. Assuming that the same wage has to be paid to all workers in horticulture, then the person who was just willing to stay in horticulture at the old wage level, because the earnings just equalled what could have been earned elsewhere, now finds the rewards they get are greater than their transfer earnings. At wage level w_2 labour will flow into horticulture from other industries, but this flow will cease at the point at which the last person who switches loses as much in terms of forgone earnings as is gained from horticultural wages. For this marginal worker, the horticultural earnings are of the same size as the earnings in the best-paid alternative employment (i.e. the person's transfer earnings) but for all the other horticultural workers whose earnings exceed their transfer earnings, part of their reward consists of economic rent (i.e. a surplus above the payment necessary to keep them in their horticultural employment).

The size of the total economic rent accruing to labour at wage level w_2 is shown by the shaded area in Fig. 6.5, and the amount of transfer earning in its reward by the unshaded area below the supply curve. From this graph it can be deduced that the shallower the supply curve (i.e. the more elastic or price sensitive is the supply of the factor) the smaller will be the proportion of economic rent in its total reward. Conversely, the more inelastic the supply, the greater the proportion which is economic rent, and if the supply were completely inelastic with the supply curve a vertical line, all the reward would take the form of economic rent.

The elasticity of supply of a factor of production can vary with the time scale under consideration and this can have consequences for the economic rent part of a factor's earnings. For example, currently in the UK doctors in general practice (GPs) as a group probably receive higher salaries than their

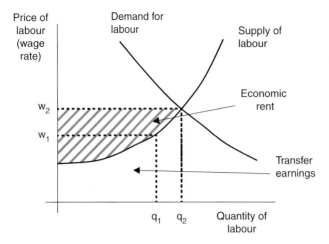

Price of labour (wage rate)

Demand for labour

Supply of labour

Economic rent

w_2

w_1

Transfer earnings

q_1 q_2 Quantity of labour

Fig. 6.5. Economic rent and transfer earnings in the reward for labour in horticulture (hypothetical).

abilities would have earned them in other professions – in other words, their actual earnings are above their transfer earnings, so that a sizeable part of their remuneration consists of economic rent. This state of affairs had arisen because of a sharp rise in demand for GP services without a corresponding increase in the number of doctors. The high earnings are attracting potential entrants to the profession but considerable time is required before they are fully trained, necessitating the expansion of university and professional teaching facilities. Under such circumstances the highly inelastic short-run supply of GPs has resulted in the profession earning considerable economic rent. The government could have taxed doctors heavily in the knowledge that, as long as they were still earning more than in other professions that were open to them, they would stay in medicine. In the longer term, economic rent could be expected to be reduced as the new entrants were trained and became qualified.

In this example, the economic rent part of GPs' earnings may be eroded by the eventual arrival of newly qualified people to increase the supply, but with some factors of production no increase in supply is possible. If a footballer earns £10,000 a week because of his unique abilities (i.e. there is only one person with his qualities), and his best-paid alternative job as a building labourer would earn him only £500 a week, the economic rent element of his present employment is £10,000 – £500 = £9500. This economic rent element is what he receives for his uniqueness; any increase in demand for his services would raise his income and the economic rent content of his earnings, while a fall in demand would have the reverse effect.

Turning to the factor land for an example, we can cite the case of the high rents paid for land in city centres. As land for office development, a city centre plot might command, say, an annual rent of £10,000/ha. If the next-best paid use of the land were for farming, let at £200/ha (i.e. its transfer earnings), then its economic rent would be £10,000 – £200 = £9800. This is an extra reward the land earns simply by being where it is – in the city centre. The number of sites in an existing city centre is absolutely fixed, so the overwhelming proportion of the land's reward will always consist of economic rent which can never be eroded by creating more sites; this is another case of an equilibrium differential which could be taxed (and has been in the form of rates paid by businesses) without changing the use to which the land is put.

Normal profits, surplus profits and monopoly profits

When economic rent and the entrepreneur are discussed, the term **normal profit** is encountered. This term has already been introduced, notably in the chapter on competition (Chapter 4). Normal profit is the level of profit which an entrepreneur requires in order to compensate themself for the uncertainties of production. If he or she is not rewarded to the level of normal profits, they will transfer their ability to some other line of production. Because the level of uncertainty differs between types of production, some being more hazardous than others, it is not surprising that the level of profit which is considered 'normal' also varies.

If the entrepreneurs in, for example, beef production are just earning normal profits and then the demand for beef goes up because of, say, an increase in the incomes of consumers, beef producers will be earning more than normal profits, and this surplus above the 'normal' level is a form of economic rent. It is the surplus above the payment necessary to keep the entrepreneur in beef production. This surplus will attract entrepreneurs from other industries, more beef will be produced, the price will fall somewhat and the entrepreneurs will each be earning only normal profits again.

The profits made by a monopolist through his manipulation of the market have something in common with the economic rent accruing to the owner of a factor of production, since the monopolist also receives a reward greater than would be necessary to keep him in his particular line of production. The distinguishing difference is, however, that while economic rent is thrown up by the market system for factors, a system which is outside the control of individual factor owners, the monopolist deliberately contrives his surplus profit by exercising his power over the market through controlling the supply of goods.

The functions of profit

Profit has been mentioned a large number of times in the text so far in addition to its discussion immediately above. To conclude this chapter on the factors of production we will summarize the functions of profit in a free enterprise, perfectly competitive capitalist model of an economy. *First*, profit induces entrepreneurs into production by acting as a compensation for the uncertainties involved. *Second*, levels of profit above or below the 'normal' level indicate to entrepreneurs which industries or product lines should expand and which should contract to reflect the changing wants of consumers. *Third*, profits in excess of the 'normal' level enable entrepreneurs to obtain the resources they require for expanding production by outbidding competitors from other industries. Production is thus expanded in these lines for which consumer demand has risen. *Fourth*, profit encourages the most efficient firms in each industry (hence benefiting the consumer) by enabling the most profitable

firms to expand by outbidding the less profitable firms. In that the more profitable firms are those which have lower costs of production, this means that the consumer will be able to purchase his or her requirements at the lowest possible prices. In such a model of the economy, profit can be seen as a useful mechanism in ensuring that productive resources are allocated in the way which serves consumers' interests best.

Within agriculture in Western economies, a good case could be made that profit performs these four functions adequately. The numbers of inefficient and unprofitable small farms have decreased rapidly since the Second World War. A striking example of the concentration of production has occurred with poultry meat, where many small units have been displaced by fewer very large producers and at the same time the real cost of this meat to consumers has fallen. However, it must be recalled that farming is unusual among industries in developed economies in that it is still largely comprised of many relatively small independent units, each of which has little power over the market.

Where some degree of monopoly power exists, or where the state plays an important part in fixing prices, the existence of profit becomes far less obviously beneficial and the role it plays in the mechanism of allocating scarce factors of production in the best interests of consumers may be considerably reduced. Nevertheless, the governments of virtually all developed economics see the pursuit of profits by competition among firms, both within countries and internationally, as a vital ingredient in promoting an efficient and growing economy and have made changes to encourage this. These include the removal of impediments to competition (such as stopping the development of monopolies and removing unnecessary administrative red tape) and the privatization of some previously nationalized industries, such as telephone, gas, water and national airlines. The former planned economic systems of countries in Central and Eastern Europe have largely been replaced by competitive market economies as a way of stimulating efficiency. Even in China, which remains essentially a socialist country, privatization and financial incentives have been introduced to the basic framework to improve productivity, to great effect.

Exercise on Material in Chapter 6

Answer the questions in the manner indicated. Answers and explanations are given in Appendix 2.

6.1. Why is the demand for a factor of production a derived demand?

6.2. Give five separate goods from which timber derives its demand.

6.3. (a) If the demand for glasshouses is elastic, will this tend to make the demand for glass elastic or inelastic?

   (b) The price of water makes little difference to the demand for it for bread-making. Why?

   (c) The demand for individual brands of fertilizer is more/less elastic than the demand for fertilizer in general. Why?

6.4. Delete as appropriate:

   An entrepreneur with a given quantity of labour available to distribute between the various enterprises of his business, must so allocate it that the *first/last* hour used in each enterprise yields *the same/different* values of output. This will of course *mean/not mean* that the same amount of labour must be devoted to each enterprise.

6.5. A farmer has only 8 days left for cultivating his land before he must plant his crop of potatoes. According to the best estimates he can make, his yield of potatoes from each of his three fields to be sown will vary depending on how much cultivation time is devoted to it according to the following table.

Days of cultivation	Potatoes (t)		
	Field A	Field B	Field C
0	63	40	65
1	78	60	75
2	88	70	82
3	96	79	88
4	100	85	89

Use the Principle of Equimarginal Returns to devise the best use of his 8 days. At this optimum, how many days will he spend on:
Field A? .... Field B? .... Field C? ....

6.6. Farmer A is being displaced by a reservoir. He is willing to take a farm anywhere else in the country but is reluctant to change from sheep farming.
   (a) Is he (i) occupationally immobile?
            (ii) geographically immobile?

   (b) Is his labour specific/non-specific?

   (c) Which is generally the more occupational immobile, specific or non-specific labour?

   (d) Which is normally the more geographically immobile, specific or non-specific labour?

   (e) Which of the following are the more specific factors?
      (i) Motor van or road roller?
      (ii) Moorland or field with medium loam soil?
      (iii) Older worker or young person just leaving school?

6.7. Economic rent: a good actor earns £100,000/year. In public relations he could earn £20,000/year.
   (a) Define economic rent.

   (b) As an actor:
      (i) What part of the payment he receives is transfer earnings?
      (ii) What part is economic rent?

(c) Give two reasons why the actor receives economic rent for his services.

(d) In the long run, training schools for actors can be established which will increase the supply of actors. Why is it likely that good actors will always be paid above their transfer earnings?

6.8. Entrepreneurs' dual functions are the organization of production and the bearing of risk and uncertainty.

(a) How may the entrepreneur remove risk from production?

(b) Give two sources of risk in agriculture.

(c) Give three sources of uncertainty in agriculture.

(d) Give four ways in which entrepreneurs can reduce the effects of uncertainty.

6.9. Profits.

(a) Cross out the inappropriate alternatives.

When prices and costs are stable an industry in perfect competition will so arrange itself that its entrepreneurs are earning *good/bad/normal* profits. If the beef-producing industry is in such an equilibrium, and the demand for beef suddenly increases because of an increase in consumers' incomes, in the short term beef producers will earn *surplus/normal* profits. New entrepreneurs will be attracted into the industry and force *up/down surplus/normal* profits until a level is restored where entrepreneurs are again earning *good/bad/normal* profits.

(b) Define 'normal profit'.

(c) Is interest on capital included in normal profit? *Yes/No*

(d) What happens if an entrepreneur earns less than normal profits in one industry?

(e) An entrepreneur would require a higher normal profit in the pop record business than he would require in the car retailing business. What is the chief reason causing this phenomenon?

7 Market Failure: Some Problems of Using the Market to Allocate Resources

Introduction

At several points in the text so far it has been pointed out that the pattern of consumption, production and resource allocation which would result from the free interaction of supply and demand for goods and services may need modifying if the best interests of society are to be served. Examples of this happening in practice are not difficult to come by. The consensus view is that addictive drugs such as opium should not be available on the open market, as would almost certainly happen if the state did not ban its possession and sale. On the other hand, the state provides some goods and services, such as education, without direct charge to the recipient, and uses legislation to compel children to 'consume' education until they reach minimum school leaving age. The cost to farmers of some forms of farm capital, such as slurry handling equipment to reduce the danger of pollution, is sometimes lowered by grants, thereby encouraging a greater level of investments, while in contrast the price of consumer 'luxury' items is frequently raised by a higher rate of tax than is applied to items such as food. Some utilities such as rural buses and country pharmacies receive subsidies out of taxation and can thereby charge a lower price than they otherwise might.

Clearly the view leading to these modifications to prices and quantities must be that the pattern of consumption and production which an unhindered price mechanism would result in, is not the optimum. A preferable pattern (preferable that is from the view of society as a whole) can be obtained by adjusting the solution provided through the market. Although the price system can be useful to an economy in indicating the pattern of consumer demand, where current production leaves shortages and surpluses, and in channelling resources into satisfying demand, it also contains imperfections which should be recognized and understood. This chapter explains why society acts to overcome the imperfections and paves the way for a study of the working of the whole economy in the next chapter.

Adam Smith and the 'invisible hand'

The father of modern economics, Adam Smith, wrote in his famous book of 1776 *An Inquiry into the Nature and Causes of the Wealth of Nations* (often just called *Wealth of Nations*) the following:

> Every individual endeavours to employ his capital so that its produce may be of greatest value. He generally neither intends to promote the public interest, nor knows how much he is promoting it. He intends only his own security, only his own gain. And he is in this led by an invisible hand to promote an end which was no part of his intention. By pursuing his own interests he frequently promotes that of society more effectually than when he really intends to promote it.
>
> (Smith, 1776)

The doctrine implied in this quotation of leaving the economy to its own devices, known as **laissez-faire**, we now realize can only result in the 'best' way of using the nation's resources under a very special set of conditions which do not exist in reality, certainly not in the sort of economy found in Western developed countries. One condition is that perfect competition should exist throughout the economy, a provision that carries the implication that information available to buyers and sellers is complete and freely available. Another is that the actions of producers or consumers should not be felt by others in the form of 'externalities'; this latter point will be developed later. A third is that the legal conditions for markets exist, including the institutions that make this possible (contract law, courts, etc.), so that all ownership rights are clearly defined and can be defended in the courts. Because these conditions are not completely fulfilled in practice and for a variety of additional reasons, it is not desirable to pursue a completely laissez-faire policy if the object is to promote the general well-being or happiness of society, commonly described as society's 'welfare'. While there are considerable differences of opinion as to how far the state should interfere with the market mechanism, most economists would agree that there are valid arguments for *some* modification to the price system's solution to the allocation problem on the grounds of improving the welfare of society.

The Value Judgements of Society

Even if the market system were totally devoid of imperfections and the 'invisible hand' of competition worked in the way Adam Smith envisaged it, there is no guarantee that society would find the resulting pattern of production and consumption to its liking. On the contrary, many people would want to change things on the grounds of improving society's welfare and would attempt to do this through the political system. The sort of goals they might envisage as being desirable characteristics of society would probably include: (i) the provision of reasonable medical care for everyone irrespective of how able they are to pay; (ii) the right for children to be educated according to their several abilities and irrespective of the size of their parents' income; (iii) the relief of unnecessary anxiety from being out of work; and (iv) everyone's right to have a roof to live under. The unhindered market system would not necessarily bring these desired goals to fruition. It might bring about 'economic efficiency', but this will not automatically result in a 'fair' or 'equitable' pattern of consumption.

The price mechanism responds to people's purchasing power, their command over the market, so that goods will be produced only for those members of society with money to spend; the greater a consumer's purchasing power, the greater will be their influence in the market for goods and services and consequently the greater will be the proportion of national resources used in satisfying their wants. The price system makes no moral judgements on the desirability of the existing pattern of purchasing power, which is a reflection of the distribution of income and wealth in the community. However, members of society do make moral judgements on the desirability of situations such as may be instanced by the ability of a rich man to buy medical care for his pet cat while the children of a poor man may be denied attention through being unable to pay; in order to provide medication for the poor children it might be necessary to tax the rich man and society might feel this to be justified on the grounds of improving the happiness of society as a whole.

In effect a redistribution of real income is being brought about and the shares of total national productive resources accounted for by the rich man and the poor man are being made a little less unequal. Attempts at quantifying improvements in social welfare resulting from such actions, when some people are made better off while others are made worse off, are beset by theoretical objections and complications (see Box 7.1). However, the general political attitude, in which society's value judgements are presumably reflected, even if imperfectly, suggests that, among other things, equality of opportunity and a basic minimal standard of living are two goals to be striven for in the process of increasing the welfare of society.

Hence governments often aim for a reasonable degree of equality of opportunity for students to

---

Box 7.1. The idea of a Pareto optimum.

Because the happiness (or utility) experienced by different individuals cannot be added, it is not possible to rank changes that make one person better off but someone else worse off as being a net improvement or deterioration in the welfare of society as a whole. An improvement can only be judged to have taken place if at least one person is made better off without anyone being made worse off. A Pareto optimum (named after the Italian economist Vilfredo Pareto, 1848–1923) is an arrangement of the economy in which the resources are allocated in such an efficient way that it is impossible to make anyone better off without making someone else worse off. Under rather restricting conditions, a perfectly competitive market will achieve this, though the outcome will be affected by how the initial spending power is distributed in society. A Pareto optimum solution might be an aim if we are looking for the most efficient use of resources, but it might not be a very fair outcome. The political preference might be for a less efficient but fairer way of using the productive resources available to society. Thus a trade-off between efficiency and equity would be encountered.

---

attend college or university irrespective of the size of parental incomes and try to achieve it through a mix of the public finance of educational institutions out of general taxation and the use of grants and state-sponsored loans to individuals (in the UK since the first edition of this book appeared the balance has shifted markedly towards the last). In rural areas basic services (education, health, transport, post, etc.) are often subsidized by the government, directly or indirectly through regulation on what suppliers can charge, especially to offset the higher average costs of providing them there and to ensure that everyone has access, a value judgement based on equity. Similarly old-age pensions, financed by a taxation structure which falls more heavily on those with higher incomes, is in reality a mechanism for transferring command over goods and services from the rich with their more ample spending power to the poor (i.e. a redistribution of the income pattern thrown up by the price system).

Because wealth is linked with income, the holders of large concentrations of wealth are liable to be subject to attempts by society to tax it away; in addition, a more even spread of wealth is seen by some to be itself a desirable goal because of the influence and social status which wealth gives. In many industrialized countries the value of farmland, determined by supply and demand where the influences are by no means restricted to those coming from the farming industry, makes the owners of farms one of the wealthiest sectors of the community. This has rendered them particularly sensitive to society's policy of striving for a more equitable wealth distribution through capital taxation, a policy which some would claim is contradictory to simultaneous attempts to promote an efficient and productive farm sector by allowing businesses to grow in size.

Imperfect Competition

If we put on one side questions of social value judgements and externalities, a situation of perfect competition among producers and among consumers can be shown to maximize utility. The case for perfect competition can be summarized as follows: if perfectly competitive conditions were to exist throughout an economy, including complete and perfect knowledge among buyers and sellers, the prices that would result would correspond to the marginal cost of producing goods. They would also correspond to the utility or benefit which consumers

attached to the marginal unit of each commodity. Because marginal costs indicate the quantity of productive factors used to produce the marginal unit, factors which could have been used to produce marginally more of other types of goods, we can say that under perfect competition the quantities produced will be such that the utility derived from the last unit of each good is just equal to the utility which might have been given by the goods which have to be forgone in its production. If we wanted to produce more apples than would result under perfect competition, the benefit in terms of extra utility derived from the additional apples would be less than the utility lost through producing fewer pears or beef or whatever has to be forgone. In this simplified model total utility will thus be maximized by the perfectly competitive solution to the problem of how much of each commodity should be produced.

Monopoly

We have already seen that this 'optimum' pattern that would be produced by perfect competition may need changing to correspond with social value judgements, and in practice perfect competition throughout the economy is not a reality. Monopoly or oligopoly is common, and under these imperfect conditions prices do not necessarily correspond to marginal costs of production (see Chapter 4). Monopoly is quite likely to develop where an industry's cost structure is dominated by fixed costs, for example railways, gas and electricity supply, where fixed equipment such as a gas distribution system is very expensive, yet the cost of producing each extra unit of product (cubic metre of gas or rail journey) is relatively small. Under such a situation, price competition tends to result in a small number of rivals, each operating at a loss with eventually only one survivor who then is in a position to exercise monopoly power. Alternatively, the rivals may come to an agreement between themselves and collectively act against the consumer by raising prices. If the reduction in competition results in a rise in price and a distortion in the pattern of production most desired by consumers, then society can be shown to be worse off.

However, as was pointed out in Chapter 4, a reduction in competition, even to the extent of the emergence of a complete monopoly, is not something to be decried automatically, since positive features may emerge, including economies of scale not considered in the simple model in which a perfectly competitive industry is suddenly replaced by a monopoly. Nevertheless, society keeps a careful watch on situations where the diminishing degree of competition *could* act against the public interest and applies a range of measures to influence the market outcome. In the EU these include legislation which restricts large-scale mergers and takes steps, under EU policy on competition, to break up organizations that appear to have too much market power.

Sometimes, in an attempt to avoid the problems associated with a privately owned monopoly, and to gain a greater control of what governments feel are key parts of the economy, the state may take ownership of an industry – it may be 'nationalized'. In many countries railways and postal services are run as nationalized industries, as often are the public utilities (electricity, gas supply, etc.). Defence forces (armies, military air cover, etc.) are almost always state-owned operations; there are strong arguments against entrusting the defence of a country to a private enterprise army hired by the state but under the control of some mercenary commander who might not have unswerving loyalty to his 'customer' if offered a bigger fee by a belligerent foreign power. However, a distinction must be made between nationalized funding and nationalized provision or production. The former is where the state takes over the responsibility for funding the service (such as by making basic health care free at the point of delivery, which happens in many countries, or for providing free education). The latter is where the state owns the productive assets and is the monopolist entrepreneur, responsible for surpluses or losses; often the product may be purchased by consumers (such as nationalized power or postal services) or used by the state on their behalf (state schools, police forces, etc.). In the UK the National Health Service (NHS) combines the two, which is often a source of confusion in argument for and against privatization or nationalization (see Box 7.2). The experience seems to be that nationalized industries with near monopolies, although set up to act in the public interest, tend to lose their efficiency and, where UK governments have sold them back to the private sector (gas, electricity, telephone, water, lorry transport, etc.) the greater competition has usually improved performance and lowered prices to consumers. This has led to pressure for greater privatization across much of the economy (health care, education, provision of information technology services to government, etc.).

Transaction Costs

Perfect competition assumes a state of complete and perfect knowledge by all those involved. Again, this is not often found in practice. Merely finding out the information about the availability of goods and services (if you are a buyer) or where and how you can sell your product (if you are a supplier to the market) involves costs, either in financial terms or time. Getting in contact, making the transaction and exchanging the goods and services all involve these **transaction costs**, of which **information costs** form one part. Where these costs are too high, the market will not operate, and below that level, the market will work less satisfactorily than if information and transaction costs were lower. Where one party to a deal is better informed than the other, a state of **asymmetry** (or inequality) of information exists, and the outcome will be different from what might happen if both sides were equally and completely informed.

Take a few examples. If a farmer wants to sell cattle at a market, he will face the costs of transporting his animals there and the costs charged by the auctioneer. If you want to buy a second-hand car, you might buy a local paper and hope that enough potential sellers had also spent money on advertising their vehicles that the prices shown in the paper are a reasonable guide to what you might have to pay. When you conclude the deal, you may prefer to pay for an inspection by a motoring organization's experts as a precaution against the seller only providing an incomplete picture of the car's history.

Anything that lowers transaction costs is likely to make the market work better. A good example is the impact of the Internet on sales of airline tickets, books and holiday and hotel accommodation. Information on what is on offer is much more readily available, and even the cost of writing a cheque and posting can be avoided by electronic purchase. Increased competition lowers prices and the volume of goods and services traded expands. Naturally, a few people suffer from this greater access to the market (those who previously exploited their special knowledge and the relatively lesser knowledge of their customers, such as some travel agents) but, overall, it is likely that a net benefit to society has taken place.

Box 7.2. Should health care provision be a state-run monopoly?

Discussions about nationalized industries frequently confuse *public funding* of goods and services with public ownership of the organizations that provide them. This is perhaps seen at its most perplexing when the future of the UK NHS is debated. For reasons explored later in this chapter (externalities of infectious diseases, judgement distortion that lead to poor planning for health spending by individuals, and the pursuit of equity) a case can be made that the state should take over the responsibility for ensuring that each member of society has access to health care free at the point of delivery. In the UK funding for the NHS comes very largely from the central government budget, the resources of which derive from taxation. Within the EU, Ireland and the Scandinavian countries broadly follow the UK funding model, but many others arrange things differently, most having an insurance-based system, which is usually compulsory (Austria, Belgium, France, Germany, Luxembourg).

A different set of arguments surrounds the case for *public provision* of health care services by a state monopoly. Should all hospitals be owned and run by the government? Should all doctors, nurses and other staff be civil servants? On the positive side are the economies of size discussed in Chapter 5 – such as the ability of the NHS to take advantages of planning the most economic provision of services (e.g. the size, numbers and locations of childbirth and Accident and Emergency units, planning of staff numbers and training to meet emerging needs, purchasing power discounts on consumables, research and development possibilities, etc.). On the other are the disadvantages of the need for a bureaucracy to coordinate the NHS's wide range of activities and the inefficiency, inflexibility and loss of staff motivation that sometimes result. Privately owned and operated hospitals and clinics often seem to be able to offer health care with shorter waiting times and greater convenience to patients, at least for routine treatments that do not demand the depth of specialized resources that are only available within an organization the size of the NHS. Added to this, in other EU countries which operate insurance-based funding, private hospitals (often non-profit bodies) are far more common than here and dominate provision in France, Germany and Belgium.

This has led to calls for the privatization of some activities in the UK. Of course, the NHS already engages with the private sector. Most GP clinics and dentist surgeries are privately owned businesses that have contracts with the NHS. Some procedures (e.g. hip replacements) for which public capacity is insufficient are contracted out to private hospitals, even to some abroad. Given the commitment of all UK political parties to the principle of a service that is free at the point of delivery and a volume of resources that is constrained by economic circumstances, the challenge is how to devise a structure (including a use of private sector suppliers) that achieves the optimum quantity and balance of high quality services for society. Other countries may prefer different solutions.
Source: adapted from European Parliament (1998).

In Chapter 6 it was stated that the market for factors of production was by no means frictionless. Labour which is unemployed in one region may not flow swiftly and easily to another area in which jobs may be available. The transaction costs may be too high. Even if information on job opportunities were freely available to those people who are unemployed (which is unlikely) there may be other factors that prevent them moving. The non-monetary costs of breaking social connections and of re-establishment in a new community may be significant. The state has become involved in the labour market through trying to remove inefficiencies in the job information system by establishing and running job centres, and attempts to direct industry by financial incentives or planning control to locations where labour and other resources are available. These actions can be seen as attempts to lower transactions costs.

Externalities Associated with Production and Consumption

A further reason why the price mechanism is imperfect as an allocator of productive resources from society's viewpoint is that it fails to take into account **externalities**. Externalities occur when actions of an individual through his or her consumption of goods and services, or a firm through its production processes, impinge on other firms or people in a way which is not reflected in market prices, so that the consumer or firm does not take these external effects into account in decision making. Externalities may be beneficial, as when a farmer installs beehives in his orchard at his own private cost to pollinate his apple trees and where the bees stray and perform the same function for adjacent orchards and gardens. The neighbouring commercial orchards are thus receiving a benefit in terms of increased yields without cost to them and society in general has its enjoyment enhanced by more productive gardens. These bees are described as bestowing an **external economy** on others, but this will not influence the decision of the farmer to install the hives. His action will be based on the cost of the bees and the resulting improvement in yield from *his* trees.

Another example of an external economy might be where a firm operates a training scheme for its employees and where these are free to leave and join rival firms which do not run such schemes. Without bearing any training costs these rival firms reap benefits of ready-trained recruits. Similarly, when a government improves a coastal road for defence purposes local businesses and hotels may benefit because of better access for trade and tourists. Yet another example is provided by the farmer who invests in a better drainage system for his land to improve its ability to be worked and to enhance yields; the land of adjacent farmers may also benefit and their profits increase. In all these cases the benefit to society in general is greater than the benefit to the individual who undertakes the initial spending. This difference between private and social benefits has implications for the optimum allocation of resources.

Take, for example, Farmer Bloggs who spends money on improving his field drainage. A range of expenditures would be possible depending on the thoroughness of the drainage system and the better the system the higher the crop yield which might be expected over a run of years. We can assume that diminishing returns apply so that the additions to yield resulting from each extra £1000 of drainage equipment installed decline as the total amount of expenditure increases. In deciding how simple or elaborate his drainage system should be, Farmer Bloggs will balance in his mind the extra cost of the more complex systems with the extra benefits they will give in terms of higher crop yields and revenues from his land over the expected life of the system; for him the optimal, most profitable system is where the marginal cost is just balanced by the marginal revenue (benefit). Both the costs and the benefits are 'private' to him since the effects on other farmers have not been taken into account; their extra yields do not benefit Farmer Bloggs and are irrelevant to his investment decision. However, if extra yield enjoyed by the surrounding farmers *is* taken into account, the wider benefit to society (marginal social benefit) of the field drainage can

be assessed, and this is greater than the private benefit to Farmer Bloggs.

Just as there are external economies that cause a divergence between private and social benefits, so there are **external diseconomies** that cause private and social costs to diverge. External diseconomies arise when the actions of firms or individuals impinge deleteriously on others and are not taken into account by the decision maker. A favourite example is the brickworks' chimney which belches dirty smoke, imposing 'external' costs on others in society by causing housewives to wash clothes and curtains more frequently, offices to be cleaned more often and necessitating more public services to be provided to cater for the damaged health of people. All these remedial actions use up productive resources and are costs to society at large which the perpetrator of the nuisance does not take into account unless compelled. He arranges his production according to his **private costs** only – the cost of labour, fuel and other inputs – with scant regard to the costs to the rest of society of his activities. However, the full cost of bricks to society (termed their **social cost** of production), on which the decision of how many bricks should be produced in the economy should be based, must take into account both the private cost of production and their associated external diseconomies.

Having recognized that, from society's viewpoint, a wider view of costs and benefits must be taken, the optimum quantity of any good or service will be where the marginal social cost of its production equals the marginal social benefit its consumption brings. Returning to our drainage example where the social benefit was greater than the private benefit, in order to balance the benefit to society (marginal social benefit) with the cost involved (social cost which we will assume for the present to be the same as the private costs borne by Farmer Bloggs) and so move towards the optimal use of resources from a whole-society view, it will be necessary to encourage Farmer Bloggs to invest in a greater quantity of drainage than he would otherwise do. This could be done by modifying the price system in the form of a grant on the cost of drainage. The external benefit of his action is thus brought into his decision-making process (or **internalized**) by the grant which lowers his private costs, so that he equates marginal (subsidized) private costs with marginal private revenue at a higher level of drainage installation.

In the EU we see the price mechanism modified in a whole range of situations where external economies are associated with production or consumption, although there is only sufficient space here to touch on a few. The UK government supports training schemes and gives grants to research establishments; part of the reason for providing grants for further education is to obtain for the rest of society the benefits of a highly educated and trained sector of the population in the form of a more rapid rate of technological development and economic growth. Grants are available to farmers willing to make holiday accommodation available in their houses; benefits accrue to the wider community, which the individual farmer does not take into consideration, through retaining employment in the countryside, preserving the appearance of the landscape for the enjoyment of townsfolk and obviating the necessity for alternative and perhaps more costly government measures to support the incomes of farmers. In the case of subsidies to rural bus services which would otherwise close down, it can be argued that the welfare of the rural community is greatly enhanced by the continued existence of the service, even if many of the benefits are not easily measurable. One such benefit might be the peace of mind given to rural non-car-owners by feeling mobile and part of the whole community rather than cut off, easing family pressure on them to move to an urban environment.

We turn next to situations where attempts are made to take external diseconomies into account, narrowing the gap between private and social costs of production. Tools used by governments include legislation to set **standards**, or the use of **taxes** to discourage activities that cause the negative externalities, or **subsidies** to encourage producers to behave in less damaging ways. Again, a few examples must suffice. In the case of the smoking brickyard chimney, where the social cost of brick production exceeded the private cost, the brick manufacturer could be compelled to take the social costs more into account by legislation forcing the brickworks to adopt methods which eliminate offensive smoke at source, such as the installation of filters on the chimneys or the use of a different fuel. This could take the form of setting 'standards' on the minimum level of permitted emissions, or requiring the use of 'best practice' technology (standards on the process) or banning inputs known to cause pollution (standards applied to the inputs to production). Inevitably bricks would then cost more, but only the buyers of bricks would feel the direct effect, not the third-party housewife or office worker. More costly bricks would mean that less would be demanded and hence less produced and the output would contract to nearer the level where marginal social cost was in line with marginal social benefit.

Another method of mitigating external diseconomies, appropriate in the case of aircraft noise, is to require the initiator to meet the cost of remedial equipment of the people who suffer the externalities, such as double glazing of domestic houses for sound insulation. As a result, air fares have to rise above the level necessitated solely by 'private' costs, but the ordinary householder who never travels by air is little affected by this. Without insulation, he or she has to bear a cost of air travel (noise) without enjoying its benefits of speed and convenience. This requirement that the creator of the external diseconomy has to face the cost of cancelling out his harmful effects on others is often called the 'Principle that the Polluter Pays'. Another method is to set up the necessary legislation to enable compensation to be exacted for external diseconomies suffered – a farmer who knows he can be made to pay for damage done to downstream fisheries or fined heavily for polluting rivers will be cautious in his use of herbicides. Private and social costs are, therefore, kept better in line although the legislative system itself will not be costless. Publicly financed advertising is used to dissuade drunken drivers from taking to the road and imposing external costs on the rest of society. Certain countries will not allow visitors to enter unless vaccinated against infectious diseases no longer endemic in the country, since the social cost of readmitting the disease is viewed as being enormous and far greater than the benefit which any visitor who refused to cooperate could bring. Action in all the above cases is more likely if the source of the external diseconomy can be pointed to precisely; it is much easier to take steps against a chemical factory which is polluting a river (a point source) than it is to combat background noise arising from an increased amount of traffic on city roads.

A major problem, associated with many negative externalities, is that the rights to clean air, a quiet environment, etc. are not well established in law. If they were, people would act to protect them. Markets act well only where property rights are well defined, and where a legal system exists whereby they can be defended – the market for cars is a good example. Things like a clean environment are trickier to handle via the market mechanism because the rights are less clearly established in law. Sometimes the act of clarifying rights by legislation is enough to enable the market to work to reduce externalities (see Box 7.3).

The production and consumption of many goods and services will have both external economies and external diseconomies associated with them. For example, the country bus service which bestows an external economy by reducing loneliness may also impose an external diseconomy by damaging country lanes developed for less weighty vehicles; neither externality is taken into account by the bus operator when fixing fares and the frequency of his service. The farmer who spoils the view from a cottage by erecting an ugly building in front of it may also be providing a windbreak which benefits the cottage's garden and lowers its heating bills.

The general picture of private and social costs and benefits, the solution produced by the market based solely on private costs and benefits, and the effects of modifying this solution in line with the wider social implications are illustrated in Fig. 7.1. This is a diagrammatic generality and is not based on a particular case.

The extent to which external diseconomies and economies are taken into account (i.e. are internalized in the cost–benefit decision) will depend on society's awareness of them, their seriousness, the practicality and administrative cost of doing something about them, and the political will to act. It is not necessary for precise estimates of externalities to be made before action can be taken: for example the potential deleterious effect of some persistent insecticides on human health is obvious enough to cause their banning from general use. In terms of Fig. 7.1 the social cost at *all* levels of use is so high that the only intersection of marginal social cost and marginal social benefit is when zero quantity is in use; furthermore, a complete ban may be the only enforceable method of control.

Society is becoming increasingly aware of the externalities of production and consumption. The action of individuals and firms is now more correctly seen as being inter-related in ways which are not reflected through market prices. Agriculture's impact on the environment rose to political prominence in the late 1980s with the realization that farming created external costs, such as: (i) pollution of water courses; (ii) reduction in wildlife diversity; (iii) changes in landscape appearance; and (iv) less easy access to land for recreation. As will be demonstrated in Chapter 10, government agricultural policy now takes externalities into account. Many major changes to the economy, such as the siting of an airport, would not now be undertaken unless some attempt had been made at a wide-ranging cost–benefit analysis. This would not be restricted to private costs and private benefits, such as the cost of the land, buildings and staff on the one hand and the anticipated revenue from flights on the other. Rather it would also take into account the externalities of the project which a private airport operator would not consider, such as on the cost side: (i) the implications of more traffic on the approach roads for the people living there; (ii) the noise disturbance to householders from aircraft; (iii) loss of earnings and disruptions to the farm workers displaced; and (iv) the deterioration of the countryside's appearance. Examples of externalities on the benefit side that may be considered include: (i) the reduction of congestion at existing airports; and (ii) the employment generated locally. On both sides of the balance there will be elements which are difficult to quantify, frequently of the environmental type, but they may be some of the most influential in the decision on whether the project goes ahead.

Box 7.3. Coase's Theorem.

This suggests a market solution to an environmental externality problem is possible if rights are defined. Rather surprisingly, it does not matter who owns the rights – the polluter or the people suffering the pollution – the outcome will be the same.

Take the example of a factory polluting a river that runs through a village, causing problems for the people living there. If the villagers are given the legal right to a clean water supply, they can take the polluting factory to court, forcing it to cut pollution. Or, if the factory has a right to put out emissions, the villagers could band together to persuade the factory to emit less by compensating the factory owner for not exercising the right to pollute (the villagers are buying at least part of the right to pollute). Of course, the shoulders on which the cost falls differ, but either initial allocation of rights would reduce pollution.

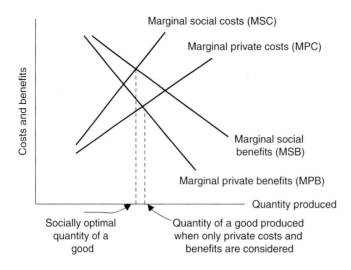

Notes: (1) In this example the socially optimal quantity of the good is less than that which would result from the consideration of private costs and benefits only. However, the socially optimal quantity under other sets of conditions can be greater, for example, if in the above diagram the MSC line were very close to the MPC line.

(2) It is possible to envisage situations where the MSB lies below MPB (e.g. where external diseconomies of consumption are involved), or where MSC lies below MPC (e.g. where unemployed productive resources exist so that more of the good can be produced without sacrificing the output of any other good).

Fig. 7.1. Divergences between private and social marginal costs and benefits arising from the production and consumption of a good.

Supporters of the centrally planned economic system would suggest that, though it had many disadvantages that led to its demise in the countries of Central and Eastern Europe, it was superior to the capitalist free-market economy in taking into account the full costs of production and consumption. Decisions for society could be made on the basis of social costs and social benefits. Prices were not the result of market forces but the result of planning, so that from the outset they could be set at levels which reflected not only 'private' costs and benefits, but also any external diseconomies or economies. Such economic systems, however, still faced the twin problems of identifying and assessing such externalities.

Public goods

There exists a group of goods called **public goods** or 'social' goods where external economies are particularly relevant. The services provided by the police force and national defence are of this type. If one man decided not to contribute to their cost, he would still benefit from the peace and security which the existence of the armed forces provides. The spending by the rest of the community on these services bestows an external economy on him and he cannot be excluded from benefiting, i.e. public goods are non-excludable. Another feature of public goods is that one person's benefit does not diminish other people's benefit from the same good. For example, the protection the British Army affords me by deterring the forces of a potential aggressor does not stop my neighbour enjoying the same protection – we are non-rivals in the consumption of defence. Public goods, then, are characterized by **non-excludability** and **non-rivalness**. In contrast, a good such as a tractor is both excludable and rival. Legal possession implies that the services of the tractor belong to its owner and everyone else can be excluded from using them (unless, of course, it is such a handsome machine that it can be considered as providing visual pleasure, and it is difficult to stop admiring glances from the rest of the neighbourhood). Tractors also

Box 7.4. Is there an optimum pollution level?

Given that in the real world pollution exists, a question arises as to how much effort should be expended in lowering it. Simple tools of economic analysis can be applied. While the environment can absorb a small amount of pollution (emissions from production or consumption) without damage, beyond this level the damage done will normally increase with the level of pollution present, and does so at an increasing rate. Thus the **marginal environmental damage** rises with higher levels of pollution. However, the more pollution there is, the easier it is to reduce emissions a little, but the cost of making further reductions tends to rise. This cost of reducing pollution from emissions can be termed the **marginal abatement cost** The concept of the **optimum level of pollution** is where the cost of abating the marginal unit of emissions just balances the value of the damage it does. To reduce pollution further would incur additional costs that are greater than the value of the damage done.

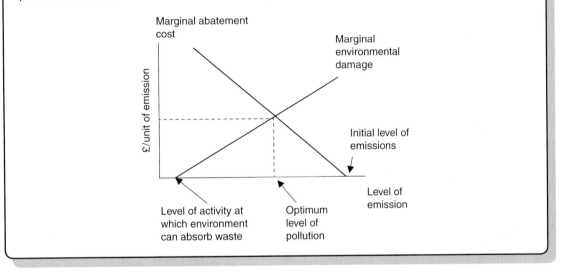

exhibit rivalness since, if a farmer buys a tractor, other farmers are denied the possibility of acquiring it, and the more tractors one farmer buys the fewer are available for the other farmers; all the farmers are rivals for the tractors.

The market mechanism is at its best when handling goods which fall into the excludable-rival category, but it cannot be relied on to provide the socially optional quantity of public goods (i.e. that quantity where their marginal social cost just balances the marginal social benefit they bring). By itself the market mechanism would result in too little public goods being provided. For example, although most people would agree that some defence force is necessary and might express the willingness to contribute to defence costs as long as everyone else paid their share, the unhindered market system would not result in an appropriate amount of defence being provided. A formal proof of this statement's validity could be made, but it is sufficient here simply to point out that there will always be a tendency for

individuals to opt out of paying for public goods like defence and become 'free riders', thereby reducing the amount spent on it below the optimum level. Society finds it necessary for the government to finance such services from tax revenue which is collected on a compulsory basis from individual members of the community.

Similarly, although all farmers benefit from the public control over the health of animals that are imported into their country, it is most unlikely that an adequate inspectorate would be provided by private enterprise with the necessary powers to refuse entry to unhealthy stock. Human health provides another important example: treatment against many infectious diseases is only effective if everyone agrees to take part in its control, which in effect means action organized by the state. As noted above with monopolies, public finance of a service should be distinguished from public production. For example, while the control of foot-and-mouth disease by slaughter can only be effectively financed as a public

> **Box 7.5. Other combinations of rivalness and excludability.**
>
> Two further categories of goods are worth mentioning, **rival non-excludable goods** and **non-rival excludable goods**. Honey bees are an example of the former, as the owner of the bees is unable to exclude his neighbours from the pollinating services of the insects. A seat in a less-than-full train demonstrates the latter; the ticket-holder can exclude other people from using his seat, but his travelling does not prevent other people being transported over the same journey (i.e. they are non-rivals). The market system finds difficulty in handling both of these categories.

policy, there is no reason why the slaughterers have to be government employees; private vets could be contracted to do the job and paid by the state.

Imperfect Knowledge by Consumers

Reference has been made above to imperfect information hindering the operation of markets. At least part of the reason why the state intervenes in two important sectors of consumption – health and education – is because it feels that individuals, even if they have access to the necessary details (though this is unlikely), are not in a position to make an informed choice. The tendency is for people to ignore the longer-term consequences of their actions in favour of short-term utility maximization. Leaving aside the arguments for intervention in the health treatment market on the grounds of externalities and that health treatment is in part a public good, if experts can show that non-experts are under-consuming health care (or over-consuming dangerous commodities) then the state may wish to intervene to correct the level of consumption. Cigarettes are a suitable example: the link between their consumption and subsequent ill health is well known by the general public but seems inadequately appreciated by smokers. Consequently the government may attempt to correct this by advertising the dangers involved, trying to change consumers' tastes and buying patterns, and raising their price through taxation so that fewer are bought.

As with many government measures, taxes on tobacco attempt to serve more than one purpose, in this case both the consumption-changing function and the traditional role of collecting revenue for other services provided collectively. Similarly, with free health care and education financed by taxation, some income redistribution from the rich to the poor is simultaneously achieved since, by and large, the wealthy pay a bigger proportion of total tax than they receive in proportion of total benefit from these services. Some people would argue, however, that the form this provision takes, with the health service and schools run as public institutions, is a relatively inefficient method of realizing the government's community health, education and income redistribution intentions. A more effective system might be simply to give the less well-off members of the community income supplements financed by taxation, accompany this with a government-financed advertising campaign aimed at improving public knowledge of health measures and the benefits of education, and allow individuals to spend their enhanced incomes as they wish. According to its supporters, such an alternative approach would be devoid of the heavy and paternalistic state direction of 'what is good for them', which they maintain is built into the present mechanism.

Macroeconomic Reasons for Intervening in the Market System

It was also noted in Chapter 6 that wages do not fall easily in times of unemployment to expand the numbers of workers taken into industry. Labour markets do not work in quite the same way as markets for, say, potatoes and wheat, in which prices respond quickly to changes in supply and demand, falling as well as rising. The labour market is apparently far more rigid, influenced by trade union strength, employment legislation, long-standing differentials and a sense of what is fair. In consequence, during a time of economic depression governments which are committed to maintaining a low level of unemployment will need to resort to more active policies to achieve their goal. These policies will be concerned with the level of economic activity in the economy, because this bears a close relationship to the level of employment.

Rather than just acting as a redistributor of the income generated within the economy from one sector to another (as with unemployment benefit or pensions paid from taxation), or as a provider of public goods again financed by taxes, or as a modifier of the pattern of production so that it matches more closely social costs and social benefits, the government is attempting to influence the *level* of national production. The factors determining the level of economic activity in a country and the manner in which a government can influence them to achieve its goals, not only that of achieving high employment but also others such as controlling inflation and stimulating economic growth, are covered in the next chapter.

Exercise on Material in Chapter 7

Answer the questions in the manner indicated. Answers and explanations are given in Appendix 2.

7.1. List three reasons why governments may wish to intervene in markets to change the allocation of resources and outcomes.

7.2. Cross out the inappropriate alternatives.

The market mechanism is best suited to handling *private/public* good and services. The pure private good displays *rivalness/non-rivalness* and *excludability/non-excludability*. In contrast the pure public good displays *rivalness/non-rivalness* and *excludability/non-excludability*.

7.3. In the table that follows, which examples of goods and services would you class as predominantly private goods and which are public? To justify your answer apply criteria of rivalness and excludability. (You may wish to specify circumstances, because sometimes these affect the classification.)

Example	Private	Public
Shoes for sale in a shoe-shop		
A country's defence forces		
Lighthouses		
Public buses		
Dental treatment financed by the public health service		

7.4. Cross out the inappropriate alternatives.

The presence of externalities to a particular production process means that, in order to achieve an optimal level of production from the point of view of society, governments will need to *reduce production/increase production/either reduce or increase production depending on circumstances.*

7.5. Cross out the inappropriate alternatives.

Taking into account the effect of external costs of private transport to society will *raise/lower* the supply curve based on private costs of production. If there are external benefits from consumption of education, the demand curve for educational services from society's viewpoint will be *above/below* that from individuals and families.

7.6. You wish to buy and import into the UK a car made in Germany.

(a) Which of the following might be regarded as transactions costs?

Costs	Transaction cost?
Cost of the car	
Currency exchange commission (conversion of £ to €)	
Risk that the cost might change because of changes in exchange rate	
Time spent in finding out details of how to make an import	

(b) What is/are the likely outcome(s) in terms of transactions costs of the introduction of a single currency operating in both the UK and Germany?

(c) What is the likely outcome on the demand for German cars in the UK of provision on the Internet of information on procedures for importing vehicles?

(d) What are the likely implications of reducing transactions costs for: (i) international trade; and (ii) the welfare of people in the UK and Germany?

8 Macroeconomics: the Workings of the Whole Economy

Introduction

Macroeconomics is the study of how the economy as a whole works. Up to now we have been looking at bits of the economy separately – offering explanations for how consumers behave, how individual firms organize themselves, how the demand for the resources available for production arises at the firm and industry levels, etc. – but we must bear in mind that all these elements are linked together and react with each other as broad aggregates (or totals), e.g. aggregate output, which is the total output of all firms in the economy. It is important to try to understand how the aggregate economy works because many of the economic problems which have to be faced by society are essentially collective both in cause and in remedy.

Some of the basic macroeconomic problems are these: (i) the problem of maintaining the value of money to prevent the harmful effects of inflation; (ii) ensuring that the country's productive resources (especially manpower) are not wasted by being unused, so that as high a living standard as is attainable is achieved; (iii) ensuring the country's stock of transport infrastructure is maintained; (iv) encouraging economic growth; and (v) encouraging beneficial trade with other countries and simultaneously achieving stable exchange rates between the different currencies. All these problems can only be tackled from a combined (or aggregate, or whole-economy) standpoint.

As was noted in the preceding chapter, the early economist Adam Smith believed that, as long as each member of society acted in his or her own business interest, the national economy would automatically take care of itself and the interests of society would also be best served. It would be as if an 'invisible hand' were directing events. Governments should therefore adopt a laissez-faire policy and not try to regulate the economy, since such meddling would almost certainly do more harm than good.

For reasons discussed in Chapter 7 we now know that Smith's theory was not correct, unless heavily qualified, and that some government interference with the economy on behalf of society is required if many of the objectives which society sets itself are to be reached. A good example is the control of inflation (fall in the value of money) which can be achieved by government action. However, too much government interference, or incorrect measures, can make problems worse ('government failure'), and some would argue that government should restrain itself rather than take actions whose full implications are not understood. An added complication is that some of the goals for which government aims are mutually incompatible – actions taken to control inflation, for example, appear often to result in a rise in the rate of unemployment. In such cases the government has to balance its priorities.

As a first step in analysing the workings of the aggregate economy and offering explanations for macroeconomic phenomena we can construct a simple model of the flow of income within the economy and use it to explain what determines the level of that flow.

The Circular Flow of Income

When an individual is asked the size of his annual income he will usually give an answer in money terms because he is paid in money. But really what he is implying is that his income gives him a certain purchasing power to obtain goods and services. Buying and enjoying them are what give him satisfaction; money is only used as a convenient common denominator. Similarly, at the national level the country's income is really the quantity of goods and services which are produced and consumed by its inhabitants in a given time period, usually 1 year. All other things being equal, an increase in the quantity of goods and services produced and consumed which had been caused, for example, by an improvement in the efficiency of manufacturing processes implies that the nation's income has gone up.

In economies where money is used, consumers buy their requirements from the market which in turn receives its supplies from firms and other producing units. There is a flow of goods and services from producers to consumers and an opposite money flow in payment from consumers to producers.

But where do consumers get their income from? They earn it by working in factories, offices, schools, etc.; they are selling the service of their labour (which is a factor of production) in exchange for wages and salaries. In a basically capitalist economy, such as those of Western industrialized countries, individuals may also own land which they can rent out to producers, or capital which they can lend directly (as loans or by buying shares or bonds) or through the banks in return for interest. Their various factors of production are put at the disposal of producers in exchange for payment to their owners – the consumers. There is thus a flow of payments from the producers of goods to consumers, which consumers then pay back to producers as they buy shoes, food, electricity, etc., forming a circular flow. A beneficial arrangement exists: the consumer sells the services of the factors of production which he controls and ends up with goods and services which he wants. Money enables the process to take place far more easily than would be possible under a primitive barter economy. This circular flow is illustrated in Fig. 8.1.

In the real world the circular flow is not so simple. As an illustration, imagine what would happen if consumers, instead of spending all their incomes on goods and services, put part away in tin boxes under their beds 'for a rainy day'. Unsold goods would tend to pile up in shops and in manufacturers' warehouses because more was being produced than was being demanded with the consumers' reduced spending. Because of their lower sales, producers would soon cut back on their output and would no longer require to employ as much labour and other factors. In short, the whole circular flow would shrink, fewer goods and services would be produced and consumed (i.e. the national income would fall) and unemployment of productive resources would arise.

Saving is a leak, a **withdrawal**, from the circular flow. Other withdrawals, which have similar effects in running down the circular flow are: (i) payments by consumers to foreigners for goods bought from abroad, such as imported food and the purchase by producers of foreign-made equipment and raw materials from abroad; (ii) payment by producers to the foreign owners of capital; and (iii) taxes taken by the state from both consumers (e.g. income tax) and producers (e.g. corporation tax). Here we are thinking of consumers as a whole and producers as a whole; the government may give back some of the taxes it raises in the form of benefits to the less well-off people and subsidies to firms it wishes to assist (both are forms of **transfer payments**), but there will be a net effect on these sectors that represents a withdrawal from the circular flow.

Fortunately a parallel set of **injections** exists to counteract the effect of the withdrawals, tending to increase the size of the circular flow and hence raising national income. Injections include: (i) investment in new factories and machinery undertaken by producers and financed partly by borrowing from individuals and from banks; (ii) goods which are exported bringing in money to producers from abroad; (iii) inhabitants in this country owning

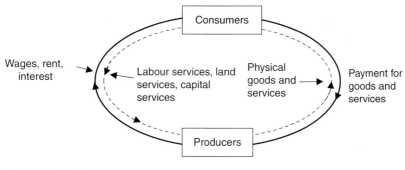

Fig. 8.1. Simple circular flow of income.

property and capital abroad receiving annual rent or interest payments; and (iv) the government spending money on buying goods such as defence equipment and drugs for the NHS and on employing civil servants. All these injections are payments into the circular flow but are originating outside it. These injections and withdrawals are combined in Fig. 8.2.

Over a given time period as long as the magnitude of total withdrawals is the same as total injections there will not be any reason for producers to expand or contract their level of output of goods and services – aggregate output will remain the same. What happens, however, if the two do *not* balance? If withdrawals exceed injections, does the production of goods contract indefinitely? The answer we observe from the world about us is, obviously, no.

The reason for this is that the size of the withdrawals which people are willing to make is a function of the level of national income, while plans for injections are very largely (and, for the sake of simplicity, taken here to be completely) independent of national income. Individuals will want to save part of any increase in income they receive, so that total intended saving (a withdrawal) will be greater at higher levels of national income; but the size of investments (injection) businessmen are willing to make depends not on the level of national income but on their confidence in the future, the rate of interest on borrowed funds, etc. As people's incomes rise they want to spend part on imported goods, so the level of imports (a withdrawal) rises;

but the amount foreigners buy of our products (injection), while being dependent on the size of *their* incomes, is not affected by the size of *our* incomes. Similarly, taxation systems are normally designed so that with higher national incomes more is raised in taxes (withdrawal); but the government can spend less than it raises (and produce a budget surplus), the same, or more (by borrowing) depending on what it considers the most appropriate level of injection for the economy at the time.

Overall, the intended level of withdrawals rises with national income, but the intended level of injections is independent of it. This is shown in Fig. 8.3 and from this diagram it can be seen that there is only one level of national income at which planned (or intended or desired) injections are of the same size as planned (or intended or desired) withdrawals. This is the equilibrium level of national income and, as long as the intentions of people and institutions who make withdrawals and injections do not alter (and ignoring for the present external influences like droughts or wars and the longer-term effect of economic growth), there is in the short term no reason why national income should change once it has become established at its equilibrium level. Furthermore, the actual amounts withdrawn and injected will coincide with the planned amounts and people will have no cause to revise their intentions.

Imagine, however, what might happen if people plan to inject less into the circular flow of income. For simplicity, let us assume that savings are the

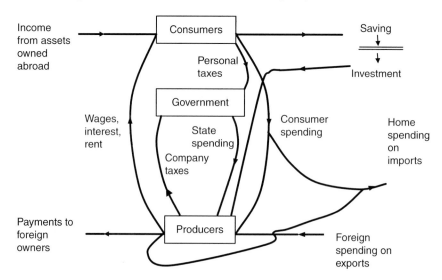

Fig. 8.2. An elaborated circular flow of income.

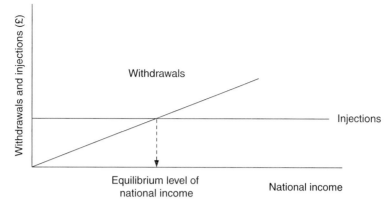

Fig. 8.3. The equilibrium level of national income.

only withdrawal and investment is the only injection (i.e. we are assuming that there is no government and no international trade). Take a situation where investors in new machinery and factories for some reason lose confidence in the future and decide to invest less than hitherto. The 'injection' line of Fig. 8.3 will fall, cutting the 'withdrawal' line at a lower level of national income. In practical terms this will mean that the demand for new machines, etc. will fall because investors will want less of them, less will be produced and fewer people needed to produce them, or less overtime worked. Less money will in total be in people's pockets, so they will want to save less, and the firms which produce machines will be less able to put money aside for later expansion. The general level of output and income falls to a level where planned savings (withdrawal) again equals planned investment (injection). As long as investors maintain this lower level of planned injection there is no reason why national income should rise from its new low level. There is little reason in the short term why any pool of unemployed labour which might result from this lower level of total production should disappear; history has shown that wages do not fall in times of unemployment by anything like the amount necessary to mop up the pool of available labour by making the factor cheaper and hence more attractive to businessmen.

If, on the other hand, investors are prompted by premonitions of a bright and confident future for the country to raise the level of their spending, the 'injection' line of Fig. 8.3 will rise, intersecting the withdrawal line at a higher level of national income. Demand for machinery and other capital

goods will be strong and their manufacturers will try to increase output by taking on more people. Output of goods will rise. In total the income of workers will increase (because more will be employed and earnings higher) so they will be able to save more and, together with the savings of firms, the total level of saving will continue to rise until it reaches the same level as investment when there will be no further tendency to rise. National income will have risen.

National income will thus rise if the intended level of investment injection increases, and fall if it falls. Similar effects would occur if people's propensity to save were to change. If at each income level they decided to save more, the savings-withdrawal line would tend to rise and a lower equilibrium national income would result; conversely, if they decided to save less, national income would rise (see Fig. 8.4). If a government is attempting to control the size of the national income, clearly it should be aware of what is happening to the intended levels of saving and investment in the economy. Why it would be interested in doing so and the actions it might subsequently take will be returned to later.

The multiplier

So far we have pictured the effect of changing levels of investment injection only on the activity of the capital goods industry – the producer of factories and factory machinery and its workers. In real life the effects of injections into the economy spread much wider. If more workers are employed in the capital goods industry, part of their extra earnings

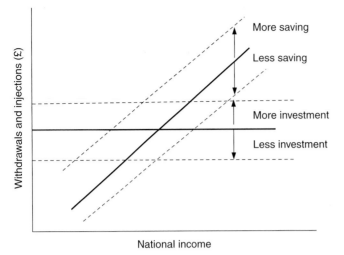

Withdrawals and injections (£)

More saving

Less saving

More investment

Less investment

National income

Fig. 8.4. The effects of changing levels of saving and investment intentions on national income.

will be saved (as already mentioned), part taken away in taxation and part spent on goods from abroad, but much of it will be spent on more goods and services from other sectors of the national economy – clothes, cars, etc., – so the demand in these consumer industries will rise and production will be expanded to meet it. In turn, workers in these expanding industries will have more in their pockets, some of which will be spent on domestically produced consumer goods. Firms producing consumer goods will probably require more plant and machinery, reinforcing the original rise in demand for capital goods.

Thus the effect of an initial amount of spending in one sector of the economy spreads throughout the whole economy, rather as a stone dropped in a pond causes ripples to spread over the whole surface. Also, like ripples, the effects of the spending diminish with distance from the initial spending because, at each stage of income which the initial spending generates, a proportion is withdrawn and not passed on. Spending an extra £1m on new machine tools (injection spending) could cause *total* increased spending to add up to £4m because of the ripple effect, of which £1m will correspond to the initial spending on machine tools and £3m to the subsequent rounds of generated spending. The relationship between the initial spending and the total amount of spending is called the **multiplier**.

In our example the multiplier would be 4. The size of this multiplier can be shown to be related to the fraction of each additional increase in spending which is *not* passed on in a second round of domestic spending because of saving, taxation or the purchase of foreign goods. This fraction of the marginal £ which is not passed on is termed the **marginal propensity to withdraw** (MPW).

The multiplier is numerically equal to the reciprocal of the MPW. For example, if in passing round the circular flow of income one-quarter of each additional pound of spending would be withdrawn through saving, taxation or spending on imports, etc. (i.e. the MPW is 1/4), then the multiplier will be 4. Similarly, if only 1/6 is withdrawn, the multiplier will be 6.

In the explanation of the equilibrium level of national income given earlier, a very simple model of the economy was taken in which saving was assumed to be the only withdrawal, and investment was assumed to be the only injection. In this case the **marginal propensity to save** (MPS) is what determines the size of the multiplier, the MPS being the fraction of each additional pound of income which is saved (the only withdrawal in the model). An MPS of 1/3 will give a multiplier of 3. The fraction which is not saved must be spent on consumption goods, and forms the **marginal propensity to consume** (MPC). The MPS and MPC must therefore sum to one, i.e. MPS + MPC = 1.

$$\text{Multiplier} = \frac{\text{Total spending generated (including initial increase in spending)}}{\text{Initial increase in spending}}$$

Box 8.1. Marginal and average propensities.

Marginal propensities must be carefully distinguished from average propensities. The **average propensity to save** (APS) is the proportion of a person's total income that is saved, i.e.

$$APS = \frac{\text{Amount saved}}{\text{Total income}}$$

Similarly the proportion of total income which is spent is the **average propensity to consume** (APC), i.e.

$$APC = \frac{\text{Amount spent on consumption}}{\text{Total income}}$$

Since what is not saved must be spent on consumption, the APS plus the APC must also sum to one, i.e.

$$APS + APC = 1$$

The existence of the multiplier effect is of great relevance to government policy. For example, if a pool of, say, 500,000 men are unemployed in the country and the government decides that, as a matter of policy, they must all be given jobs, it can achieve this by placing orders to build roads with construction firms who will employ more men. However, it is not necessary to place sufficient new road contracts to absorb *all* the available labour. The government knows that much of the money it pays to construction firms will be spent by their employees and shareholders and generate expansion of other sectors of the economy which will in turn demand more labour. All the government has to do is to place sufficient road orders, which may themselves only need 100,000 men, to generate a *total* demand for 500,000 men. If it goes beyond this a shortage of labour will eventually result. Because the exact size of the employment multiplier is unlikely to be known in advance, the government may in reality under- or over-achieve its aim.

Another example might be where the government encourages a firm, looking for a site to establish a new factory, to set up in a region where incomes and living standards are lagging behind the rest of the country. It knows that the wages paid by the factory will generate spending and incomes right through that region far greater than the initial amount paid out in wages. Again, quite by how much may not be known in advance.

The multiplier effect has often been described as a two-edged sword. This is because, just as stimulating increases in spending (injections) can spread throughout the economy, so can depressing withdrawals. If, for example, a large company were to reduce the amount it normally invested in new machinery, firms making machines would face reduced demand and would lay off workers who would then cut down their spending on consumer goods, and so on. Total output (and incomes) of the commodity would fall, as might well the level of employment. In such circumstances the government would be under political pressure to counteract the tendency for the economy to run down by spending more itself.

The Level of Aggregate Demand

An alternative way of analysing the circular flow of income is to consider the level of aggregate demand, alternatively called aggregate expenditure. The term **aggregate demand** is used to mean the combined demands of the several sectors of the economy.

Referring back to Fig. 8.2 shows that the demand for goods and services comes to producers from three sources: (i) consumer spending (C); (ii) spending on capital goods in the form of investment (I); and (iii) spending by the government (G). For simplicity we will assume that spending on exports and spending on imports equal each other, cancel out and thus can be ignored at this stage. As investment and government expenditure are both injections, it is possible to think of aggregate demand as consumer expenditure (C) + injections (I + G). As long as aggregate demand is just sufficient to absorb the quantity of goods and services that the nation is producing, the economy will be in equilibrium. Put another way, national income, which is the goods and services produced in a year expressed in terms of money, will have no tendency to rise or fall if it is equalled by the value of total expenditure; if expenditure is greater or less than national income, then national income must rise or fall before equilibrium is restored.

Figure 8.5 shows all the levels of national income which are potential equilibrium combinations of income and expenditure, i.e. where income (Y) and expenditure (E) equal each other – this is simply a line at 45° between the income and expenditure axes. Any national income on this line is maintainable indefinitely, up to the limit set by the country's productive resources. Only a certain quantity of labour, capital and the other factors of production are available at one time, and once these are all fully employed no further increase in national income is possible, at least in the short term. Below this full-employment level of income a whole range of equilibrium national

incomes is possible, although some unemployment of resources will then occur.

Determining the equilibrium level of national income

Expenditure within the economy comprises, as already mentioned, expenditure by consumers (C), expenditure on investment (I) and expenditure by government (G). We have also already noted that the latter two (I + G) can, as a first approximation, be assumed to be independent of the level of national income. Consumer expenditure is, however, not so divorced, and relates to national income in the way shown in Fig. 8.6. As national income rises, so does the amount consumers desire to spend, but not at the same rate; if it did it would have the same slope as the E = Y 45° line. At incomes greater than N planned spending is *less* than the national income, and so some is saved. At levels of Y below N consumers will want to spend *more* than the national income in order to maintain their living standards, implying that

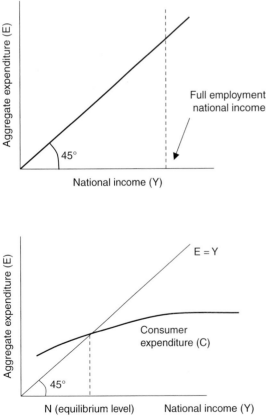

Fig. 8.5. The range of possible equilibrium combinations of national income and expenditure.

Fig. 8.6. The equilibrium level of national income.

they must dig into past savings to do so. If there were no investment going on in the economy and no government spending (admittedly an unreal situation), national income would be in equilibrium at N because, at that level, intended aggregate expenditure and national income would be equal. There would be just enough demand to take up everything that producers were putting on the market. Any attempts by producers to produce more would simply result in stock accumulating on shelves, and production would be cut back to its original equilibrium level.

It is worth noting that consumer expenditure can also be affected by their confidence in the future. If the 'feel good' factor is increased, they will tend to spend more, raising the C curve, whereas if they become more apprehensive they will be less inclined to spend, lowering the C curve.

Introducing investment (I) and government spending (G), the other two components of aggregate demand, into the model results in Fig. 8.7. Note that, if I and G are independent of Y, then adding them to consumer expenditure (C) gives lines for total national expenditure parallel to but higher than the original curve of consumer expenditure. The equilibrium level of national income is where the combined level of expenditure (C + I + G) intersects the E = Y 45° line. At this level intended aggregate expenditure just balances the value of goods and services produced so people have no reason to revise their intentions and national income will be stable at this level.

It is worth noting that the equilibrium level of national income shown in Fig. 8.7 could also have been prescribed by the method described earlier – the balance of intended injections with intended withdrawals. The same equilibrium income level would have resulted in either case. Only the ways of analysing the situation differ.

Policies to control unemployment

After the Second World War governments in the UK accepted responsibility for preventing high levels of unemployment. This they attempted to achieve principally by encouraging national income to rise until any serious underutilization of the country's labour resources was wiped out. Policy, developed from the thinking of John Maynard Keynes (1883–1946) was built on the belief that, without government manipulation of aggregate demand, there is little reason why the equilibrium level of national income should correspond to high or full employment levels. Keynes pointed to this as the explanation for the prolonged period of low national income and high unemployment in the late 1920s and 1930s; less enlightened economists had thought that full employment was the norm to which the economy would eventually return.

In Fig. 8.7 the position at which the C + I + G line (aggregate expenditure) cuts the 45° line indicates an equilibrium level of national income corresponding to less than that which could be generated when all resources are fully employed. To raise the level of employment aggregate expenditure must be

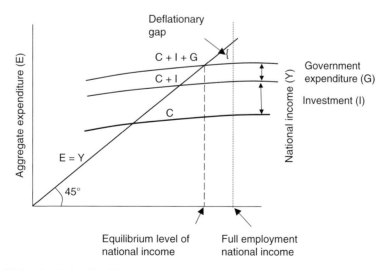

Fig. 8.7. The equilibrium level of national income including investment (I) and government spending (G).

increased to make good the shortfall between the level of expenditure which would occur with the present C + I + G line in Fig. 8.7 and the national income at full employment. This short-fall is termed the **deflationary gap.**

If businessmen can be encouraged to invest more, perhaps by politicians giving them confidence in future prosperity and guaranteeing access to loans at attractive interest rates, the I component will increase, and the C + I + G line will shift upwards. Or, with a higher degree of certainty of affecting demand, the government can spend more on road and rail infrastructure, hospitals, social services, etc. increasing the G component and raising the C + I + G line, creating a higher level of equilibrium national income corresponding with a higher employment level. They must be careful, however, not to raise C + I + G to such a level that the equilibrium national income is beyond that which is attainable with the full employment of all available factors (see Fig. 8.8). Because this is unattainable, intended aggregate expenditure will exceed the value at existing prices of the quantity of goods and services which can be produced (i.e. national income). This 'excess demand' will result in continuing inflation as the prices of the full-employment output of goods and services are bid up, assuming that the supply of money is allowed to increase to accommodate the higher prices and higher money wages which follow. In such a situation the gap between the level of expenditure at full employment and the value of output (national income) at full employment is termed the **inflationary gap.** Inflation will be returned to later.

Similarly, if consumers gain confidence in the economy, the security of their jobs and the prospect of rising income levels and decide to increase the proportion of their incomes which they spend, the level of consumption spending in the economy will increase and the C curve in Fig. 8.7 will rise, taking the C + I and C + I + G curves with it and a higher level of national income will result, assuming there is capacity in the economy to deliver it. However, increased consumption spending might also happen if consumers come to believe that saving is futile because anticipated inflation will greatly reduce the value of their savings. The raised C + I + G line might well produce a situation of 'excess demand', that is where the equilibrium level of national income is unattainable, itself producing the inflation which consumers fear. To counteract this, the government could reduce its spending, lowering the G component until the inflationary gap disappeared and an attainable equilibrium national income was produced.

Full Employment and the Longer Term

The economic theory presented so far in this chapter is generally accepted as valid. But the precise workings of the national economy are by no means fully understood. In particular, there is a range of views on the extent to which governments should attempt to manipulate the economy, and on the means by which this should be done.

From the early 1970s some economists and politicians, in the UK and internationally, started to express serious doubts on the advisability of striving for full employment through 'demand management'.

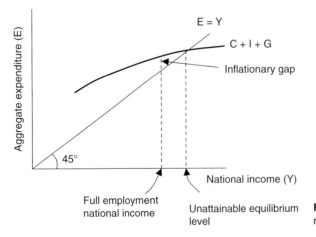

Fig. 8.8. An unattainable level of equilibrium national income.

Its pursuit by successive administrations along Keynesian lines was blamed for the inflation which had increasingly dogged the UK post-war economy. The argument is that, while expanding government expenditure will eliminate unemployment and increase output to the full-employment level in the short term (i.e. a boom period will be created), growth in the economy will be insufficient to sustain the necessary demand for this extra output permanently, and production and employment will fall back. Further expansion of spending by the government will be required to sustain employment and this will produce inflation because, although the volume of money in the economy is increasing (due to the government borrowing to finance a continual budget deficit), the quantity of goods on which it may be spent is not rising to the same extent, so their prices rise. Each time the level of output falls back, unemployment reaches a new peak.

Moreover, high levels of employment sustained by government demand management are believed to lead to: (i) a loss of dynamic drive within industries; (ii) poor output per worker; (iii) an unwillingness to innovate; (iv) unresponsiveness to changing patterns of demand; and (v) a failure to be competitive with suppliers from overseas. Contrary to the views of many followers of Keynes, his critics thought the economy has significant self-regulating properties if only the free market were allowed to operate more in the way that industries organized themselves, implying more competition among producers and greater productivity and efficiency. These ideas are associated with the American economist Milton Friedman. They emphasize the supply side of the national economy ('supply side economics').

The proponents of this view of the economy advocate an abandonment of a full-employment policy. They would argue that jobs are best created by industries becoming internationally competitive, and this is more likely to happen under private control than if run by the state. An example of this thinking being put into practice was the return to private ownership (privatization) of many nationalized industries in the UK by Conservative governments in the 1980s and 1990s.

A hands-off approach to the economy may well mean having to accept levels of unemployment, at least in the short term and before the benefits of a better supply side are reaped, that would have been strongly resisted in earlier post-war decades. However, higher levels may now be more tolerable for a variety of reasons. With the improved Social Security system the amount of overall hardship resulting from a relatively high rate of unemployment could well be no greater than that which formerly resulted from a much lower rate. With increasing awareness of job opportunities and job satisfaction, people may be more careful about choosing jobs and do not necessarily take the first available, so the period during which they are prepared to be unemployed may well be greater and hence the level of unemployment in the economy may be higher. From the national viewpoint higher unemployment probably also means that labour can be switched from declining to expanding industries more easily, speeding the modernization of the economy. In addition, if full-employment policy can be shown to have been a major contributor to past inflation, many members of society would be willing to accept a small degree of unemployment as preferable to rapid falls in the value of money.

Supporters of this school would not claim that a government should *never* use its spending power to stimulate national income. There may be exceptional circumstances in which aggregate demand management may be needed (though not always possible because of conflict with other aims, such as to reduce public borrowing). But normally government should restrict its role to creating the framework of economic stability, notably by controlling inflation, within which private industry can operate. And this can be best achieved, so they believe, not by demand management but, as we will see later, by policies which relate to the nation's supply of money and rates of interest.

Inflation

Any economy that has progressed past the stage of barter requires some common denominator against which the wide variety of goods and services it generates can be exchanged – in other words, a money system. Wages are paid in the form of money and spent on items priced in money units. Great difficulty would be experienced if a farm worker were paid solely by being given a leg of a cow each week, which he had to barter with the grocer and the electricity supplier. In most developed countries the monetary unit is of little or no intrinsic worth – a £5 note is just a scrap of paper – but it has exchange value because it is accepted in exchange for goods that *have* intrinsic value by all members of a country. As long as the exchange value of the monetary unit is maintained, people will not feel compelled to spend all their earnings immediately because money acts as a store of value. They will save in terms of money with banks and building societies and be willing to make loans, accepting repayment in the future. However, when the value of money changes rapidly, and this usually means downwards with rising prices, money becomes less acceptable as a store of value and the smooth running of a developed economy is jeopardized.

Falling money values and rising prices erode the value of savings and reduce the living standards of those people dependent on interest from savings or on fixed incomes, notably the elderly living from pensions. Borrowers benefit and lenders suffer because the nominal money size of loans, say for house purchase, does not rise with inflation, but the nominal value of houses and most incomes usually do rise. With the passing of time, repayments constitute a diminishing part of the borrower's income and become worth progressively less in terms of purchasing power to the lender. The general effect is that spending on borrowed funds is encouraged and saving discouraged, tending to exacerbate the inflation as will be shown later. Inflation, then, redistributes real income and wealth in a somewhat arbitrary and generally undesirable manner.

Frequent wage claims will be made to compensate for rising prices, bringing with them industrial unrest especially if some groups, such as those employed by the public sector, have wage control strictly applied to them in an attempt to halt inflation, while others manage to secure larger increases. With inflation the price of exports will rise (unless an exactly compensating depreciation in the international exchange rate for the currency is allowed, see Chapter 9) making them less competitive abroad, causing an adverse movement in the balance of payments; in turn this may lead to depreciation and more expensive imports and hence higher living costs at home. Recession starting in the exporting industries can spread by the multiplier

effect to other industries. This can stunt the rate of economic growth through businesses losing the confidence to invest and having to face the high interest rates which lenders will try to obtain as a way of counteracting the fall in the purchasing power of the currency. House ownership is also put beyond the aspirations of many by these high borrowing rates.

This formidable list is by no means exhaustive, yet illustrates why reasonable stability in the value of money is a highly desirable feature, almost a prerequisite, of a smooth functioning exchange economy.

The Quantity Theory of Money

Money consists not only of notes and coins but also bank deposits and loans where cheques, or some other type of purchasing-power transfer such as credit cards, are involved. If a bank were to give all its customers an increased overdraft, they would spend that money in much the same way as if the bank handed them notes. However, the simple Quantity Theory of Money states that the number of times a piece of money changes hands (called the **velocity of circulation**) is also vitally important and that the real quantity of money is better represented by the size of the money supply times its velocity of circulation. For example, a single £1 coin changing hands in a market ten times will finance just as much total transaction (£10-worth) as ten £1 coins changing hands once. The theory begins with the tautological equation (see equation at bottom of page).

Sometimes Y is given in the equation rather than T, Y standing directly for the flow of goods and services over the given period.

The number of trade transactions (T) will depend on how many goods and services the country is producing and putting on the market. The equation is a tautology because both sides show the same thing; the left-hand side shows the total money value of transactions over a given period while the

$$MV = PT$$

Amount of **M**oney × **V**elocity of circulation = General level of **P**rices × Total number of trade **T**ransactions
(bank notes and
deposits including
overdrafts)

right-hand side measures the total money value of goods sold.

From the equation several predictions can be made. Taking first the variable M, if the government allows banks to lend more to customers, increasing aggregate demand, or if it spends more itself by borrowing in a way which increases the reserves held by banks, thereby enabling them to create further credit, the money supply (M) will increase.

Assuming that the velocity of circulation (V) is constant, as long as the quantity of goods and services (T) the nation produces increases in step, there is no reason why the general level of prices should change (i.e. no inflation will occur). This can happen if the country has unused productive resources (as happens in a slump) which can be used to boost output. If, however, all the resources are fully employed, increasing aggregate demand and the money supply cannot be accommodated by changes in T, and P must rise instead (i.e. inflation will result as an increasing quantity of money chases a non-increasing quantity of goods). This is equivalent to the situation referred to back in Fig. 8.8 where an inflationary gap was identified. This type of inflation is termed **demand pull** because it is caused by excess demand in the economy.

There is much dispute over the ability of an expansion in the money supply to *cause* a rise in demand; some economists (the 'monetarists') believe that money has this active property. They point out that increasing the amount of money causes interest rates in the economy to fall. They further believe that these lower interest rates cause businessmen to invest substantially more heavily and consumers to borrow and spend more, thereby increasing the level of aggregate demand. If there is unused productive capacity in the economy, production (and national income) will rise. If not, inflation will result. This thinking has been behind the decision to increase the money supply (termed **quantitative easing** in the UK (and EU and USA) to deal with the recession that started in 2008. Others (the neo-Keynesians) hold the view that investment and consumer spending is not very sensitive to interest rates, being mainly influenced by other factors such as expectations of future prosperity; a lack of confidence appears to have been a major constraint on the post-2008 UK recovery. These economists deny the ability of an expanding money supply to affect the real level of economic activity, and conclude that it simply enables excess demand

resulting from the intended level of government and private spending being greater than the value (at existing prices) of what is available for purchase to be realized in higher prices. Most exponents of either view would agree, however, that a money supply which consistently increases faster than the growth in output of goods and services will be associated with rising prices.

Increases in productivity have a role to play in mitigating the inflationary effect of a rising money supply as they increase the quantity of output, from a given stock of factors, on which the rising spending power can be dissipated. Even if all productive resources are fully employed, they can often be used more efficiently by eliminating wasteful production methods, developing better types of equipment and other forms of innovation, reducing over-manning, etc. In practice, however, such attempts to 'grow' out of an inflationary situation in the UK have not been successful at the national level; increases in productivity have been exceeded by concurrent increases in the money supply (as was illustrated by the policy of the Conservative Government in the UK from 1970). On a regional basis, however, the general principle may be found working, for example: (i) seaside holiday areas receive large seasonal influxes of spending power which would bid up the price of food enormously if it were not for the increased supplies of food which are channelled to them; and (ii) expanding towns often try to ensure an adequate supply of housing so that the arrival of big firms with staff does not result in gross inflation in house prices.

The velocity of the circulation of money (V) is often taken as constant, but it will be affected by people's expectation of future inflation rates. If inflation is expected to worsen, then consumers will be encouraged to spend rather than save because: (i) the value of savings can only be expected to fall; (ii) it is known that the money values of many durable items like houses rise with inflation, maintaining their real value, and so making their purchase highly attractive; and (iii) waiting to purchase goods will often put them out of reach because of price rises. This urge to spend increases the velocity of circulation, which, without further real production, aggravates the inflation that the spenders were hoping to circumvent. In inflationary times the velocity should, if possible, be slowed by giving consumers greater confidence in the future value of money and a willingness to save.

Until the mid-1960s **demand pull** was the traditional explanation for UK inflation; such inflation could be cured, it was believed, simply by reducing the level of aggregate demand until a pool of unemployment was created. Once this was created inflationary pressure was removed because workers would not be in a strong bargaining position; if they did not accept the existing level of rewards, other people were supposed to be waiting for their jobs. A.W. Phillips showed that, from the 1860s up to the 1950s, a relationship held between the level of wage and price rises and the level of unemployment, more unemployment being associated with low inflation. The general relationship was of the type shown in Fig. 8.9.

However, the late 1960s and the 1970s saw rates of inflation far greater than the Phillips curve, based on historical data, would lead one to expect. Both inflation *and* the level of unemployment were high whereas the simple demand-pull theory would have regarded the two as incompatible. This gave rise to a range of alternative explanations for the inflation, mainly of the **cost-push** type. These pointed out that, under the conditions of industrial organization and unionization then found in the UK, increased costs of production, notably successful wage demands by 'pushy' trade unions, were passed on as price rises which again caused further wage demands as the cost of living increased, and so the spiral continued. Cost increases which arose outside a country's economy, such as an increase in the price of imported oil, or by chance factors like a crop failure were obviously unavoidable, but there seems little reason why internally generated,

self-sustaining cost-price spirals should be tolerated, although the supporting evidence that increased trade union strength or strike activity was the cause of inflation was not strong. The increasing failure of real wages to keep up with expectations was cited, as was also the insistence of workers on maintaining parity or established differentials between those industries where increases in productivity had occurred and those where none took place.

The most likely explanation for the inflation of the 1970s in the UK would appear to be one based on the demand-pull idea, but modified. The argument runs that, once inflation was established, it was built into any wage bargaining procedure as a basic demand, on top of which negotiations for 'real' wage increases were superimposed. With an anticipated and 'allowed-for' inflation established, and a government monetary policy which permitted the money supply to expand to accommodate higher levels of costs and prices, there was no reason why inflation should slow down when the economy was close to or at the full-employment level of activity.

We have already shown that, while high levels of employment and high output can be stimulated in the short term by employing unused and potentially productive resources, in the longer term aggregate expenditure cannot rise faster than the sustainable increase in output of the economy (in other words, the economy's rate of growth) without producing inflation. If the government creates an 'excess-demand' situation by allowing too much expenditure and striving for an unattainable level of national income growth and high levels of employment, inflation will accelerate from its expected base level. According to the proponents of this theory, to slow inflation aggregate expenditure must be cut, and in particular government borrowing from the banks to cover its budget deficit spending. This creates a pool of unemployment the size of which depends on how fast inflation is to be checked. At the same time people's *expectations* of the future rate of inflation must be reduced. For this latter purpose a prices-and-incomes policy of restraint may be a useful tool; historically they have had little if any long-term effect but may be useful as part of a package of policy measures including a reduction in aggregate expenditure. However, they carry the danger that, because they tend to be applied more vigorously in the public sector than in privately owned industry, they can result in industrial unrest.

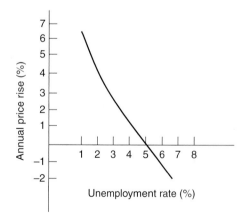

Fig. 8.9. 'Phillips curve': a trade-off relationship between inflation and unemployment.

Concurrently the aim of monetary policy must be to slow down the growth of the money supply so that eventually it is expanding in line with rising national output. This is the type of policy that was followed by Conservative governments in the UK over the 1980s. Money supply was controlled and public spending cut back. High interest rates caused many less-competitive firms to close down and a great number of jobs to be lost. The power of trade unions was curtailed both by legislation (on closed-shop agreements and on strike ballots) and by their failures in disputes with determined employers (most notably the year-long miners' strike against their public-sector employer, the National Coal Board). High unemployment was seen as a necessary part of restructuring industry to make it profitable and internationally competitive and to achieve an economy which was less prone to inflation.

More recently the Labour Government (1997–2010) in 1998 passed responsibility for monetary policy to the Bank of England, an independent public organization (though nominally owned for the government by the Treasury Solicitor). The Bank was given the aim of maintaining inflation below a low target rate. One purpose was to take this major lever on the economy out of the realm of party politics; previously governments had tended to regard the manipulation of interest rates as something to be used to attract voters, lowering interest rates and thus the costs of borrowing in the run up to an election, thereby boosting consumer spending and creating a (generally short-lived) economic boom. The longer-term consequences of the inflation and sharp rise in interest rates that followed were not considered conducive to longer-term sustainable economic growth.

Economic Growth

The promotion of economic growth is another of the goals of government, although growth can go on quite happily without government action to stimulate it. Growth occurs when the amount which can be produced and consumed per head of the population increases. As will be discussed later, this does not necessarily mean that the inhabitants are happier, simply that they can buy more goods and services. If we assume that the average European, because his or her income is higher than the average Indian's, is thereby 'better off', we are making a subjective judgement about what brings happiness.

If some of the country's productive resources are unemployed it will be possible in the short term to increase national income by increasing expenditure to take up the slack of unused potential. These short-run changes are *not* what is meant strictly by economic growth. Economic growth occurs when a country's **productive capacity per head** increases. Productive capacity is the output, measured by **gross national product** (GNP – see Box 8.2), of a country that is achieved when all of its productive resources are fully employed. However, in practice, this full-employment qualification is often dropped and growth is taken as the change in GNP per head, after allowing for changes in the value of money. If a country's total output is expanding, but its population is rising even faster, the quantity of goods and services available for each person on average will be declining. We will refer again to the race between output and population for the goal of economic development.

A country's rate of growth will depend on a variety of factors that are related to those that determine its present output per head. Principally they concern capital, labour and the state of technology. Growth will occur rapidly if the amount of **capital per head** of the population can be increased rapidly. This in turn will allow more consumer goods and services to be produced per head because each worker on average will have a greater quantity of machinery, etc. to assist them. An obvious example is the man on the powerful tractor who can do so much more cultivation per day than a man with a spade. The number of men needed on farms has fallen dramatically over the last 100 years, yet farm output has also increased; production per head is much greater because of all the capital now employed.

When the Russians wished to turn their country from its largely peasant state in the 1920s into a modern industrial nation, they embarked on two 5-year plans of building up their stock of capital, first (1928–1933) in the heavy industries (e.g. iron and steel) and later the lighter manufacturing ones. This was achieved by reallocating resources, especially from agriculture, into the production of capital items and by exporting cereals in exchange for capital from abroad (such as railway equipment). As was stated in the chapter on factors of production (Chapter 6), countries with high outputs per head (so-called developed economies) find it much easier to build up further their stocks of capital than do low-output countries. Increasing capital

Box 8.2. Gross national product – GNP.

GNP is the value of the final production of goods and services measured at factor cost (i.e. with the market prices of goods and services adjusted for price-rising effects of taxes on them and the price-lowering effects of subsidies). GNP is calculated by adding together the **value added** by each industry; this approach avoids double counting when the output of one productive activity uses that of other activities as inputs. For example, agriculture's value added is its sales of crops, livestock and livestock products less the costs of inputs such as fuel and fertilizers bought from other industries. In GNP, due allowances are made for property income received by nationals from abroad (it is added) and payments to non-nationals abroad (it is deducted) – ignoring these would give **gross domestic product** (GDP) (which relates to the activity taking place within a territory irrespective of who gets the rewards). Gross implies that capital depreciation allowances have not been removed; this is because international comparisons are often made with GNP of countries in which information on capital consumption is missing or weak. Sometimes net figures (after deducting estimates of capital consumption) are quoted, and this is probably preferable as an indicator of national income.

GNP is an indicator of what an economy produces, not of the well-being, welfare, satisfaction or happiness of the nation to which that production gives rise. Moreover, the measurement of GNP relates to activities that lie within a boundary set for the country's national accounts; most use the conventions of the United Nations' System of National Accounts (SNA). Included within the boundary is the production of goods and services sold on the market plus the cost of the armed forces, police and similar publicly provided services (as they are not sold on markets it has to be assumed that their values to the nation equal their costs). The work done unpaid by housewives and carers is not treated as part of the GNP (though when paid staff do the work, they are included). Environmental benefits of agricultural production are not included when measuring the value of farming's output though, perversely, the costs of farmers clearing up pollution are. Various attempts have been made to redraw the boundary of what is included in national accounting, in particular to integrate the environmental benefits by attaching money values to them. However, most internationally available GNP estimates stick to the conventional SNA coverage. As long as users are aware of what is not included, and interpret appropriately, GNP and related concepts drawn from national accounts form very useful indicators for what is happening to the economy.

accumulation means diverting resources from the production of consumer goods to capital goods; in a poor country all the production may consist of necessities, so the margin available for capital build-up may be very limited. Aid in the form of gifts or loans of capital from rich countries can be an important growth stimulant.

A moment's thought should raise the question of diminishing returns. Given a population of a certain size and a stock of natural resources, is not capital subject to diminishing returns, so that each successive unit of capital added to a country's stock of assets generates progressively less extra output? This is illustrated in Fig. 8.10 where the marginal product of capital (MPC) is shown as falling as the capital stock is built up. And could not the situation arise where more capital added nothing to production, or even reduced it? While such a situation is possible, it appears that, for the richer countries, new techniques of production and new products are discovered at about the same rate as extra capital becomes available. This has the effect of shifting the MPC curve to the right (Fig. 8.10) so

that marginal blocks of capital continue *to increase* national output by about the same extent.

Clearly another important determinant of the rate of economic growth is the rate of *invention and innovation*, both that undertaken by private firms and that supported by the state. A good example of an invention is the desk-top computer, which enables so many business operations to be performed more rapidly than hitherto, and hence makes the workforce more productive. New plastics, improved alloys, more fuel-efficient aircraft, etc. all provide new investment opportunities and permit greater production, while new consumer goods (smartphones, electronic games, etc.) and entertainments provide an ever-widening range in which entrepreneurs can exercise their production abilities. In the UK for much of the second half of the 20th century the state actively supported research and development, though now the responsibility has mostly been passed back to the industry itself. Copying good ideas from abroad is an age-old practice. In the period following the Second World War the Russians and Japanese were particularly

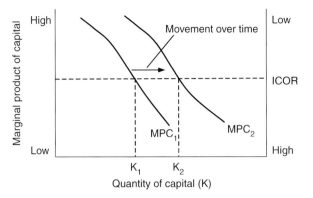

Fig. 8.10. Movement in the marginal product of capital (MPC) when new technologies are being invented (ICOR = incremental capital:output ratio).

successful in aiding their own growth by 'borrowing' ideas from the West; this was readily observable in motor cars, where their products tended to resemble out-of-date American and European models. However, this 'borrowing' of ideas is increasingly difficult once the gap between the 'borrowing' and the 'lending' countries has narrowed. Furthermore, the 'borrowed' techniques may well be more appropriate to situations where labour is scarce than where pools of unemployed labour already exist, as is the case in many less developed countries.

The *quality of the human capital* also has a bearing on growth. Spending money on education and training is at least in part an investment – scarce resources must be devoted to it and an educated and trained labour force can be more skilful, more productive and more adaptable to changing conditions. Literacy and job-related skills training certainly aid productivity, but many would argue that certain sorts of esoteric higher education add nothing to a country's ability to produce and little or nothing to the quality of life of the individual or the community. Far from promoting growth, an education which encourages students to question growth's desirability may inhibit it.

Changes in the *quantity* of labour have an important relationship with growth, as has already been mentioned. If the productive capacity of a country is increasing fast, yet its population is increasing faster, growth in the sense of an expansion of productive capacity *per head* will not occur. Rapidly expanding populations are a major problem of poor countries. On the other hand, increases in the size of the population can contribute to growth under some circumstances.

In Fig. 8.11 a production function is shown, with labour taken as the variable input, all others being held constant. The standard of living can be represented by the average product curve. It will then be seen that, with low labour usage, increasing quantities of labour will raise average product. Such a country could well be labelled 'underpopulated'. An example might have been Australia in the early 20th century. Increasing the labour force, however, causes output per head to reach a peak, beyond which living standards decline with further increases in the labour force because of diminishing returns. This was the sort of thing the early British economist Thomas Malthus (1776–1834) foresaw in his *Essay on the Principles of Population as It Affects the Future Improvement of Society* (first edition 1798). Malthus predicted what would happen to the supply of food with a rapidly rising population; because of diminishing returns, food production per head would decline as the population increased until only a quantity just sufficient to maintain the population was available. If the population were to go beyond this, he believed that it would be cut back automatically by famine, disease and other miseries. With the rapid increases in population that were occurring in the UK at the time of his writing, such catastrophes seemed unavoidable. Fortunately, the unforeseen development of North America and the British Empire as suppliers of food and the possibilities of trade meant that the UK had both a larger population *and* higher living standards in 1900 than in 1800. New production techniques raised (and continue to raise) the average product curve to offset the effect of falling average product which might have been expected with a rising population. Increasing numbers are no longer a matter of serious concern to

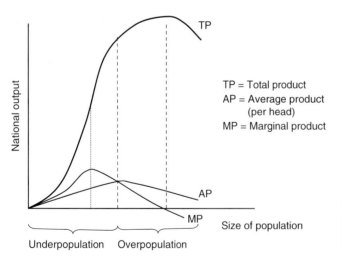

Fig. 8.11. The relationship between population growth and national output (assuming a strong positive correlation between population and workforce size).

developed economies (with some seeing declining populations), but developing countries still face problems of population expansion and the predictions of Malthus are by no means irrelevant to them.

The types of *economic, social and legal institutions* with which a country is equipped are further influences on its rate of growth. A centrally planned socialist economic system has shown itself to be capable of rapid growth (for example in the early period of the former USSR, etc.) because productive resources could be directed towards the build-up of capital more readily than in capitalist economies, although this does not preclude the possibility of remarkable capitalist growth rates, as exhibited in the post-war recovery of West Germany and Japan. China's spectacularly rapid growth since 1992 (typically about 10%/year, about three times that of the UK even before the UK's recent recession) has been essentially market driven and fuelled by privatization and incentives, but backed by strong central strategic decisions on things like housing provision, low taxes and building good transport infrastructure. Within agriculture, land tenure is often a barrier to growth where land is fragmented into many small units, making mechanization impractical. Landlords, often absentees, who are content with a low rent because of the large areas they own and who operate restrictive policies on their tenants are a hindrance to the achievement of higher productivity. Economic growth has often necessitated a major revision of the land tenure system. Religion has on occasion been quoted as a brake on the rate of growth, especially where it is mixed with inflexible social norms. Religions that place little emphasis on materialism in this life or that restrict the availability of female labour by stressing the home-making role of the mother are not likely to encourage growth.

International trade can speed the role of growth by allowing specialization and exchange, removing the necessity for a country to develop all its lines of production. For example, growth in Britain in the 19th century was facilitated by its ability to import food from abroad to feed its rising and increasingly urban population. In exchange it could trade exports of manufactured goods which it was best suited to produce and which were in demand by the food-exporting countries. More recently, China's growth has been driven by its ability to sell exports of manufactured goods. Growth requires increasing a country's stock of capital, and trade has a role to play here. Accumulation at home may be a slow and difficult process, but trade can provide the means by which capital can be borrowed from abroad in exchange for a future flow of exports. The size of the repayments must be viewed in light of the extent to which the loans allow the country to produce and export more. For example, loans from the USA aided post-war recovery in Britain and had to be repaid by currency earned through exports. The developing countries are in the unfortunate position that many of their exports are agricultural and are not in high and increasing demand by richer countries. This limits their opportunity for growth emanating from trade in these commodities.

In developed economies sustained growth seems primarily constrained by the rate of build-up of capital (both investment in goods and human capital) and the rate of invention and innovation. The developing economies frequently face the additional hindrances of: (i) unhelpful legal, religious and social structures and institutions; (ii) poor nutrition; (iii) poor roads and communications; (iv) crude banking systems; (v) high levels of unemployment which may be aggravated by copying labour-saving but capital-demanding production techniques from the rich countries; (vi) a bleak trading position; and (vii) often few natural resources which can be exploited for industrial development. Furthermore, the gap between slow-growing poor countries and faster-growing rich countries is an ever-widening one. Many people, particularly in the poor countries, find the disparities in material living standards increasingly objectionable as the greater ease of communication makes their existence more obvious.

The Costs of Economic Growth

Not everyone would agree that economic growth is desirable. While the poor usually strive for increased possessions and consumption of more goods and services, the lure of materialism has palled for many in the developed world. Personal satisfaction is only in part derived from easily identifiable goods and services and a major source is the heterogeneous group comprising job satisfaction, stability in society, absence of excessive noise and pollution, and many others which are contained in the diffuse term 'quality of life'. Attaining higher material standards of living through economic growth imposes costs which, upon reflection, at least partly offset and may completely nullify increases in satisfaction gained from having more clothes, housing, consumer durables, etc.

Growth derived from increasing a country's stock of capital can be more costly than is immediately apparent. Capital build-up involves the diversion of productive resources from making consumer goods to making producer goods, implying a sacrifice of living standards in the short run in return for higher standards in the future. This sacrifice is a cost of promoting growth. While consumption may return fairly soon to its original level, the net advantage of growth is only enjoyed after sufficient extra consumption goods have been produced to offset the original sacrifice (see Fig. 8.12). If less value is ascribed to goods which can only be enjoyed many years hence than on goods which are available now (i.e. a process of discounting is employed) the attractiveness of growth derived from capital accumulation diminishes further. As most countries encourage investment as a means of promoting growth, although probably it is only responsible for about half of the total in developed countries, we must assume that on balance such short-term sacrifices are considered worthwhile in the national view (or rather, in the views of the politicians who make the decisions).

Higher levels of national production, it is alleged, imply costs in the form of a faster pace of life, more heart attacks, more stress and frustration, more suicides and a less caring society. Externalities of production such as air and water pollution become of increasing importance and an increasing proportion of the nation's resources has to be used in their control. Because of the difficulty of dealing with the less tangible forms of external diseconomy such as ugliness, the quality of the landscape is frequently thought of as deteriorating through the housing and industrial development which accompany growth, despite attempts at planning control. High technology societies, as well as permitting increasing amounts of leisure time, run the risk of reducing some sections of the population to becoming machine minders with little interest in the product of their labours. Added to this, the increasing amount of mechanization is likely to reduce the number of unskilled jobs which are available. Rapid technological advances make old skills obsolete and necessitate retraining if their owners are not to become redundant. Consumers are asked to make spurious choices between brands of goods, like soap powders, where contents are identical and only packaging differs. At the opposite pole, adequate products are withdrawn from the market when new models are introduced, reducing in this case the ability of consumers to choose between the familiar and the 'improved' version. Advertising seems set on convincing people that only material goods are important, that they are dissatisfied with what they have and lures them into buying more in the hope of satisfying themselves.

In the EU, and especially in the UK, economic growth has brought changes in farming practice and the appearance of the countryside which many observers find objectionable from an aesthetic or ethical viewpoint. A decline in the number of people working on the land has been accompanied by increased mechanization, the greater use of herbicides, pesticides and fertilizers, and (particularly in the UK) field enlargement and the removal of hedgerows. Changes in the size and the composition of the rural community have taken place (by no means all being detrimental) and problems have been experienced with services such as rural education and public transport. The ability of European agriculture to provide environmental and social services of a

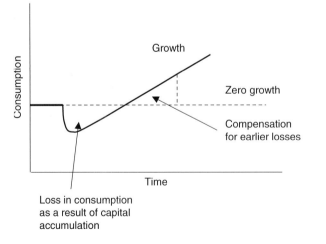

Note: A similar diagram could have been given when a growth rate of, say, 2% is accelerated to a faster growth rate by the accumulation of capital.

Fig. 8.12. Losses and gains in consumption over time when growth is pursued through capital accumulation.

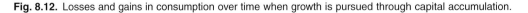

public good nature at the same time as producing market goods (mainly farm commodities) has become the centre of a debate about the 'multifunctional' role that this industry plays in society and the best way to encourage and reward it for producing non-market services (see Chapter 10). Here it is simply necessary to point out that increased levels of production either by the whole economy or by the farming industry alone should not automatically be welcomed without also examining their wider consequences.

The critics of economic growth are not short of ammunition with which to bombard the policies of governments aimed at its promotion. Their salvos have not succeeded in halting growth, nor has this been their considered aim in most cases. They have achieved considerable success, however, in dragging before public attention the fact that satisfaction is not derived solely from the fruits of production – goods and services – and that measuring changes in the standard of living by ascribing money values omits many of the factors which determine the quality of life. The economist has to take into account that most changes in the economy – such as the building of a new airport or the restructuring of farming – have implications for many sectors of society. Some implications will be beneficial, others deleterious. Only some are readily quantifiable and an assessment of the desirability of growth, or indeed any change in society, must in the last resort be a subjective judgement.

This section has emphasized the interrelationships of individuals and firms, both singly and collectively. Such interactions are not confined to within the national boundary and a natural extension is the notion of international trade, considered next.

Exercise on Material in Chapter 8

Answer the questions in the manner indicated. Answers and explanations are given in Appendix 2.

8.1. Which of the following are injections into the circular flow of income, which are withdrawals and which are neither?
 (a) Saving of wages by workers
 (b) Wages paid to civil servants by the government
 (c) Spending by foreigners on our exports
 (d) Income tax
 (e) Investment by a firm in new machinery

 (f) A gift of £5 from a father out of his spending money to his son
 (g) Spending of wages by workers

8.2. Cross out the inappropriate alternatives:

The marginal propensity to withdraw is the *proportion/amount* of any increase in *average/total* income which is withdrawn by saving, taxation, purchase of imports, etc.

8.3. If saving is the only withdrawal from the circular flow of income, and the marginal propensity to save is 0.25, what will be the size of the multiplier?

8.4. Cross out the inappropriate alternatives:

National income will be in equilibrium when the intended level of injections is *greater than/less than/equal to* the intended level of withdrawals. The equilibrium level *must always/need not* correspond to one at which all the nation's productive resources are fully employed.

8.5. Examine the following diagram.

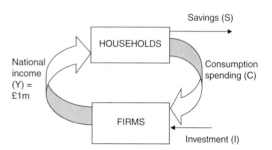

National income is the sum of wages, rent, interest and profit. The level of national income is in equilibrium. The marginal propensity to consume (MPC) is constant at all income levels. Ignore other forms of injections and withdrawals.
 (a) If spending on consumption is £600,000 and the national income is £1m, what will be the level of saving (S)? .... and investment (I)? ....

 (b) Assuming that MPC is constant at all income levels and that MPC = APC (average propensity to consume), what size of MPC will be required to give the consumption level given in (a) above?

(c) If the MPC rises to 0.8, and assuming that MPC = APC, what will be the new levels of C, S and I which are appropriate to give a national income of £1m?

(d) If I is in reality not the figure in (c) above, but is £400,000 and MPC = 0.8, to what level will national income have to rise before equilibrium is restored? At this level what will consumption spending (C) be?

8.6. Study the following diagram. The line C–C shows the level of intended consumption spending. Are the statements below *true* or *false*?

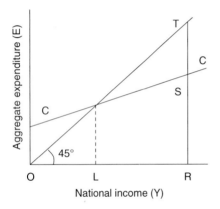

(a) Intended consumption spending increases as income increases.
(b) At some levels of income, intended consumption spending is greater than the level of income.
(c) Saving is positive at all income levels.
(d) Consumption is a fixed proportion of income (i.e. the average propensity to consume (APC) is constant at all income levels).
(e) There is a diminishing marginal propensity to consume (MPC).
(f) For the level of saving to be TS, the level of income would need to be OR.
(g) If OR is the level of income, the level of consumption would be SR.

8.7. In the next diagram state whether the following will have the effect of raising or lowering the C + I + G line:

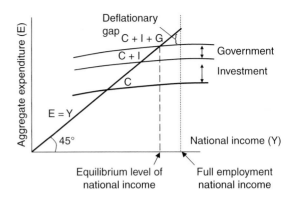

(a) Increasing consumers' spending intentions by a relaxation of restrictions on their borrowings.
(b) An increase in government taxation without more government spending.
(c) Additional government spending without more taxation.
(d) A government programme to encourage personal saving.
(e) Raising the rate of interest on bank loans and hire-purchase agreements.
(f) Making credit from banks easier to obtain.
(g) Increasing retirement pensions and family allowances without an increase in taxation.

8.8. Examine the following diagram. I, II and III represent three levels of aggregate demand in an economy.

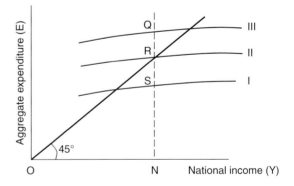

If ON is the full-employment level of national income:
(a) What is the deflationary gap by which the government may wish to raise the aggregate demand of Curve I to achieve full employment?

(b) Conversely, what is the inflationary gap which must be closed if level of aggregate demand III is operating?

(c) If aggregate demand III is allowed to continue, what will be the consequences for the level of prices?

8.9. If there is an inflationary gap, which of the following is the most promising way of closing it?

(a) Increase government expenditures on goods and services, leaving tax collections unchanged.

(b) Increase taxes, leaving government expenditures unchanged.

(c) Increase both government expenditures on goods and services and tax collections by the same amount.

(d) Increase government transfer payments from the rich to the poor.

8.10. If M = money supply, V = velocity of circulation, P = average price level, and T = number of transactions, which arrangement of the above letters represents the tautology which is the basis of the crude Quantity Theory of Money? ....

8.11. Inflation is conventionally described as having two main types of causes. What are these two types?

8.12. Cross out the inappropriate alternatives:

In the Phillips curve the *level of prices/annual percentage change in the level of prices* is shown as being *inversely related to/proportional to* the level of unemployment, so that the higher the level of unemployment, the *higher/lower* the level of inflation.

8.13. Cross out the inappropriate alternatives:

UK gross national product (GNP) is a measure of *all/only some* of the production activity under the control of UK nationals. A rise in GNP *always/never/often* implies a greater level of well-being of the population.

9 International Trade

Introduction

International trade is an extension of the phenomenon of specialization and exchange, which can be seen operating at all levels of society. No individual in a developed economy attempts to be wholly self-sufficient; no one will attempt to grow all their own food, make all their own shoes and clothes, build their own house, make their own car and mine their own coal. Rather, they specialize in one process, where their abilities are relatively most pronounced, using the money earned to purchase the output of other people who themselves have specialized. This process of specialization allows a greater quantity of goods and services to be produced and consumed, and hence a higher material standard of living to be enjoyed, than would be possible if every individual were to be completely self-sufficient.

Specialization must always be accompanied by the exchange of goods and services between people. At the individual level in a market economy this involves earning and spending money – a very useful medium of exchange. Regions or countries can also be seen to specialize according to the abilities of their inhabitants and the natural resources of the area. Rural areas produce agricultural commodities (though much else in addition) whereas the bulk of manufacturing is concentrated in urban areas. Evidence of regional trade resulting from specialization is obvious; simply look at any motorway in any country and see goods being shifted from one area to another.

International trade can be considered as an extension of regional trade. Indeed, in an association of states such as the EU that share a single large market, trade between Member States is essentially like that between regions of one country, especially those that share the same currency. Specialization and trade result in higher material standards of living for their inhabitants than self-sufficiency provides. In market-based economies international trade will occur simply through the natural inclination of

businessmen and individuals to buy from the cheapest source. If imported Israeli oranges were cheaper in England than oranges grown in glasshouses in Britain, a trade in oranges would soon be set up. On the other hand, British cars might be cheaper in Israel than Israeli-built vehicles, so trade in cars might also occur. Trade between Britain and Israel in oranges and cars would benefit both countries.

Trade involving the movement of goods can be seen as an alternative to the movement of factors of production in pursuit of their optimal allocation. If the marginal product of a factor is greater in one area than another it will be beneficial in terms of maximizing output to switch units of the factor from the low productivity area to the high productivity one until the marginal products are the same in each. This is one more example of the use of the Principle of Equimarginal Returns. The price system will encourage movement in this direction because factor earnings will be higher in the higher productivity area.

In practice the geographical movement of factors is limited. Despite publicity given to the 'brain-drain' of scientists and technicians from Europe to North America and migration of workers from the EU's newer Member States to the UK to take relatively low-paid jobs in the health sector, most labour is reluctant to move far and frequently within a country, and is even less willing to take the major step of moving abroad, with its inherent unfamiliarity and possible cultural and language problems. Most physical capital is immobile in the short term, although the reallocation of capital over a number of years between countries is within the powers of large multinational firms. Added to this, governments may impose restrictions both on the international movement of labour and of capital, though between EU countries such barriers have largely been dismantled. Land surface and climate, both particularly important in agricultural production, are obviously geographically immobile. While the shifting of naturally occurring substances such as oil or iron ore, where a country sells part of its stock of productive resources in exchange for immediate purchasing power, is an important element in international trade, to a large extent countries have to put up with the resources with which they find themselves. The problem then becomes one of rebalancing their production in such a way that each country specializes in those goods in which it has a comparative advantage and cuts back its production of the others. Trade in goods and services accompanies this process of specialization, and the ease with which factors of production can be switched between industries – occupational mobility – becomes an important determinant of the advantages flowing from trade.

Diversity of resources between countries

No two countries are identical in the productive resources they possess. It follows that the ease with which they can produce the various items which their people want also varies; in other words their production possibilities differ. It is conceivable that a country *could* produce internally some of almost every commodity that its inhabitants demanded – grapes under glass in Finland, petrol from coal in Germany, etc. – but such self-sufficiency denies the great advantages that can be earned through specialization and trade.

Countries differ in the *natural resources* they contain. For example, coal is mined in the USA, Russia, Germany, etc., while other countries may have none. Large iron ore deposits occur in the USA, France, Sweden and Russia, diamonds in South Africa and copper in Zambia. Climates differ, so that citrus fruits can be readily grown in Mediterranean countries, coffee in Brazil and rubber in Malaysia. It is obviously beneficial for Britain to trade with those countries that possess natural resources or climates that enable them to produce things Britain cannot or can only with extreme difficulty; hence we sell cars to Zambia in exchange for copper, defence equipment to Israel in exchange for oranges, manufactured goods to the West Indies in exchange for tourism.

Countries may have acquired certain *skills* that act as the basis for trade – the Swiss watch trade or wine-making skills of France are examples – or for historical reasons have built up stocks of *capital*, such as the shipyards of Scotland or the concentration of banking and insurance institutions to be found in the City of London.

Theory of Comparative Advantage

What is less obvious is that trade between countries which are basically similar in the resources they embrace can be beneficial. For example, trade between the various Members States of the EU has raised living standards, yet they are similar (though by no means identical) – Belgium and the Netherlands for example are much more alike than England and Saudi Arabia, yet trade is beneficial to both.

The benefits from specialization and trade can be illustrated by taking a simple model of a world which consists of only two countries (A and B), each of which has its resources fully employed in producing two commodities, bicycles and oranges, and examining the **opportunity costs** of increasing production of each of these goods in the two countries. The opportunity cost of producing one extra case of oranges in terms of the *number of bicycles* forgone is as follows:

In Country A	0.2 bicycles
In Country B	0.5 bicycles

Country A is said to have a **comparative advantage** in the production of oranges, since its opportunity cost of expanding its production of oranges is the lower.

Alternatively we can examine the output of oranges that must be lost if bicycle production is expanded in the two countries. *The number of cases of oranges* forgone if one extra bicycle is produced is:

In Country A	5 cases of oranges
In Country B	2 cases of oranges

(These figures are the reciprocals of those above.)

Country B has a *comparative advantage* in the production of bicycles since expansion in that product involves a lower opportunity cost in terms of oranges forgone. Note that one country cannot have a comparative advantage in *both* commodities.

Specialization

Specialization in the goods in which a country has a comparative advantage will permit a greater total quantity of both goods to be produced. Let us assume that *before trade* Countries A and B produce and consume the following:

	Cases of oranges	Bicycles
Country A	500	300
Country B	800	700
A and B together	1300	1000

The figures are the quantities involved in both production and consumption since, without trade, the inhabitants can only consume what they produce.

Let us assume that A expands its production of oranges, the good in which it has a comparative advantage, by 30 cases. By doing so it will have to forgo the production of six bicycles (remember the opportunity cost of production was 0.2 bicycles per one case of oranges). Similarly, let B expand its production of bicycles, the good in which it has a comparative advantage, by ten bicycles, reducing its output of oranges by 20 cases to do so. The new production schedule will be as follows:

	Cases of oranges	Bicycles
Country A produces	530	294
Country B produces	780	710
A and B together	1310	1004

Note that specialization has resulted in the combined output of the two countries expanding so that ten extra cases of oranges are produced (1310 as opposed to 1300) and four extra bicycles (1004 as opposed to 1000).

Trade in surpluses

When countries specialize the balance of production between the two goods will be wrong for domestic consumption – in Country A too many oranges will be produced and too few bicycles. However, this balance can be restored by the export of excess oranges in exchange for bicycles produced in Country B, which will find itself initially with a deficiency of oranges and a surplus of bicycles. Obviously some exchange ratio between oranges and cycles will have to be arranged between the countries. It will be found that this international exchange rate will settle by bargaining at a level somewhere between the internal opportunity cost ratios, i.e. somewhere between five cases of oranges per bicycle (A's opportunity cost of production) and two cases per bicycle

(B's opportunity cost of production). Let us assume that the international exchange rate settles at three cases of oranges for one bicycle.

At this ratio Country B will be able to trade seven out of the ten extra bicycles it produces after specialization in exchange for 21 cases of oranges. Country A will, reciprocally, trade 21 of its extra 30 cases for seven bicycles. After trading has occurred the quantities of the goods which each will be able to consume (i.e. the consumption schedules) are as follows:

	Cases of oranges	Bicycles
Country A consumes	509	301
Country B consumes	801	703
A and B together	1310	1004

Comparing this schedule with the situation before specialization and trade shows that both countries now consume greater quantities of both commodities. Both countries have benefited. Combined production is increased. When countries are not forced to be self-sufficient (i.e. no longer compelled to produce themselves what they wish to consume) they can specialize in those goods in which they have a comparative advantage, trading part of their excess production for goods which are the speciality of other countries. Thus the advocates of freer trade point to the economic advantages that flow from the removal of impediments to trade, especially those that have been introduced by countries in the form of taxes or other barriers. In the second half of the 20th century many countries agreed to lower such hindrances by internationally coordinated action – formerly through the General Agreement on Tariffs and Trade (GATT) and, since January 1995, using the World Trade Organization (WTO), the institution that superseded it.

Comparative and Absolute Advantage

Note that no reference has been made to the level of productivity or efficiency of either country. One could have been a place where output per man, and hence wages, were high and the other a low-output country. Output per man (or per some other unit of resource) indicates which country has an **absolute advantage** – Country A may produce both more oranges and more bicycles per man than Country B – but this is of no relevance to the advantages to be gained from trade. What determines the advantages from trade is comparative advantage, which means differences in the opportunity costs of production.

The irrelevance of absolute advantage can be illustrated by an example of specialization at the individual level. A solicitor employs a secretary to type business letters. If the solicitor discovers that, as well as being a much better lawyer than the secretary could ever be, he or she is also a marginally better typist, will it pay the solicitor to sack their typist and spend half the day typing letters, cutting back the legal practice to allow time? The answer is no: the lawyer should keep the secretary so that all the professional's available time can be spent in doing the job in which lawyer has a comparative advantage – law. By doing so the solicitor could, on balance, earn more, even though the cost of the secretary has to be met. The solicitor has an *absolute* advantage both in law and in typing, but this is irrelevant to specialization. What is important is that the solicitor has a *comparative* advantage in law (or, alternatively, a comparative disadvantage in typing – an hour of typing has a higher opportunity cost in terms of earnings forgone for the lawyer than for the secretary), and it is comparative advantage that leads to benefits through specialization and exchange at personal, regional or national levels.

Countries do not carry specialization to extreme for a variety of economic and political reasons. To illustrate the fundamental economic reason we must return to our simple model.

Limits of specialization

Country A cannot expand its production of oranges indefinitely, nor B its production of bicycles. While initial expansion in the direction indicated by comparative advantage may be easy, further specialization will be increasingly difficult. Country A will find that, to produce successive additional cases of oranges, it has to give up producing an increasing number of bicycles (i.e. the opportunity cost of orange production will rise but the opportunity cost of cycle production will fall). In Country B expanding cycle production will cause its opportunity cost in terms of oranges forgone to rise, but the opportunity cost of orange production will fall. Expansion of orange production may bring with it diseconomies of scale or the land, labour and other factors switched to orange production may tend to be less suitable for orange production than they were for cycle production.

The quality of factors is not uniform, and the grades giving the greatest benefit in terms of oranges gained per cycle forgone will tend to be transferred first, the less suitable later. This can also be illustrated at the farm level. If a farmer wishes to increase his area of wheat at the expense of another crop, potatoes, he will tend to select the most suitable of his land first; this is not necessarily the land which gives the highest wheat yield but that which gives him most of the expanding crop per unit of the crop which is being contracted (i.e. the largest ratio between tonnes of wheat gained to tonnes of potatoes lost). Continuing wheat expansion will involve transferring land which gives a progressively diminishing benefit in terms of wheat yield for every tonne of potatoes forgone.

You will recall that the reason why specialization and trade occurred was that opportunity costs of production differed between countries. However, the very process of specialization causes opportunity costs to change and converge, and a point will be reached at which the opportunity costs are the same in both countries. Beyond this point further specialization will not occur.

Geometrical representation

A geometrical representation may be helpful. Figure 9.1 shows the production possibility curves (boundaries) for bicycles and oranges for the two countries, A and B. These convex curves imply that the marginal rate of substitution (MRS) of products (i.e. the slope of the curve) is not constant; increasing the production of either good by 1 unit requires progressively larger quantities of the other to be sacrificed. (For a further illustration of this point see the product–product relationship described in Chapter 5.)

Before trade, each country is at point X on their respective production possibility curves. At these points the MRS of products (i.e. the opportunity costs indicated by the slopes of curves at point X) differ. With the opening of trade, each country adjusts its balance of production until it arrives at point Y, at which the MRS of products (or opportunity costs of production) are the same; this MRS will also correspond to the rate of exchange, bicycles for oranges, between the two countries. By trading, countries can exchange their surplus production and consume combination Z; for each country point Z lies further out

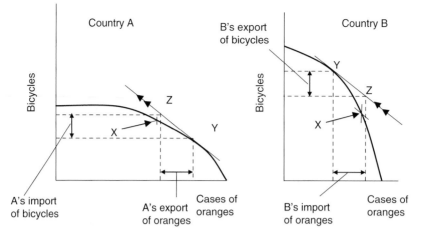

Fig. 9.1. A two-country, two-commodity model of the gains from specialization and exchange.

from the origin than X, showing that specialization and exchange has enabled them to enjoy both more oranges and more bicycles than they could before trade. (Indifference curves relating to the whole populations of Countries A and B are being implied here although not shown. Also, because the geometrical approach was required to fit the earlier numerical model, the possibility that after specialization and exchange a country need not consume more of *both* commodities to reap the greatest rewards from trade has not been explored.)

Extension of the simple model

So far we have dealt with only two countries and two commodities and have shown the benefits which can flow from specialization and trade. In the real world, however, many countries and many goods are involved.

Many countries

Trade between pairs of countries is termed **bilateral trade** (such as is met in bilateral trade agreements in which two countries agree to buy a certain amount from each other) and between many states is called **multilateral trade**. Unhindered multilateral trade is more beneficial than bilateral trade. This is easily illustrated by Fig. 9.2 where trade flows between three countries.

If trade could only go on between pairs of countries in this example, perhaps no trade at all would occur. The UK might not want anything produced

by Chile in exchange for the mining equipment. Similarly New Zealand might be unable to sell anything to Chile in exchange for fertilizers. By allowing multilateral trade, in which the value of trade between any two countries need not balance but combined imports and combined exports for any one country will balance, the benefits of specialization and exchange can be more fully realized.

Many commodities

Within a country it should be possible to estimate the opportunity costs of the whole range of its products in terms of a 'reference' product. These opportunity costs can then be compared with a similar set from another country. For example, Table 9.1 shows the opportunity costs of a range of goods in two countries in terms of food production. In Country A far less food has to be sacrificed in order to produce one extra unit of clothes than in Country B. By comparing the opportunity costs of each good in terms of food forgone in the two countries, it is possible to 'rank' comparative advantages; in Table 9.1, Country A's comparative advantage is most pronounced in clothes production but least pronounced in cigarettes. Alternatively one could conclude that Country B's comparative advantage is greatest in cigarettes and least in clothes. Where this 'rank' is split for specialization will depend on the total demand and supply situation. Note that it is not absolute advantage that determines the ranking, but comparative advantage.

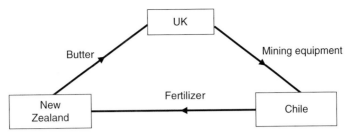

Fig. 9.2. Multilateral trade.

Table 9.1. Ranking of comparative advantage.

Goods	Opportunity costs in terms of food production[a]		Ratio of opportunity costs (B/A)	Ranking of opportunity costs
	Country A	Country B		
Wine	10	20	2.0	3
Cigarettes	6	2	0.3	5
Beer	1	3	3.0	2
Clothes	2	10	5.0	1
Food	1	1	1.0	4

[a]Increasing production by 1 unit causes the production of food to be cut back by these numbers of units.

Transactions Involving Currencies

So far in this section, trade has been treated as if it were conducted on behalf of nations by their governments. But this is, of course, not what happens. International trade, at least between market or mixed economies, is basically undertaken by individuals and firms who find that they can buy and sell goods across national boundaries with financial advantage. The reason why a businessman in England buys a Dutch-made lorry is probably because the foreign vehicle is cheaper or better or more available than one made in Britain. Americans buy Scotch whisky for similar reasons. Although the role of government in trade is essentially a secondary one – encouraging trade or restricting it for economic, political or social purposes – it also enters directly into trade by purchasing supplies (such as armaments) from abroad, using foreign contractors to provide services, maintaining troops in foreign stations, etc.

Trade across national boundaries is complicated by the fact that each country generally has its own currency (though the introduction of the euro (€) has rather changed this for the 17 EU Member States that have adopted it as their shared single currency). If a British manufacturer sells a machine to a Japanese businessman, the Briton will not welcome a payment in yen, because workers in his British factory will not accept their wages in yen. There must be some mechanism whereby yen can be exchanged for pounds sterling. In other words, for trade to occur there must also be the facility of international currency exchange; multilateral trade will require the facility to exchange units of one currency for any other currency. Currency exchange is undertaken in the Foreign Exchange Market, in which commercial banks and central banks (like the Bank of England) are active. It is a market because, for each currency, both a demand and a supply exist; the interaction produces a price for each currency which can be expressed with reference to any other of the currencies in the market. Hence we can talk of the exchange rate between pounds sterling and US dollars or between sterling and yen or between dollars and yen; each of these three exchange rates will be closely linked to the other two.

For the moment we will concentrate attention on the way the market for currencies is related to trade, although later the effects of international capital movements and the interventions by central banks will have to be considered too. Take the example of the exchange rate between pounds (£) and US dollars ($); the demand for pounds (and also the supply of dollars) arises from American businessmen going to their local banks in America and arranging to buy pounds sterling in exchange for dollars so that they can purchase goods from

British firms. The supply of pounds (and the demand for dollars) arises from British business-men wishing to sell pounds in exchange for dollars so that they can import American goods. There will be some rate of exchange between pounds and dollars where the supply and demand of each is in equilibrium. If this exchange rate is allowed to alter, and is not held rigid by government order for political or social reasons, its flexibility will ensure that the trade in goods and services normally flows in the direction suggested by comparative advantage. A simple model will show this.

Flexible exchange rates

Take the example of a world consisting of only two countries (the UK and the USA) and with only two commodities (food and clothes). It can be shown that, under perfect competition, the relative prices of commodities are indicative of the quantities of resources used in their production. The prices of home-produced food and clothes in the two countries before trade starts up are as follows:

	USA	UK
Food	$1/unit	£1/unit
Clothes	$3.30/unit	£1.25/unit

In the USA a unit of clothes is 3.3 times the price of a unit of food, indicating that in the USA pro-ducing a unit of clothes takes 3.3 times the quantity of resources required to produce a unit of food. In the UK it only takes 1.25 times as much; the UK therefore has a comparative advantage in clothes production since producing one extra unit of clothes involves sacrificing a smaller quantity of food in the UK than in the USA.

If on the international currency exchange market the rate between the two currencies is initially set at $10 = £1, the Americans will be able to undercut the UK home prices of both food and clothes. Ignoring transport costs, the cost in the UK of an American good can be easily calculated using the exchange rate to be as follows:

Price of a unit of US food in the UK = 10p
  (in UK currency)

Price of a unit of US clothes in the UK = 33p
  (in UK currency)

UK citizens will naturally try to buy the cheaper imported goods. To buy them they must obtain dollars to exchange for the food and clothes. This demand for dollars will bid up the price of dollars in the currency exchange market (i.e. the exchange rate will alter, with the dollar appreciating). Eventually the value of the dollar will rise so that US goods only undercut the home market in one commodity. If this rate is $2 = £1, the cost of American goods in the UK will be as follows:

Price of a unit of US food in the UK = 50p
  (in UK currency)

Price of a unit of US clothes in the UK = £1.65
  (in UK currency)

At this level of prices UK citizens will find imported food cheaper than home-produced food, but foreign clothes more expensive. Similarly, US citizens will find imported clothes cheaper and imported food dearer than that produced at home. This situation will not only produce a two-way flow of goods but also, as a by-product, there will be both a supply and a demand for each currency. In this model the price and exchange rate mechanisms have ensured that each country specializes in the commodity in which it has a comparative advantage. The absolute level of prices before trade was irrelevant.

If UK citizens suddenly take an increased liking to American goods and try to buy more, the demand for dollars will increase (and the supply of pounds will also increase) so that the value of the dollar will be bid up, i.e. the exchange rate will alter, with the dollar appreciating (or, which is the same, the pound will depreciate). This will make American goods in Britain more expensive and reduce the demand for them, tending to restore the balance between the total value of imports and exports.

Domestic inflation will also be reflected in changes in the exchange rate. As long as the exchange rate is allowed to move freely, inflation will not affect the directions in which traded goods flow. This is illustrated simply: assume the level of all UK prices goes up ten times but the *ratios* of prices in each country remain unaltered. The new level of prices will be:

	USA	UK
Food	$1/unit	£10/unit
Clothes	$3.30/unit	£12.50/unit

At $2 = £1$, which was the exchange rate at which trade had been occurring before UK inflation, UK citizens find both US food and clothes cheaper than the home-produced items. The demand for dollars rises and the exchange rate is bid up to a level where only US food is cheaper than UK food, US clothes becoming more expensive than UK clothes, and trade is re-established. The new rate will value the dollar ten times higher than previously, i.e. $2 = £10$. (This can also be thought of as a fall in the exchange value of the pound, to a tenth of its pre-inflation level.)

This model again emphasizes the point that the absolute price levels are irrelevant to specialization and trade, which depend on differences in comparative costs.

Monetary Union

The introduction in 1999 of Monetary Union for 11 of the (then) 15 countries of the EU led to the adoption of a single currency (the euro) for this group (known as the Eurozone) in 2002 and the issue of euro notes and coins. Countries were required to pass certain tests (a budget deficit less than 3% of GDP, national debt below 60% of GDP, low inflation, and interest rates close to the EU average). The UK did not participate, nor did Denmark and Sweden (Greece adopted Monetary Union in 2000, but in retrospect the evidence by which it met the tests was flawed). Subsequently the Eurozone has grown to 17 countries (out of EU-28) with Latvia joining in 2014, but with the UK still retaining its own currency.

The sharing of a common currency does not change the validity of the economic case for international specialization and exchange. Indeed, in several ways a single currency encourages trade. The avoidance of the necessity of currency exchange makes the differences in prices of goods originating in different countries more apparent and certain. The danger is removed that fluctuations in rates between currencies can change the price of imported goods between when the decision to buy from abroad is taken and when payment is due. Also, the costs of the act of currency exchange itself disappear; there is no longer any need to pay banks and bureaux to provide this service. These advantages of a single currency have existed for long periods in, for example, the UK (where the constituent countries all use the pound sterling) and, before the euro, in Belgium and Luxembourg (these countries shared their currencies, with valid use of either).

However, there are drawbacks. A single currency implies that there is no possibility of inflation being handled automatically by a fall in the value of the currency of the country suffering from it (as outlined above). There are major implications for the way in which macroeconomic policy is put into practice (see Chapter 8). Interest rates can no longer be used by governments of single countries as a way of influencing their national economies, as the same rates (set in the eurozone by the European Central Bank) will in practice apply across all those sharing the currency. Consequently the economy has to be managed in other ways, which in effect means policy on spending (including national and EU regional policy) and (national) taxation (fiscal policy). To help ensure that this remained possible, EU Member States sharing Monetary Union agreed to accept a set of rules on the way they handle their own economies (a 'stability pact' relating to factors such as the extent of government borrowing). If acute problems arose in a country it might (in theory) withdraw from the eurozone, but this would be practically difficult and politically unwelcome to those who see greater economic and political cooperation as the way forward.

The UK may at some date join the Eurozone when a set of economic conditions deemed favourable comes about (meeting certain 'tests'), though this is unlikely in the medium term. The decision is seen as much more than just an economic one, with political overtones of loss of independence, only some of which can stand up to scrutiny.

Assumptions in the Theory of Comparative Advantage

The Theory of Comparative Advantage shows that higher material living standards can be attained through trade. However, the theory makes a number of assumptions which are not entirely borne out by the real world, and acknowledging these assumptions helps to explain why, despite the apparent advantages to be gained from unhindered trade, most countries in the past have used devices such as tariffs or quotas to restrict it. (A **tariff** is a tax on imported goods that, in effect, raises their price to the importer, and a **quota** is a restriction on the quantity of imports permitted.)

Static or dynamic conditions

We have assumed that a country with a comparative advantage will always possess it. But in reality things are dynamic: for example, less developed countries may not yet have a comparative advantage in, say, car manufacturing, but if such industries could get established they might develop a real comparative advantage. Protection of such 'infant' industries may well be justified in the long run as, when young, they will usually be undercut by the lower prices of the established industries in other countries. However, countries must beware of protecting infants which never develop and require permanent protection.

Technical advances also tend to shift comparative advantage. In the 19th century Britain had a comparative advantage in textile production, but the development of artificial fibres and new machinery, together with changes in other sectors of the economy, shifted comparative advantage to countries such as Taiwan. More recently China has developed comparative advantage in clothing manufacturing, and telephone call centres for European customers are now often situated in India, where there is a comparative advantage in activities that are labour intensive and use information technology.

Full employment of resources

Specialization and exchange involves the contraction of those industries in which the country *has not a* comparative advantage. This means that factors of production – labour, capital, land and business ability – have to be switched from declining to expanding industries.

Much depends on the rapidity with which such switches of factors are made. A sudden decline in, say, the ship-building industry will simply swell the ranks of unemployed shipyard workers. The potential advantages from trade will be largely lost unless this labour is taught new skills and is absorbed by the expanding industries. Indeed, without this re-absorption it is possible that the nation may be worse off, particularly if unrest caused by redundancies leads to strikes. The loss to national income by such action can easily be greater than the potential gains from trade. Similarly, capital goods in industries shrinking rapidly under pressure from overseas competitors frequently becomes unemployed because of their specific nature (e.g. textile machinery) and do nothing to aid the expansion of the industries in which the UK possesses a comparative advantage.

The moral is that factors need time to adjust – labour to be retrained, capital to wear out and be replaced by other types, land to be switched to other uses – and a too-rapid change caused by opening up trade may cause factor unemployment which more than offsets any potential gains from freer trade. To prevent this waste and social distress, continuing short-term restrictions on trade may be justified to give factors time to be switched. Thus the process of lowering trade barriers, as agreed between countries, is usually a gradual one. For example, in the multilateral negotiation for reductions of tariffs that took place formerly within GATT and now under the WTO, each round of talks aimed for only modest reductions rather than complete removal. Even these agreements tend to be implemented in phases over a period lasting (typically) 5 years.

Distribution of the gains from trade

While trade can be shown to be beneficial to nations as a whole, given certain assumptions, not everyone benefits to the same degree, and some will suffer, at least in the short term.

The manager and owner of a business which could suffer from competition from abroad may take a poor view of government policies to promote international trade – their firm's prices will be undercut and they will need to look for other opportunities of using their resources. They may well advocate more tariffs, quotas, etc., to protect their own interests. They may support a Buy British campaign and attempt to make buyers of foreign-produced goods feel unpatriotic. They will attempt to extract subsidies from the government by persuading it that an industry vital to the national interest is being threatened.

The specialization and exchange that trade involves have been said to benefit many people a little but hurt a few people a lot. The few sufferers will readily band themselves into pressure groups to protect their own interests, but it is far more difficult to organize the rest of the population, each of whom benefits only marginally, into an effective lobby to press for more trade. However, if at the whole-economy level, trade is beneficial, it must be at least theoretically possible to adequately compensate the sufferers from the gains of the beneficiaries, with something remaining left over as a net gain.

Similarly, there is no guarantee that the broad factor groups will receive benefits in the same proportion. For example, the owners of capital may find themselves much better off if comparative advantage indicates that a country should expand its production of goods which require much capital. On the other hand the rewards to labour may not increase as much. Specialization and trade generally work so that the factors which were initially the most scarce become less scarce as specialization proceeds, so that the share of the nation's income they receive will tend to fall. If this produces a distribution of income which society considers unjust, incomes can be redistributed by differential taxation and benefit payments.

Arguments put in support of trade restrictions

If, in principle, trade results in an improvement in incomes and living standards, why throughout history, has there been a tendency to impose restrictions? Several strands of argument can be identified, many of dubious validity.

To protect against competition from cheap labour countries

The arguments put forward for protection often place emphasis on the harm the economy can suffer through one industry being undercut by foreign competition from countries where labour is cheap. Such competition is said to be 'unfair'. This argument is not valid. It is true that specialization and trade may reduce the *relative* earnings of labour in 'dear labour' countries, i.e. its share of national income may decline and the absolute level of income of some individuals may fall. However, as the levels of national income in both the 'cheap' and the 'dear' labour countries will increase through trade (after re-allocation and adjustment), it is quite possible, although not necessarily done, to arrange via the tax and benefits system in each country that everyone benefits.

To prevent dumping

Taxes on imports (tariffs) are often advocated to prevent dumping on the domestic market, that is the sale at very low prices of a surplus from abroad which the exporting country might have accumulated as the result of some domestic policy. For example, before Britain joined the EU in 1972 (then the European Economic Community) there were occasions when French 'Cheddar-type' cheese was sold to British shopkeepers for retailing at prices considerably below that of the equivalent British product, displacing British cheese and causing difficulties for the UK dairy industry. This dumping was only temporary, yet its possible effect on the British dairy industry could have been longer term if serious disruption and bankruptcies had occurred. UK consumers might have found themselves eventually paying more for cheese once French dumping had ceased. Tariffs or quotas, especially those that can be brought in quickly, are tools which can be used against such disruptive dumping.

On the other hand, if Korea and Malaysia are willing to sell their motor cars in Europe at prices which are less than the costs of production in order to earn sterling and euros, and they are prepared to do this on a long-term basis, would it not be foolish to reject their generous donation of the fruits of their production resources?

To raise revenue

Governments need funds to operate their policies. Taxes on goods where no significant domestic production is possible, such as on imports into the UK of tobacco and French brandy, have long been used as a source of state revenue. But the amount of revenue which a tariff raises depends on the price elasticity of demand of the good on which the tariff is imposed. Increasing the tariff on goods with high price elasticities of demand could result in a *fall* in tax revenue as the rise in price will severely choke off the demand for imports. In performing their revenue-raising role, tariffs protect home industries from competition from cheaper foreign goods and hence prevent, *ceteris paribus*, the optimum international allocation of resources. To prevent such a distorting effect, some would suggest that a tax equal to the tariff should also be placed on home-produced goods. With tobacco and French brandy no distortion between domestic and foreign production of each particular good arises since domestic output will always be constrained by climatic factors to zero, or to very low levels.

For retaliation

An understandable emotional reaction to, say, the USA erecting a tariff wall against EU goods is for the EU to place similar tariffs on US goods coming to Europe. Yet fundamentally this is about as sensible as digging holes in Danish roads to damage the cars of Belgian holiday-makers visiting Denmark because Danish cars are sometimes damaged by poor Belgian roads; other people suffer too. However, the *threat* of a retaliatory tariff may, however, sometimes be sufficient a deterrent to prevent a country contemplating such a restriction from implementing their action. Threats of this nature have been used on several occasions in the trade relations of the EU and USA.

To 'export' unemployment

We have agreed that a temporary restriction on trade may be necessary to prevent the unemployment of factors in the short term, where this is caused by their inability to shift from declining to expanding industries. However, where the whole country is in a slump, with widespread unemployment caused by a general deficiency in demand, countries have occasionally been tempted to help home employment by imposing import quotas or by erecting a tariff fence, thereby cutting down imports and hopefully expanding home production and home employment to take advantage of the drop in imports. But this ultimately causes unemployment abroad and excites retaliatory action by other countries.

The 'export' of unemployment was practised during the 1920s and 1930s and resulted in tariff walls being erected and a rundown in international trade. In order to discourage such a situation from arising again, after the Second World War a large number of countries (including the UK and the USA) signed the GATT. Rounds of negotiation continue to be held, now under GATT's successor, the WTO, with the aim of dismantling existing tariffs and other trade barriers and of preventing additional distortions; the Doha Development Round started in 2001 but has not yet (2014) culminated in a set of agreements.

For social or security reasons

History shows that, during periods of 'conventional' warfare, a home food supply has been a valuable asset. For this strategic defence reason, measures to protect agriculture against international competition in peacetime may be justified. However, there is room for debate on the target level of home supply, on the need to consider not just output but also inputs bought from overseas (particularly energy), and the changed nature of threats (terrorism rather than conflicts between states).

Restrictions on imports can be justified to guard against the accidental introduction of diseases that could be a threat to human health or of agents that could infect animals or plants. Sanitary and phytosanitary measures are considered as legitimate by WTO as long as they meet certain conditions and are based on scientific evidence, though there is a danger that the real reason why a government may want them is to protect an industry that is not internationally competitive but which carries political weight. Also restriction may be legitimate for social reasons, such as those placed on the imports of addictive drugs – in the case of tobacco the underlying reason seems to have moved towards that of improving the standard of the nation's health from one purely of raising revenue.

On a more general level, agricultural protection might be used to maintain a prosperous rural population with many small farms, since a depopulated countryside and large-scale farming practices may be neither politically nor socially acceptable. However, it is highly unlikely that trade restrictions are the cheapest way of achieving the desired results. Within the EU, where unrestricted trade between member countries and regions in agricultural products was intended to apply under the Common Agricultural Policy (CAP), regional specialization has not been allowed free reign. For example, farming in hill areas would largely disappear if subject to undiluted market forces. For social (and political) reasons these producers are protected, using special payments to farmers in Less Favoured Areas (from 2014 called Areas of Natural Constraint).

Summarizing restrictions on trade, it can be concluded that many of the arguments put forward for 'protectionism' cannot bear close scrutiny. While short-term trade restrictions can be justified where waste would be caused by attempts at sudden adjustments in the economy, long-term restrictions are likely to be harmful to the national economy unless any economic benefits to be gained are more than offset by social, security or environmental deteriorations. It is necessary to examine closely the real motives behind the case being made for any trade restriction. Self-interest rather than national interest may be the dominant factor.

As was pointed out earlier, at an international level the second half of the 20th century has seen a gradual move to the freeing up of trade (or 'trade liberalization') by the dismantling of impediments such as tariffs, quotas and other barriers (which may take the form of differences in national regulations on product safety or trading standards). However, this is generally not a quick or easy process, not least because, while more trade benefits the economy as a whole, there are always some individuals, firms or organizations that are made worse off, at least in the short term and if they choose not to accept that adjustment will be necessary. Consequently, the path towards trade liberalization has many interim stages that take different forms. These range from simple bilateral agreements between pairs of countries to full Economic and Monetary Union among groups of them; in the latter all tariffs and non-tariff barriers to trade between countries are removed, there is a common set of import customs duties with countries in the rest of the world, and a single currency ensures that there are no exchange rate issues (see Box 9.2). These various arrangements should not be seen as one of natural progression, though in some cases countries that have started with a more simple arrangement between each other have moved to more a complex one. The UK's relationship with other EU Member States has shifted from one of a Common Market to a Single Market, though there is little likelihood that this will progress to Economic and Monetary Union as the UK does not currently wish to adopt the euro as its currency.

Balance of Payments

Deficits or surpluses in the Balance of Payments figures within national trade statistics form the material for newspaper headlines and strong political debate. The preservation of an acceptable balance is an important objective of national economic policy, and economic growth and low unemployment have frequently been sacrificed in its interest. Before discussing the reason for its prominence we must briefly describe the elements of the balance.

The Balance of Payments is usually divided into four categories: (i) the trade in 'visibles'; (ii) the trade in 'invisibles'; (iii) flows of capital; and (iv) monetary movements. **Visible trade** is comprised of items which literally can be seen (e.g. cars, television sets, food, etc.). Imports into the UK increase the demand for foreign currency (because we need the currency to pay the foreign exporters) and increase the supply of sterling. Our exports increase the demand for our currency (because foreigners need it to pay for the goods they buy from us) and increase the supply of foreign currency. The balance of the value of visible imports and exports is called the **Balance of Trade**. It is negative if the value of visible imports exceeds the value of visible exports. If this negative figure becomes greater (i.e. grows from –£10m to –£20m) the equilibrium between supply and demand for currencies will tend to alter, with the exchange value of the pound falling.

The **Terms of Trade** is not to be confused with the Balance of Trade. The *Terms of Trade* compares relative movements in indices of the *prices* of exports and imports.

Box 9.2. Degrees of liberalization of trade between countries.

There is a bewilderingly complex set of relationships between countries set up to facilitate trade, though often also carrying economic and political connotations. The main broad types are, in ascending order of liberalization:

- **Bilateral Trade Agreement** between pairs of countries – where two countries agree to give preferred status to each other, typically by lowering tariffs and other trade restrictions, though these may only apply to a small range of goods. Example: the EU has concluded (in 2012) an agreement with Singapore that will open up trade in services in banking, insurance and other financial services.
- **Multilateral Trade Agreement** – where three or more countries agree to lower the barriers to trade between them, thereby increasing trade. Example: the GATT Uruguay Round agreement made in 1993 between major trading nations (including the EU and USA) that *inter alia* committed countries to convert all import restrictions on agricultural products to tariffs (tariffication) and to a programme of reducing them by 36% over 6 years, with a minimum cut of 15% per product.
- **Free Trade Area (FTA)** – where there are no taxes or quotas on trade between two or more participating countries who form a Free Trade Agreement, though there is no common set of trading arrangements or tariffs with other countries. To prevent goods imported into a country that has a low import tariff simply being exported to another country in the FTA, bypassing the latter's higher import tariff, various rules of origin are applied, and only those goods that meet them are allowed to be traded freely within the FTA. Example: North American Free Trade Area (NAFTA) consisting of the USA, Canada and Mexico.
- **Customs Union** – like a FTA but also with a common external tariff. This avoids the need for rules of origin restrictions on trade between participants. Example: the Customs Union involving the EU and Turkey, set up in 1996.
- **Common Market** – comprises a Customs Union (FTA in goods and services and common regulations on products and a common external tariff) and the relatively free movement of capital and services, though there remains other restrictions and barriers to trade (such as sanitary and phyto-sanitary regulations, relating to animal and plant health and safety). Example: the European Common Market, later called the European Economic Community and forerunner of the EU.
- **Single Market** – comprises not only a Common Market in goods and services and common regulations on products, but also allows the free movement of factors of production (such as labour and capital) across national borders and gives the rights for entrepreneurs and professionals to conduct their business in any place. Technical and legal (taxation) barriers to trade are removed as far as possible, and it is as easy to buy from another country as from one's own country. Example: the EU as currently constituted.
- **Economic and Monetary Union** – comprises a Single Market but also uses a single currency that facilitates trade between member countries. However, economic union may be incomplete, with only some coordination of economic policies rather than a single, unified one, and separate national budgets. Example: the members of the EU's eurozone.

$$\text{Terms of Trade} = \frac{\text{Index of export prices}}{\text{Index of import prices}}$$

The Terms of Trade are said to 'improve' if the ratio increases, i.e. if the price per unit of our exports increases or if imports become cheaper. On the other hand, the terms deteriorate if import prices increase with no equivalent rise in export prices, as happened when the price of oil on the international market was suddenly increased in 1973/74 by the oil exporting countries acting collectively. Following the deterioration a greater volume of goods had to be exported by oil-importing countries to pay for each barrel of imported oil than before.

Invisible trade comprises the following: (i) banking and insurance services; (ii) tourism and travel expenditure; (iii) government spending abroad to maintain troops on foreign stations and embassies, etc.; (iv) shipping and civil aviation services; (v) income from investments abroad and payment to foreigners arising from investments they hold here; (vi) private gifts such as occur when foreign workers send money back home to relatives; and (vii) other miscellaneous payments for the hire of foreign films, etc. While this group of 'invisibles' is rather a hotch-potch, they have the following characteristics in common:

1. No tangible goods are *directly* involved (even with the hire of foreign films, what is being bought is the right to see the film, not the ownership of the film itself).

2. The items are essentially for current consumption as opposed to investment.

3. Spending on them increases the demand for foreign currency and hence tends to lower the value of the home currency (or increase value of foreign currency, which amounts to the same) while spending by foreigners, say, by visiting Britain, tends to increase the value of sterling.

Taken together, the Balance of Trade plus the balance in invisible earnings comprise the Balance of Payments on Current Account. This is illustrated in Table 9.2. It is this Current Account Balance of Payments which is the subject of much political controversy. A deficit, with the value of imports exceeding exports, will tend to cause the value of the pound to fall, although this may be counteracted by what is happening to flows in the Capital Account and by monetary movements.

Capital flows refer to spending on foreign goods that is not for current consumption, but is for investment. For example, if a British man buys a farm in France he is hopefully making an investment in an asset which will yield him an income. (Incidentally, this income will subsequently appear as an invisible earning in the Current Balance of Payments.) Similarly a Chinese investor may buy shares in a British firm. The purchase of foreign assets by Britons will increase the demand for foreign currency and increase the supply of sterling (i.e. it will have a depreciating effect on the value of sterling similar to that of the purchase of imports). Investment by foreigners in British assets will, of course, have the reverse effect. In a time when the Balance of Payments on Current Account is negative, an outflow of capital will tend to aggravate the downward pressure on the value of sterling and so investment abroad may be discouraged by the government. It must not be forgotten, however, that this means the *income* from such investments in future years is being forgone. The combined balance on Current and Capital Accounts produces the Total Balance of Payments.

Monetary Matters

In the short term imbalances in a country's Total Balance of Payments are covered by **monetary movements**, meaning changes in the national reserves of gold and foreign currencies or (increasingly) official borrowings from international financial organizations and banks (see Table 9.2). But this is an accounting mechanism and is not an answer to the underlying imbalance in trade.

As was shown earlier, if a country were to try to import a greater value of goods than it exported, the exchange rate of its currency would fall, making imports more expensive, choking off demand for them, and making its exports more attractive in foreign markets. Thus the imbalance would tend to be corrected. However, because trade is not an instantaneous process, and because of complications involving the demand elasticities of exports and imports, a depreciation in the exchange rate would not completely correct a Balance of Payments deficit, at least not in the short term – monetary movements would still be needed. Nevertheless, over a few years such a depreciation could be reasonably expected to contribute to a fall in the trade deficit.

The determination of exchange rates by supply and demand for currencies, and a market in which rates can change according the underlying economic forces, are now the norm. There have been periods in which governments have attempted to set 'fixed rates', either in terms of a reference currency (such as the US dollar) or gold and have been active in currency markets to maintain these rates. For example, when the value of sterling was 'fixed', if the value on the real currency market showed signs of falling, demand for it was increased by the Bank of England buying sterling using the nation's reserves of gold and foreign currency which it held. Conversely, if the value of sterling tended to rise above the 'fixed rate' (a less common experience), gold and foreign currencies were bought into reserve by selling sterling. For the UK things were made more complex by sterling itself acting as a reserve currency, that is, some other countries (the 'sterling area', mainly but not exclusively former British colonies) tended to keep their national reserves as sterling as an alternative to gold because much of their trade was with Britain and because of sterling's traditional stability. Imbalances in the trade position of these other countries could result in them buying or selling sterling and thus impacting on exchange rates. These sterling balances were largely invested as short-term loans to the British Government.

Fortunately for an introduction to economics such as this, a detailed study of fixed exchange rates is no longer necessary as both the pound sterling

Table 9.2. Balance of Payments (simplified format).

	Debits (−)	Credits (+)	Balance
1. Current Account	Visible imports (cars, TV, etc.)	Visible exports (clothes, tractors, etc.)	Balance of Trade (a)
	Invisible imports (holidays abroad, interest to foreign investors, govt. spending abroad, etc.)	Invisible exports (banking and insurance, foreign tourists, sea and air transport, etc.)	Balance of invisible earnings (b) (+ or −)
			Balance of Payments on Current Account (x = a + b)
2. Capital Account	Capital flow out (−) Official and private investment abroad	Capital flow in (+) Official and private investment by foreigners in this country	Net capital flow (y)
Combined Accounts	Current Account balance + Capital Account balance		Total Balance of Payments (x + y)
Monetary movements to cover Balance of Payments deficit or surplus	(a) Increase (−) or decrease (+) in national reserves of gold and foreign currencies (b) Increase (+) or decrease (−) in borrowing from International Monetary Fund (c) Increase (+) or decrease (−) in borrowing from foreign banks, etc.		
Combined monetary movements			(x + y)

and the euro 'float' (i.e. their values are determined by the supply and demand for them on currency markets). Under such circumstance the Bank of England and the European Central Bank, respectively, only intervene in the exchange markets on those occasions when values are *temporarily* affected, in the case of sterling such as at the time of a general election. This policy could be called a **managed float**.

Confidence is clearly an important issue in trade and monetary movements. Loans by foreigners to governments or to fund investment will only be made if they have confidence that trade deficits are temporary and that internal policy (such as inflation control) and external policy (such as international relations) are so set that stable exchange rates are feasible. Confidence will be seriously undermined if the exchange rate of a currency is persistently under downward pressure through a fundamental imbalance in trade, and this loss of confidence may do more harm than the original imbalance. Governments, then, are intensely interested in their country's trading position and have tried many methods, some effective and some largely cosmetic, to maintain international confidence in their currencies and economies.

Government manipulation of the trade balance and exchange rates

While a government will not feel excessively concerned over short-term and minor imbalances in trade, particularly if a floating exchange rate policy is adopted, it may wish to engage in countering policies if the fundamental trade trend is adverse. This it does by measures that stimulate export earnings and discourage spending on imports.

In the past a government might restrict imports by imposing quotas, tariffs, etc. Or it could stimulate producers of commodities that could substitute for imports (such as agriculture) or that were orientated towards exporting by giving them financial assistance and tax concessions. It might place a limit on how much cash tourists going abroad could take. If it controlled banking and credit institutions, the government could restrict personal loans and other forms of consumer credit so that fewer imports were purchased and more goods were available for export.

Many of these instruments are no longer available. Countries that form the EU have agreed, as part of community membership, to operate a Single Market for almost all goods and services and to

Box 9.3. European Central Bank and the euro.

Following the establishment of Monetary Union and the Single Currency, the European Central Bank has played a role in respect of the exchange rate of the euro similar to that of the Bank of England in respect of sterling. After the beginning of 1999, when the system was introduced, the value of the euro relative to other international currencies (especially the dollar) and to the pound declined. The European Central Bank tried to moderate this by coordinating the activities of the national central banks of Eurozone countries, buying euros in exchange for the other currencies. Managing the euro's value in terms of other currencies has been made more complex because of the financial difficulties experience by some Member States in the economic downturn associated with the international banking crisis triggered in 2008 by the collapse of the USA investment bank Lehman Brothers. However, by-and-large the exchange rate of the euro against other currencies is allowed to 'float' in the same way as is sterling.

avoid using trade restrictions. As part of GATT/WTO agreements (in which the EU negotiates as a bloc) increases in trade restrictions are not allowed, and the movement is towards dismantling existing ones. Banking and credit are also now international activities. The steps national governments can still take tend to be things like encouraging visiting tourists by promotion abroad, and increasing the competitive performance of home industries by encouraging skills training, supporting networking between entrepreneurs and their better management of risks. However, even such actions are only allowable if they do not significantly interfere with the EU policy of fair competition throughout the community. More generally, a domestic fiscal policy that aims to be deflationary by using higher taxes on personal incomes to cut back home demand will tend to divert UK produced goods towards export markets and lower the prices of UK goods abroad relative to foreign goods. This will encourage exports and discourage imports. Personal taxation is something on which Member States still have freedom of action, and on which they are keen to maintain their autonomy.

It must be remembered, however, that imports are the exports of other countries, earnings from which are used to buy our exports; this is particularly true of developing country trading partners. Cutting down imports is likely eventually to harm a country's exports, so, while import restriction may be a temporary expedient, it is probably against the national interest if carried on for too long.

These measures to influence trade are primarily concerned with the Current Account items in the Balance of Payments. As Capital and Current Accounts are linked, a restriction on investment abroad by Britons will eventually curtail the invisible

earnings from such investments, and increasing the rate of interest to be earned on foreign monies deposited in a country will increase the invisible payments to foreign investors on the Current Account. In the past making investments abroad could be discouraged or restricted by controlling currency exchanges for investment purposes. However, under the free movement of capital in the EU that is a feature of its Single Market, this is no longer the case, and the growth of electronic financial transfers has meant that controls of capital movements anywhere around the world are difficult. Inward investment can be encouraged by giving incentives, such as tax concessions for Japanese car firms setting up factories in Europe or making sites readily available and so on. However, probably the best way of attracting foreign investment is to ensure that the country's domestic economy is in good health, well managed and is growing.

Up to the 1980s and 1990s official national interest rates had a marked effect on exchange rates. An increase in the Bank of England minimum lending rate made the deposit of money in Britain in the form of sterling look attractive, hence increasing the demand for it. But the UK's freedom to pursue an independent policy on interest rates is in practice now quite limited, as borrowing from elsewhere in the EU is available to British residents and firms. For countries that form part of Monetary Union, with the euro its single currency, there is virtually no opportunity to set interest rates nationally. EU rates (and the quantity of money in the Eurozone) are determined by decisions taken collectively by the European Central Bank, and these may not suit conditions in all Member States. For example, an increase in the central bank's base rate may result in domestic deflation in some countries which may not be politically acceptable if, for example, domestic unemployment is already high. This lack of national independence in monetary policy has to be set against the perceived benefits flowing from Monetary Union.

Where a government still has some control over its currency exchange rate (such as in the UK), it may attempt to correct a persistently adverse Balance of Payments by allowing the exchange rate to fall. Indeed, it may have no option. At first glance this appears a cure-all, since the price of imports will rise and the price of exports will fall. Imports should appear less attractive and exports more attractive. This is undoubtedly true, but the Balance of Payments is not a *ratio* of prices, but the *difference* between values of exports and imports – that is, price times quantity.

If the price of a country's exports drops by 10% as the result of a currency depreciation but only a 5% increase in quantity of exports sold abroad takes place, the *value* of exports will have risen by 5% (measured in the domestic currency). However, if the same volume of goods continues to be imported at the lower exchange rate, the total value of imports (again measured in the country's own currency) will rise by 10% and the Balance of Payments deficit will worsen. This is almost inevitably the sort of thing which happens in the short term after a currency depreciation but before buyers

have had time to adjust to the new levels of prices. In the longer term, say 2 years, the deficit may show some improvement, but to be successful a fall in the exchange rate must be accompanied by elastic demand (i.e. price-sensitive demand) for exports and/or imports, preferably both.

Exercise on Material in Chapter 9

Answer the questions in the manner indicated. Answers and explanations are given in Appendix 2.

9.1. If a country has to give up the production of 5 t of wheat when it wishes to expand its production of cheese by 1 t, what is the opportunity cost of the extra cheese?

9.2. Cross out inappropriate alternatives in the following statement:

> A country is said to possess *comparative/absolute* advantage vis-à-vis another country in the production of a good 'X' if the expansion of production of good 'X' by 1 unit involves forgoing the *consumption/production* of a *greater/lesser* quantity of other goods in the first country than in the second.

9.3. If it can be shown conclusively that UK agriculture is the most efficient in the world, can we say that it would never pay for the UK to import food?

Box 9.4. A case study of currency exchange rates and trade – the UK.

For a mature economy such as the UK it is far from certain that the price elasticities of demand for its exports are high enough for a downward movement in exchange rates to be of great help in correcting a Balance of Payments deficit. Added to this problem is a longer term one arising from the low income elasticity of demand which, at least until recently, the rest of the world exhibited for British exports in general. For historical reasons the UK appeared to produce an array of goods which were not in much greater demand as world incomes rose. This implied that the UK fell behind other Western countries with higher income elasticities for their exports in terms of growth of exports and consequently in national income per head. A parallel can be drawn with the decline in relative incomes experienced by the agricultural sector of a developed country as national income continues to rise (Chapter 3); this decline can be explained by the low income elasticity of demand for agricultural products.

In the 1960s and 1970s the static demand for UK goods abroad was in contrast with the demand by UK citizens for imported goods as British incomes rose. Attempts to stimulate the UK economy into growth led to Balance of Payments problems, caused primarily by rapidly increasing imports. These problems necessitated rapid changes of policy to correct the deficit, thereby ending the economic expansion. The appropriate long-term action to overcome these problems appeared to be to encourage the production and export of goods which had expanding demands on world markets as incomes grew. The 1980s saw a move in this direction, in part the result of the happy discovery of oil in Britain's North Sea, a restructuring of industry away from traditional manufacturing and towards lighter, hi-tech industries, and a growth in the service sector, including banking, insurance and other financial services for which there is strong international demand.

9.4. Examine the following two-country, two-commodity model:

Maxitry and Minitry are two islands, similar in the productive resources they contain. The standards of living in Maxitry are higher than in Minitry because the inhabitants of Maxitry are better workers. Initially there is no trade, but an enterprising Maxitry merchant acquires a boat and trade starts in two commodities, peaches and coconuts. It is found that to expand production of peaches by 1 t in Maxitry, 10 t of coconuts have to be given up, and in Minitry 15 t of coconuts have to be given up.

(a) Which country has the comparative advantage in peach production? (Cross out inappropriate alternatives.) *Maxitry/ Minitry/neither/both*

(b) Which country has the comparative advantage in coconut production?

(c) Which country is likely to be an exporter of coconuts?

(d) Which country(ies) benefits from specialization and trade, assuming problems of adjustment to the new patterns of production can be successfully overcome?

9.5. With a given quantity of land, labour and capital Country A can produce 12 t of wheat or 12 t of maize, but Country B can, with the same given quantity of resources, only produce 4 t of wheat or 6 t of maize. Assuming that these figures remain constant, specialization and trade should be in such a direction that Country *A/B* exports wheat and *A/B* exports maize. (Cross out the inappropriate letters.)

9.6. In Country C and Country D the prices of beef and butter before trade starts are as follows:

	Country C	Country D
Beef	£1/kg	$1/kg
Butter	£0.33/kg	$0.50/kg

Assuming perfect competition exists in both countries (cross out the inappropriate letters):

(a) Which country will specialize in and export beef? *C/D*

(b) Which country will specialize in and export butter? *C/D*

9.7. Using the information given in Question 9.6 above, can we predict the relative prices of butter and beef after specialization and trade have occurred?

Yes/No   Level/possible range

9.8. Given that we can easily prove that some trade is better than no trade, give three reasons for tariffs or other trade regulators that could be supported by advocates of the theory of unfettered trade:

9.9. Cross out the inappropriate alternatives in the following statement:

The Terms of Trade are said to 'improve' when the *value of exports/index of export prices rises/falls* relative to the *value of imports/index of import prices*. Such an 'improvement' will have an effect on the Balance of Payments which *is always beneficial/is always deleterious/can be either beneficial or deleterious depending on circumstances.*

The following list relates to Questions 9.10–9.13.

(a) Value of goods imported.

(b) Interest payments to foreigners on investments they previously made here.

(c) Spending abroad by our tourists.

(d) Value of goods exported.

(e) The investments made by our people abroad during the year.

(f) Earnings by our transport companies in shipping goods around the world.

(g) A loan to our government from the International Monetary Fund.

(h) Spending by foreign tourists here.

9.10. Which of the above are included when calculating our country's Balance of Payments on Current Account?

9.11. Which items would be included when calculating its Balance of Trade?

9.12. Which items are treated as 'invisibles' in the Current Account?

9.13. Which items form part of the Capital Account?

9.14. A country is allowing the exchange rate of its currency to 'float' against other currencies. Which of the following items are likely to put downward pressure on its exchange rate?

(a) A rise in the value of exports.
(b) Domestic inflation.
(c) Investment abroad by its nationals.
(d) A rise in earnings from overseas investments owned by its nationals.
(e) A rise in the value of imports.

10 Government Policy for Agriculture and Rural Areas

Introduction

This book has concentrated on the underlying principles of economics, which, while they are of great importance to those who wish to understand the agricultural industry and the rural economy, are applicable in many other contexts. While the nature of the problems that students will face in the future are not predictable in detail, they can feel confident that the basic toolbox of economic concepts provided here will be of substantial benefit in sorting out the causes and implications of these problems. Agriculture has been drawn on to provide examples, but this was never intended to be a book solely about the farming industry or even the rural world. That would have been too narrow.

Nevertheless, there is a good argument for bringing together the various characteristics of agriculture which have emerged in the course of this text and to assemble them into a coherent section dealing with the problems it and the rural economy faces and the policies used to tackle them. This is particularly important when so much debate goes on about the costs of the action which government (national or in the form of the collective action of the EU) takes towards agriculture. Why government is involved demands our attention. This means that we must examine the objectives of policy and the ways that the various actions taken meet, or fail to meet, the aims set for them.

Why do governments get involved with the economy?

Let us start from a general review of why governments wish to interfere with the running of the economy. Later we will move on to examine the special case for intervention in agriculture and rural areas.

In market-based economies the price mechanism is the main way in which the fundamental decisions of what is produced, how it is produced and who gets what is produced are settled (Chapter 1). The interaction of supply and demand has many admirable properties for this purpose (Chapter 4). It operates as a way of reflecting willingness to purchase and to produce without the need for a mass of data collection and planning. Those entrepreneurs who respond to meet consumer demands sell their output and prosper, serving the interests of both themselves and their customers. Industries where demand is growing will grow and those where demand is falling will contract.

Competition between producers means that resources go to those who can use them in the most effective way. Efficient producers grow, while the inefficient are squeezed out and new technologies which make production more efficient get taken up (Chapter 5). Comparative advantage leads to specialization and exchange, resulting in trade both within countries and across national boundaries (Chapter 9). And because the market mechanism involves stable equilibria, such as between supply and demand, the system is largely automatic and can for the most part be left to itself.

However, a completely free market is not a faultless way of solving the basic economic problems. Flaws occur in various forms and governments will wish to intervene to modify the outcomes, 'correcting' for what they perceive to be the failures of the system and achieving a better overall result in terms of the welfare of society as a whole. The main deficiencies can be classed broadly as follows:

1. *Imperfections in the market mechanism (leading to 'market failure')*. We have seen that the market mechanism is not good at taking into account externalities, such as environmental pollution or disfigurements of the landscape (Chapter 7). It does not ensure that marginal social cost equates to marginal social benefit; in a completely free market (and with only poorly defined legal rights to clean water and air) polluters would lack curbs on their activities. Public goods, such as an adequate

defence system, would not be provided. Consumers are not always in the best position to know what is in their personal interest, as in the cases of smoking and car seat belts. Nor are they likely to take strategic decision on, for example, organizing national reserves of food and other essentials that are needed in emergencies. The mechanism is also subject to the growth of market power by monopolies and other forms of imperfect competition. We also know that, at aggregate level, the whole economy does not automatically operate simultaneously at high levels of employment, low inflation and sustained moderate growth rates (Chapter 8). While there are some self-correcting mechanisms at work, and economists vary in their view of how strong they are, in practice some steering of the economy seems to be necessary.

2. *Failure at reflecting non-economic goals.* The market is sensitive to purchasing power, and those without this power will not have a direct impact on what gets produced and on who consumes it. The outcome of market-determined solutions may not reflect what society prefers, as reflected by the political system, including the way that people vote in elections. For example, it may be agreed to be in the national interest that young people should have equal opportunities of access to education irrespective of where they live or their type of home background. In a completely free market system, where education had to be paid for by the person receiving it, there are grounds to think that those from poorer backgrounds would be disadvantaged. Issues to do with achieving what is considered fair, such as education or job opportunities or the avoidance of wide disparities of income in society (particularly the avoidance of poverty) are described as to do with **equity**.

For these sorts of reasons government will wish to intervene in the economic system. Its goals will include those listed earlier in Chapter 1, and include: (i) the maintenance of national security; (ii) stability in the currency (avoidance of both domestic inflation and erratic exchange rates); (iii) equity in terms of access to health and other basic services; and (iv) protection of the environment. Box 10.1 illustrates the potential trade off between achieving these goals and economic efficiency. In practice there are a large number of interventions going on at any one time – an array of policies with economic, social, environmental, defence and (because in a democracy governments wish to be re-elected) political ends in mind. Agriculture and the rural economy will be affected, directly or indirectly, by each of these broad categories.

The Policy Process

It is useful, before moving to policies which are of immediate concern to agriculture and rural areas, to consider the process by which policy is formed and applied. This process will be of widespread application in many other situations and its study is in line with the stance of this book of examining the principles first and then choosing rural and agricultural examples. The process has four basic stages, shown in Fig. 10.1. These in essence ask the questions: (i) Why is the policy necessary?; (ii) What objective is the policy trying to achieve?; (iii) What mechanism is used to bring about the desired results?; and (iv) What is the outcome of using the mechanism, and how does this compare with the intended effect? As part of this there should be feedback to policy makers so that they can learn from the experience. In this last respect it is important to recognize that this process operates over time, so that the decision of one round of policy making is influenced by what went on in previous rounds (an example of **path dependency**), and that today's decisions will affect future ones.

The four basic stages of the policy process will now be discussed in a little more detail.

1. Problem formulation

The first stage is the recognition that there is a need for some government intervention. This may sound surprising, but the need for a policy may not be self-evident. In some cases it is blindingly obvious. For example, if people are becoming ill because the nation's food supply is contaminated by inadequate hygiene in slaughterhouses, any competent government will recognize that it faces a problem and will wish to do something about it. In many other cases, though, the recognition of a problem will be less immediate. For example, the awareness that something should be done about the way that farming has been encroaching on wildlife and reducing biodiversity to a serious extent required the recognition that there was a danger in continuing in the old ways. Questions had to be asked, such as: 'What is the real extent of the loss of hedgerows and birds?' and 'Does it matter?' Society was forced to recognize that changes were taking place and to sort out its priorities.

Box 10.1. Policy to correct for market failure and to improve equity.

If the government wishes to correct for market failure, for example to reduce environmental pollution by reducing the production (and consumption) of one type of transport (e.g. private cars) and expanding another (e.g. public transport) it is, in effect, wanting to shift from A towards C along society's production possibility boundary (curve) in the left-hand section of the diagram below. It can do this, for example, by taxing cars and/or subsidizing trains and buses, altering their relative prices, to which manufacturers respond.

Similarly, if the government judges that the living standards of poor people should be increased to some minimum in order to create a fairer society, it can do so by the system of tax and benefits that transfers income from better-off groups to the less well off. In effect, it is arranging a shift along the social welfare function from A to C, the latter corresponding to a higher level of well-being for society as a whole (see the social indifference curve in the right-hand section of the diagram). Though these social concepts must be treated with caution because of the problems of making interpersonal comparisons (and are thus drawn as dashed lines), this thinking clearly lies behind government action.

Getting from A to C is not always easy and sometimes may even be impossible for practical reasons. Progress is typically through point B, which represents a less efficient use of resources than A. In the first example, firms need time to adjust to new prices, and there may be a waste of resources in the short term as they cut back production on goods now in reduced demand. Society suffers through the output forgone; B is well inside the production possibility boundary. However, there may also be a more permanent loss of efficiency. Systems of taxation and subsidies absorb resources to run them, and maybe the production levels at C are never completely reached. Similarly, when transferring income from rich to poor, the administrative system absorbs resources that otherwise might have gone to the poorer group in society, added to which the change to the tax arrangements may blunt incentives and thus lower efficiency.

In both examples a balance has to be struck by politicians between, on the one hand, the potential loss in efficiency with which resources are used that accompanies intervention, at least in the short term, and, on the other, the environmental and social benefits that the intervention brings. Sometimes it may be preferable to do nothing.

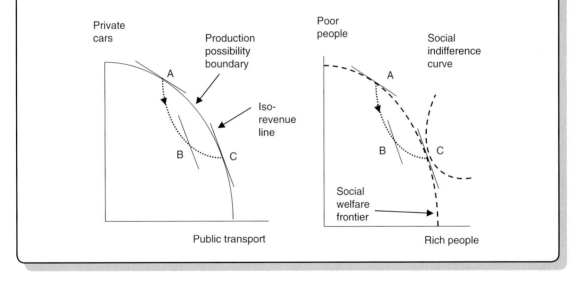

Priorities change. In times of severe food shortage, such as in a war, environmental considerations would not count highly, but in a situation where food is plentiful there will be the opportunity to take into account the broader aims of society when considering how the nation's rural land should be used, including its contribution to the environment. We now think that we should attach a larger value to preserving part of the national stock of old buildings than was the case 20 years or so ago. On the other hand, the idea that everyone should have a job has been given a lower priority in the attempt

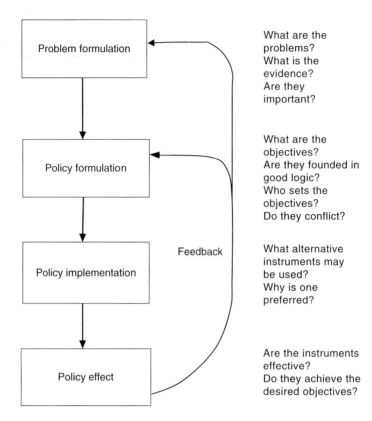

Fig. 10.1. The policy process.

to achieve other policy aims, such as the control of inflation and the need to maintain a competitive and flexible economy that can adapt to new consumer demands. Thus what is not a problem worth government intervention at one time and in one set of circumstances may be viewed as a real problem at another time, and vice versa.

In a democratic society, the process of setting broad policy aims and achieving a balance between them is left to the political system. A dictator (individual or small group) could perhaps make such decisions more quickly, but history suggests that, in the longer term, such systems operate against the welfare of society. In a democracy decision making involves politicians (and political parties), voters, interest groups and the bureaucracy (government departments necessary to put policies into operation). Individuals and groups can still be very influential in steering the system. For example, in the case of the impact of farming practice on the environment, it needed some articulate critics and pressure groups to bring to the attention of the general public and politicians the extent of the changes that were taking place. Under such a system there is no guarantee that the view which predominates is necessarily the most accurate. Pressure groups often have to overstate their case in order to attract attention and it is easy to lose perspective. Sometimes individual government ministers become preoccupied with a particular line of argument and can exert disproportionate influence. Particular government departments may encourage policies that keep their budgets growing. As was noted in Chapter 1, the way that a policy decision gets taken will influence the shape of that policy (another example of *path dependency*), and it is common to find in capitalist market economies that farmers, their unions, and politicians and civil servants who have an interest in encouraging public support of agriculture are able to steer decisions in their favour. Nevertheless, despite its drawbacks, the democratic political system and the diversity of competing views in a pluralist society do seem to work generally in a way which enables real problems to be brought to the surface (sometimes belatedly) as a prerequisite to doing something about them.

2. Policy formulation

The next stage is to decide on the type of action which is necessary to meet the problem in hand. The responsibility for formulating policy lies primarily with government ministers and civil servants. The critical part of policy formulation is to decide on the objectives which government intervention is designed to achieve. These hold the key, since if they are not set out clearly and specifically it will be impossible later to judge adequately whether the actions taken have turned out to be successful or not. For example, if the aim is to ensure that children living in the countryside can receive schooling without excessive travelling, an objective which could be tested might be to reduce the percentage of children living in rural areas who have to travel for more than an hour each day from, say, 20% to 10%. However, clear and testable objectives may not always be welcomed by ministers as they leave little flexibility to respond as problems change, and it becomes relatively easy for their critics to point to a lack of complete success.

3. Policy implementation

This consists of the design of programmes of action by which the objectives can be brought about and putting them into practice. There is a wide range of types of policy programme which can be used. Let us take, for example, the general policy aim of conserving areas of natural wildlife habitat in the countryside undisturbed by commercial agriculture. In pursuit of this aim the government may determine, as a specific objective of policy, that land containing certain types of wildlife or environmental features should not be disturbed and not ploughed. What alternative types of scheme might they use? They might consider the following:

- Passing laws on what land can be used for, with or without some compensation for the occupiers of land who find their options now limited.
- Purchase of the land, so that the government can directly influence what it is used for.
- Offering farmers and owners financial incentives to encourage them to enter into management agreements not to disturb wildlife features.
- Educating farmers and the general public on the importance of conserving the nation's environmental heritage. Part of this might involve designating certain areas as being of outstanding natural beauty.

Which method, or combination of methods, is actually used will depend on a number of factors, including: (i) the total cost; (ii) the administrative burden involved; (iii) the political acceptability of the alternatives; and (iv) the value put on the wildlife to be protected. All four approaches are in use for different types of environmental protection in Britain.

Often there are alternative ways of administering policy programmes. For example, should they be controlled from central government or should local bodies, such as regional, county or district councils, be given the responsibility of running them and the finance to carry them out? Some programmes, such as the regulation of markets for agricultural commodities, can be best organized centrally but others, such as grants for encouraging farmers to repair old stone walls to make upland areas appear more attractive, are probably best left to local bodies to work out the details and organize.

4. Policy effect

Some policy programmes achieve their effects directly, such as banning the sale of polluted food to protect health. But many others rely on a chain of causality (or **intervention logic**), such as providing people with education and training with the intention of raising an industry's productivity and competitiveness. If the logic is sound and programmes meet the objectives set for them, they can be considered as being effective or, to use another technical term, they have had **impact**. To return to the rural schooling example, if grants were made to local bus companies to enable them to run more services, and the percentage of children having to spend an hour or more travelling fell from 20% to 10% (or below), the objective had been reached.

If, however, the percentage had not fallen at all, then the policy was not successful. If it fell to, say, 15%, the policy had been partly successful. The process of judging the outcome in relation to the objectives and the resources used is termed **evaluation**. This is rarely a simple and clear-cut process. As the term implies, more often it is a matter of subjective judgement, though taking into account as much objective information on the performance as is possible.

Evaluation has some difficult problems to face. The first is to have available adequate data on policy programmes to enable their performance to be assessed. To make this possible the decision has to be taken early in the life of a policy programme to set up the administration necessary to collect it. This systematic data collection to assist evaluation is known as **monitoring**. Let us take as a first example a government programme to encourage elderly people to insulate the lofts of their homes by offering them 50% grants on the cost, the objective being to achieve '*x*' per cent of insulated houses. (We will assume, at least initially, that this is a reasonable activity in pursuit of broader goals such as reducing the emission of carbon dioxide that plays a part in global warming and in countering poverty among the elderly, though a full evaluation should probe the underlying rationale.) The sort of information needed to evaluate this programme will include the amount of public resources that have gone into the programme, such as the expenditure under the policy and administration costs. Also required will be statistics on the results of the programme, such as the number of people who applied and the number who actually went ahead with insulation. We would also want to know the number of people who were eligible to apply but did not, and reasons for their non-uptake of incentives (e.g. they were too small, the application form too complicated, etc.).

Apart from data availability, evaluation has to try to determine what would have gone on in the absence of the policy programme (called establishing the **counterfactual**). Continuing the insulation example, the evaluator would want to know also how many would have gone ahead without the grant anyway; if most of the elderly people intended to insulate even without the grant, the spending by the government may have been largely wasted. The real impact of the policy is the additional amount of insulation activity which was caused; this aspect is often described as attempting to assess the programme's **additionality**. Payments to people who would have insulated anyway are **deadweight**, in that this achieved nothing, a term also used about any benefits that go to people other than the target group. And there may be **side effects** or **spin-offs** that have to be considered as part of the overall impact. Some may be intentional but others may be unforeseen; some may be so minor that they can be ignored while others will be of importance economically, environmentally or politically. Perhaps the availability of the grant, by creating additional demand for insulation material and the services of contractors, will create extra employment and incomes through the **multiplier** effect (see Chapter 8).

Some of these questions on additionality can be answered by the use of **control** areas, for example parts of the country in which grants are not offered, and by drawing comparisons with what happened before the programme was introduced. A **baseline study** is one that tries to establish what a situation is before a policy programme tries to change things. However, in many cases this is difficult or impossible to carry out.

The workings of other policy programmes, such as special payments to the elderly in times of severe cold weather to assist with their heating costs, may be affected by the installation of the insulation – there may be fewer claims for help because houses will be warmer and therefore the public expenditure may be lower than planned. This grant scheme, by generating some jobs, may also reinforce other policy programmes which lower unemployment, and the full evaluation of insulation grants should take this into account. But, if the elasticity of supply of these insulation materials and services of installers is low, in the short term the extra demand may bid up the price of insulating a house to the extent that the net cost to the householder after receiving the grant may be little different from the initial cost without the grant (Chapter 3), so that much of the benefit going to the suppliers, which was not part of the government's intention. Drawing a reasonable boundary around what an evaluation tries to encompass is a key task for the people carrying it out.

It is important that there is **feedback** from evaluation, for without it there is little point in it being undertaken. Evaluation may throw up more information about the basic problems the policy is aimed at (its nature and extent), and thereby help shape objectives, affect the choice and design of instruments, and influence how these are used and monitored.

Box 10.2. Evaluation of a training programme intended to improve farm productivity.

- The training programme is intended to raise the competitiveness of the agricultural industry by increasing the level of skills of people working on farms. These skills might be of a technical nature (e.g. the prevention of diseases among animals) or to do with business management (e.g. the interpretation of business accounts) or marketing. It is assumed that, if people undergo training courses, there will be an impact on the productivity of the farms to which they return after training. This may not always happen: for example, the farm may not be set up for their extra skills to be used, or the skills may be inappropriate.
- The training programme will use up public resources to provide courses of training. The intermediate output is a number of trained people over a period; this is a relatively easy thing to measure and monitor.
- An evaluator will have to assess the actual impact on the productivity of the farms to which the trained workers return. This is more difficult to measure: for example, the relationship between the new skills and productivity may be indirect and take a long time to be fully realized. Other factors may be also be causing changes in the productivity of farms at the same time, such as greater investments in machinery. There may be an impact on other farms as they copy the improvements they see on the farms that have benefitted from training, though this trickle down may take years.
- In addition, the evaluator may wish to include spin-offs (or side effects), such as the economic benefits to firms that provide the training, and the impacts on non-agricultural firms if the people with training decide to move on to other industries in which their skills are valued.
- The evaluator has to choose where to draw the boundary of the outcomes of the training programme, which will depend on circumstances such as the significance of spin-offs and the practicality of measurement.
- A judgement will need to made on the size of the benefits in relation to the quantity of public resources used.

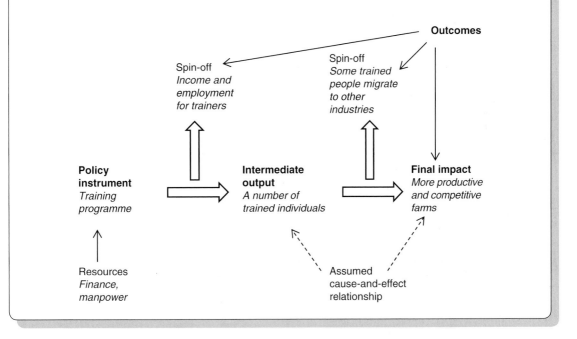

A mechanism has to be set up to ensure that decision makers take note of what previous evaluations of similar interventions have concluded, not always an easy step to ensure when civil servants are under pressure to come up with a quick response to a policy problem. And when staff shift post and are replaced with others from different backgrounds there is a tendency for awareness of previous work to decline.

As a second example, Box 10.2 illustrates some of the issues that evaluators face when assessing the performance of a training programme intended to increase the productivity (and thus the competitiveness) of farms. Such programmes are commonly found throughout the EU, including in each of the constituent countries of the UK. These include the choice of where to draw the boundary around the outcomes the programme can have and questions about the reliability of assumptions about the way in which training impacts farm performance.

In view of the problems, is this evaluation stage of the policy process worthwhile? The answer must be, in most cases, yes. This is because even though clear-cut answers cannot always be given and inevitably an element of judgement is involved, attempting to undertake evaluation will expose the more blatant examples of failure or of weakness in the logic behind policy. It will lead to a questioning of why certain policies are undertaken and the weeding out of the least effective programmes. Policy is a process that takes place over time, and there are often opportunities to avoid mistakes in policy decisions in subsequent rounds. Often a policy programme, once introduced, tends to keep going without anyone asking why, or whether it is still necessary. Unless this examination goes on, it will be difficult to stop programmes that are ineffective or which are no longer relevant to the problems of today; provisions for terminating a policy or replacing it are often not built into its design. This also implies that policy rarely starts with a blank sheet of paper. In many situations, certainly most concerning agriculture, today's policy is heavily influenced by that of years gone by (*path dependency* once again). However, without rationalization that results from evaluation it is difficult to obtain finance for new government policies in areas where new and pressing needs have arisen.

As will become evident, there are many examples within agricultural and rural policy of government programmes that were set up when the aims of policy were very different and have outlived their usefulness. In some cases they now positively act against the public interest. Funds used to support agriculture, if released, could be used in many other ways within the economy, perhaps even in supporting the non-agricultural aspects of the rural economy such as in the creation of non-agricultural jobs in villages, in the support of rural services (grants to rural buses and post offices) or in other ways of improving the welfare of

rural society (subsidies to village halls?). But, naturally, the people who are currently benefiting from the present policies are reluctant to see change. As was noted in Chapter 9 in relation to the benefits from trade, it is easier for a relatively small number of individuals who would be harmed significantly by a change to band together to influence policy makers, than it is to organize large numbers who individually would benefit only a small amount, even if on balance the welfare of society would be improved by the change. It might be possible for the beneficiaries to compensate the losers, but unless the mechanism exists by which compensation is actually paid, the vested interests may prevent any change occurring.

Government, Agriculture and Rural Areas

In virtually all countries, whether high income and mature or developing, free market or centrally planned, the government is in some way involved with its agricultural industry. High- and middle-income market economies, the focus of this book, share many reasons why there are government policies directed towards their farming industries and, increasingly, for rural areas (in which agriculture plays a shrinking economic role). The problems they address flow essentially from the economic characteristics of agricultural products and of the structures of the farms that produce them (including the incomes they generate), though there are differences stemming from particular geographical, social and historical circumstances.

Earlier in this chapter the general problems of a market economy and the rationale for government policy to address them were reviewed. Here we are concerned with specific problems and policies as they affect agriculture. These fall into two groups. First there are policies of government to tackle general problems and which use agriculture in an *instrumental* way to achieve a policy aim. Second, there are policies to deal with problems arising from the economic characteristics of the agricultural industry itself – the *intrinsic* problems of agriculture. Often both sorts of aim are set out together in some fundamental legislation, such as the EU's 1957 Treaty of Rome (see Box 10.3) or the sequence of Farm Acts in the USA. But such statements are not all-embracing and in practice policy aims evolve over time. More recent objectives for the EU are articulated in official documents agreed at high level such as the *Agenda 2000* and the

Box 10.3. The Objectives of the EU Common Agricultural Policy (CAP).

Treaty of Rome (1957). The foundation is Article 39 of the Treaty of Rome, the wording of which has been carried forward into the Treaty on the Functioning of the European Union, the consolidated text of which was agreed in 2010. The Article states that the aims of the CAP are:

- to increase agricultural productivity by promoting technical progress and by ensuring the rational development of agricultural production and the optimum utilization of the factors of production, in particular labour;
- thus to ensure a fair standard of living for the agricultural community, in particular by increasing the individual earnings of persons engaged in agriculture;
- to stabilize markets;
- to assure the availability of supplies; and
- to ensure that supplies reach consumers at reasonable prices.

Agenda 2000. The original objectives were reformulated in *Agenda 2000*, a document with less legal weight than a Treaty but agreed by the European Council (comprising the heads of government of the Member States), and are as follows:

- to increase competitiveness internally and externally in order to ensure that Union producers take full advantage of positive world market developments;
- food safety and food quality, which are both fundamental obligations towards consumers;
- ensuring a fair standard of living for the agricultural community and contributing to the stability of farm incomes;
- the integration of environmental goals into the CAP;
- promotion of sustainable agriculture;
- the creation of alternative job and income opportunities for farmers and their families; and
- simplification of Union legislation.

Political settlement of 2013. The two key EU institutions, the Council and the European Parliament, agreed the CAP's long-term objectives for the period 2014–2020 as:

- delivering viable food production;
- sustainable management of natural resources and climate action; and
- balanced territorial development (which implies assisting rural areas that face social and economic problems).

European Commission's 2010 vision for the CAP to 2020 (agreed politically in 2013 and implemented by a group of Regulations of that year). But official documents are only part of the story, and the real aims of policy are probably best understood by looking at what governments actually do.

It should also be remembered that, in an economy which consists of interrelated parts which are connected through the price mechanism, policies which are aimed primarily elsewhere will have some impact on agriculture. For example, the use of official interest rates in an attempt to control aggregate demand and inflation (see Chapter 8) will have an impact on farming through affecting its borrowing costs. Taxation policies designed to redistribute wealth may influence the sizes of farms; because agricultural land counts as an asset on which tax may have to be paid when wealth is passed from one generation to another, parts of farms may have to be sold off by heirs in order to raise the sums required. In both cases governments may wish to introduce

special treatment for agriculture as part of general policies. What can be called 'agricultural policy' is therefore not easy to define precisely. Here we will restrict attention to those government interventions which appear to have the agricultural industry as the main economic activity on which the policy is meant to have an impact or which has farmers and their households as the group who are meant to benefit.

Agriculture and general policy aims

Agriculture has an important role to play in achieving many of the wider aims which governments have for their countries. The most important of these are as follows:

1. As a provider of a *secure food supply* for the population, assisting in the defence strategy for the country.
2. As a source of *economic growth*, such as by becoming more productive, releasing resources that can be used to develop other industries.

3. As a way of assisting the country's *trade* situation, such as by improving the balance of payments and international currency exchange rate by replacing imported food or by generating export earnings.

4. As a way of *developing rural areas* by ensuring that jobs are available in the countryside, so that the rural economy is kept viable, and that a mixture of types of people will still live there.

5. As a way of *protecting the environment* by maintaining a landscape which is attractive to visitors from the urban areas for recreational purposes, and which conserves the wildlife and other natural features which the population as a whole considers as part of the national heritage. Agriculture (and forestry) can play important roles in improving water and air quality and in meeting international commitments to the reduction of greenhouse gas emissions.

In each of these cases the intention is to provide benefits which are shared widely throughout society, not just going to the agricultural *population* or even *those who live in rural areas*. For example, the defence offered by a secure food supply, so that the country cannot be starved into submission by a potential aggressor state cutting off food supplies from abroad, is shared by the whole of society. Some of these aims, it could be argued, mainly concern the urban population. It might be thought that an attractive countryside mainly benefits those people who do not normally live there but who like to visit it.

Historically the most important of these general policies is that of ensuring a *food supply*. The reason why Britain and many other European countries had agricultural policies before the EU's CAP was adopted in 1957 can be traced back to the Second World War and the post-war recovery period. During the war the UK's food production had to be expanded dramatically to compensate for cuts in supplies from abroad as the result of hostilities. In other European countries, which suffered more directly from the fighting, the vivid memory of starvation in the towns and cities embedded the notion of food security firmly in the minds of politicians in the post-war period.

As with many government policies, there are various ways in which food security can be brought about. One would be to keep large stockpiles of food reserves, but these may be expensive to maintain (e.g. the refrigeration of meat) and can only partly meet needs, and only until the reserves run out.

Or some land and machinery can be kept in reserve, so that output can be expanded fairly quickly. But often governments have decided that, in order to safeguard food supplies in times of emergency, they wished to have a high degree of self-sufficiency in peacetime. Usually this has meant having an agricultural industry bigger than would happen if market forces were allowed to operate freely.

In Britain, the economic recovery of the late 1940s and early 1950s needed agricultural expansion to assist the country's poor *trading position*, initially with the objective of replacing imports at almost any cost but then, once the critical period had passed and food supplies had expanded, more selectively. Even in the 1960s and 1970s there was a feeling that agricultural expansion should be encouraged if there was room to replace imports. Again, this meant having an enlarged agricultural industry.

The main way that this was achieved in the UK was to offer farmers higher prices for their products, though there were also subsidies on the factors of production which farmers used (e.g. subsidies on the costs of fertilizers and on buildings) and encouragements to improve the technology of production by providing a free advisory service and the government finance of scientific research related to agriculture. The impacts of these interventions in the free market are not difficult to predict, knowing the basic demand and supply theory described in Chapter 3. Higher prices caused agricultural producers to expand up their supply curves, and these were shifted to the right by lower input costs or improvements in technology. All had the effect of increasing the amount coming on to the market. But support systems, once in place, are notoriously difficult to dismantle as they build up vested interests (higher incomes among farmers, higher wealth among landowners, prosperity among businesses supplying goods and services to farmers) that will try to exert political opposition to protect themselves and oppose the change, even if it makes economic sense.

Agriculture has often made major contributions to *economic growth*. Because of the economic characteristics of its products, to which we will return later, resources in agriculture often have lower marginal value products than if they were used elsewhere. Consequently, governments will be on the look-out for ways in which resources (particularly labour) can be transferred to other industries, with an overall net benefit. This is in line with the Principle of Equimarginal Returns experienced in many sections of this book. If there are problems to the free flow of productive factors, resulting in occupational immobility (see Chapter 6), governments will wish to ease them.

Looking back in history, the rapid economic development in Russia following the Revolution in 1917 was seen to hinge on improving the performance of agriculture so that both labour and capital could be shifted to the heavy industries where major expansion was needed. The reorganization of farming along large-scale lines, with collective farms, and the forced relocation of labour, made possible by central economic planning and an oppressive political regime, was part of this. More recently, a major contribution to the rapid growth of many continental countries of the EU in the first two decades following its establishment in 1957 was the substantial transfer of labour out of agriculture to other industries in which it could be more productive. This source of growth was largely denied to Britain, where there was a far smaller percentage of the population engaged in farming and where output per man in agriculture was already relatively high by the start of this period.

In the 21st century further improvements in agriculture's productivity are sought not only as a way of promoting general economic growth but also as ways of reducing waste, with a positive impact on the nation's use of energy and levels of greenhouse gas emissions.

Policy for rural areas

Though agriculture has a role in promoting growth in the economy in general, in particular it is seen as one channel through which some of the economic and social problems of rural areas can be tackled. Rural areas are not simple to define and differ greatly in their characteristics. Though land use by agriculture and/or forestry and relatively low population densities are common characteristics, in some regions large stretches of countryside are often devoid of substantial towns or villages (e.g. the Highlands of Scotland) whereas elsewhere (e.g. in southern England and in the Netherlands) rural land use separates closely spaced conurbations. Consequently characteristics like remoteness, availability of basic services (education, health care and so on) will vary between locations. This means that the problems that people who live in rural areas face will differ from place to place. And what is a problem to one rural household (e.g. access to public transport) may be important to those on low incomes but not to the better-off. Policies for *rural development* attempt to generally improve economic prospects (jobs and incomes) and social conditions. They also have a strong environmental element.

Agriculture has often been seen as an instrument for tackling development problems of the rural economy. Some remoter parts of the EU face severe depopulation problems as farming declines. Support of farming is seen as one way of preventing the loss of rural jobs, keep a viable size of population which then can sustain rural services (schools, shops, etc.) and a lively local society. The reality is more complex. Many rural areas in the more densely populated countries and regions are gaining population as people move out from towns. Often farming has already shrunk to only a very small sector of the local economy, so that support to agricultural producers is unlikely to have much effect overall. Farmers can be encouraged to diversify into non-agricultural activities (tourism, crafts, food processing, etc.), something that will help them to be viable. In practice a more effective way of stimulating rural jobs is by encouraging non-agricultural industries based in or near rural areas to grow, achieving this by mechanisms such as improving communications (including broadband ICT) and providing education and training in skills, both technical and generic business management ones. Rather than giving support to agriculture, a subsidy to the village school or post office, or a grant to enable new businesses to start up, may be a far more effective and efficient use of public funds in pursuit of rural development goals.

Agriculture's role in environmental issues is, however, still important as (with forestry) it represents the greatest user of land. Governments increasingly recognize that agriculture, in addition to its production of food and fibre, also generates non-market outputs which are valued by the rest of society. They are aware of the importance that the public attaches to environmental quality, the appearance of the countryside and to the wildlife it contains. Safeguarding of soil, air and water quality, preservation of landscape and nature conservation can be regarded as other types of output from the agricultural industry, since the land-using activities of farmers and others determine what the countryside looks like and the wildlife features it contains. Some market services, like tourist accommodation on farms, fall into the category of 'private' goods and can be handled by the market mechanism but are dependent on countryside quality – a public good – though farmers may need to be made aware of the potential and trained to exploit it (the information gap is a form of market failure).

Agriculture is described as 'multifunctional'; in addition to its market output it is providing these non-market services for society in general. For some this comes from active consumption (e.g. recreation activities in rural areas) while others may derive well-being from simply knowing that the countryside is 'there' and being looked after and is available to all (it is seen as a 'merit good'). Because of the 'public good' aspects of such services, government intervention is necessary to enable the wishes of society to be implemented, even if only partially. Reflecting the greater affluence of the population in general, we can expect to witness an increasing role for policies which encourage farmers to use their land in ways which are consonant with emerging ideas of what the countryside should look like and the activities which should take place there.

Policies designed specifically for the problems of agriculture

The second group of reasons why governments have policies for their agricultural industries stem from the economic characteristics of agriculture itself. First we must summarize these characteristics that have been mentioned in previous chapters. For simplicity we assume that food is the only output from agriculture.

1. Basic food products typically have low price elasticities of demand.
2. Basic food products typically have low income elasticities of demand, so that little of any extra income is spent on the output of agriculture.
3. Agricultural production is subject to unpredictable variations from year to year, mainly caused by weather. The result of this, and the low price elasticity of demand, means that the revenues from sales by farmers can vary greatly over adjacent years.
4. The industry is composed of an atomistic structure, and this carries substantial implications. Though there is great diversity in terms of farm size, virtually all individual farmers will be price-takers. There is a strong possibility of suffering from a poor bargaining position when dealing with

large suppliers of agricultural inputs or buyers of farm commodities. Another characteristic of an atomistic structure is that the uncoordinated production decisions that it implies are conducive to the establishment of price cycles. The disadvantages of many small independent units can be avoided to some extent by forming farmer-controlled cooperatives or other group organizations to buy and sell and represent farmers in the increasingly integrated food chain. A central problem of cooperation, though, is the conflict it can produce between what the individual perceives as best for him and what is best for farmers as a group.

5. As technological improvements are taken up by individual farmers, starting with a few innovators but spreading to the majority in time, the industry is characterized by a gradually rising output. This expansion, when faced by an inelastic demand, has the effect of pushing down agricultural prices over the long term. The structure of the industry means that it is in the interest of individual producers to expand even though they know that it is not necessarily in the interest of all producers.

6. Falling prices squeeze incomes both of the industry as a whole and at the farm level. Though the volume of output may increase over time, the income left to reward its producers is less. Farmers look for ways of countering this, such as further expansion of area, replacing labour with machinery where this enables costs to be cut, and diversifying into non-agricultural activities. But the smallest farms can no longer provide an adequate reward for their operators and are in the weakest position when competing with other farms for extra land. Many of them have to give up. Their land is taken over by other farmers, usually the medium or large operators, to increase their farm size.

The combination of the above factors means that agriculture is a declining industry in all developed countries (and in most developing ones). There is a persistent fall in the total number of people engaged in agriculture and a shrinkage in its share of the national labour force and of the proportion of the national income generated by agriculture. But there are additional factors that must also be considered.

7. Farms are typically small businesses run mainly by families, often with the family providing much of the labour, and with the farm providing a place to live as well as a way of making a living. Personal and business wealth is closely mixed. Thus when small farms are no longer viable, there will be social

problems as well as business ones, such as the need to find a new place to live as well as a new occupation for the farmer and family. Often this means moving to an urban area, which may not be something which comes naturally to a person used to living in the countryside.

8. The transfer of resources out of agriculture, especially labour, is often difficult and slow. The occupiers of small, unviable, uncompetitive farms tend to be elderly and with little experience of other ways of earning a living. There may well be an emotional attachment to the land that makes them reluctant to leave. Whatever the cause, they suffer from occupational (and often geographical) immobility. Rather than transfer to other occupations, small farmers may struggle on, perhaps inefficient in their use of factors of production but nevertheless helping to generate the output which in turn keeps product prices low.

9. Both among independent farmers and in the hired labour force, the most likely to shift to non-agricultural occupations are the younger and better educated. This means that agriculture may contain a disproportionately large number of elderly people, especially in the more remote rural areas where leaving agriculture means moving away to a town job. Changes in the composition of the rural society may carry implications for the viability of rural communities.

Taken together, it is hardly surprising that there have been cries for governments of the more developed countries to do something to counter the pressures on farming brought about by the fundamental economic forces at work. Few people relish change, especially the sorts of structural adjustments which are necessary to accommodate the technical advances which have been going on. The inevitability of the changes has not stopped demands coming from farmers and their representatives for public support for farm product prices, which they see always falling in real terms and in relation to the costs which farmers face. The hardship caused to the small farmer has been pointed out, often by the larger farmers who stand to benefit most if the government accedes to demands for support. The threat to rural communities is highlighted. The need for a secure food supply and an attractive countryside is emphasized. Each assertion and claim for support should be examined with care. Is the logic sound? Are there alternatives to the support of agriculture which could achieve the ends more effectively?

From our analysis of the characteristics of the agricultural industry we can conclude the following:

1. That the government may wish, as a policy objective, to *stabilize the markets* for agricultural commodities and to prevent the damaging effects of large random changes of income from year to year. Some market risks fall to the farmer to deal with (e.g. by diversifying into several contrasting crop or livestock enterprises) but others may be more appropriately handled by policy interventions, particularly where they affect the whole industry. Also relating to markets, there may be the need for the government to keep a watch on the possibility of farmers being exploited by monopolists and monopsonists (big supermarket purchasers are sometimes suspected of abusing their market power), and to do this by the encouragement of farmer-controlled cooperatives and marketing agencies/companies.

2. That a policy may be necessary to assist *structural adjustment* in agriculture, by helping some farmers and their family members to leave the industry (thereby helping others increase their farm sizes to viable proportions) and to provide training in alternative types of economic activity (e.g. the setting up of craft industries using farm buildings or the development of tourist facilities on farms to give additional income sources).

3. That there is the possibility that, among some sectors – typically among small farms where resources are immobile so that structural adjustment cannot take place with sufficient speed – incomes will be so reduced that the farmers and their families cannot enjoy a standard of living which the rest of society would consider a bare minimum. Governments may set as a policy objective *income support* aimed at ensuring that the agricultural population can have a 'fair' standard of living.

It should not go unnoticed that the last two objectives potentially conflict with each other. Structural adjustment in agriculture is ultimately a necessity, but it can be fended off for a time if governments are willing to support farmers' incomes.

Policy in Practice

Each of the above aims of policy should properly be considered separately. It could well be that, for example, income support would be best achieved using policy programmes which are quite separate from those which aim to improve agricultural productivity. Environmental objectives need instruments that focus on land use rather than the welfare of farm households. But in practice many of the EU agricultural policy programmes were set up in the post-war period when it was thought that several, if not all, the aims could be achieved with the same types of intervention in the markets. For example, if prices for farm commodities were raised in order to induce farmers to produce more, backed up with grants to encourage investment in extra machinery to increase productivity, it was assumed that this would simultaneously improve the incomes from farming and, if the country were a food importer, lower the import bill and assist the balance of payments. When the belief predominated that prosperous farming produced an attractive countryside, and that farmers would only protect wildlife if they had enough income to allow them to leave parts of their fields undisturbed, support of farm product prices would also be expected to realize these environmental goals.

The ability to achieve all these aims of policy mainly through a single mechanism – support of farm product prices – is no longer seen to be possible. Price support leads to substantial conflicts between agricultural and environmental goals: farmers are encouraged to use their land more intensively and to bring into cultivation marginal land which for environmental purposes would be better left as rough grazing. Expansion of domestic production has implications for the pattern of international trade and stability of commodity prices, especially if EU surpluses are disposed of on world markets using export subsidies. Other ways need to be found of achieving income aid for farmers. Furthermore, the support of prices means that most of the benefit goes to the larger farmer whose living standards are likely already to be good, but the low-income farmer on the smallholding at the margin of viability gets relatively little help. Price support, by blunting the spur of economic hardship, also works directly against any policies which are aimed at improving the size structure of agriculture by helping the operators of too-small farms to leave the industry. This may store up even greater policy problems for the future when the costs of product price support rise to levels which the government is not willing to face.

Recognizing these difficulties, EU policy for agriculture has moved away from support by market intervention, though not entirely. Starting in 1992 (with the MacSharry Reforms) the system of product price support was gradually dismantled (though not completely for all commodities) and replaced by one of direct income payments; these payments, together with the remaining elements of market intervention, comprise what is known as Pillar 1 of the CAP. Direct payments were introduced initially as 'compensation' for the lost market support of the main types of crops and livestock, but they soon took on a life of their own as income support. A decision of 2003 combined the separate direct payments into a Single Payment Scheme (in the form of Single Farm Payments, or Single Area Payments, depending on whether these were calculated on what the individual farms previously received or on an average per hectare; the formula varied between countries), and this became operational from 2005. From 2014 these direct payments have been relabelled and repackaged as a mix of 'Basic Payments' and very similar additional payments that are conditional on certain environmental requirements being met (termed 'greening' the payment). An important point is that a farm receives this payment irrespective of whether or not it produces anything for sale (anything that is sold only receives the market price), though certain conditions apply, such as the need to keep their land in a good agricultural and environmental condition and to respect statutory management regulations on things like the pollution of watercourses – known as 'cross-compliance'. At the level of the farm, the sum received in direct payments broadly continues the pattern of support given under the previous arrangements (though with some gainers and losers), so the largest farms still receive the biggest payments and the smallest ones the least. In addition there are some residual payments that are still linked to production. Thus these direct payments fall short of a complete change to a system in which the largest amounts of income support would go to those in most need, but represent an intermediate step which is politically feasible in the present circumstances.

In the second decade of the 21st century, direct income payments account for almost three-quarters of all spending under the EU's CAP. Though these payments are, in principle, 'decoupled' from decisions on production, there are some indirect impacts that should not be ignored. For example, the fact that farm operators get a reliable payment irrespective of whether or not they produce anything means that they may continue in farming rather than be forced to exit from the industry by the lower levels of prices for their output, which will have some implication for the aggregate production level and the rate of structural change in the industry. The farmer's attitude to risk may be affected, as the cushion of the direct payments may enable a farmer to switch to farm enterprises that are inherently less safe (e.g. soft fruit), or to lower the degree of diversification (e.g. reducing the number of crop types grown), or to explore types of activity with which they are less familiar (e.g. the provision of tourist facilities).

Some tools of market intervention

Though the principal method of support to agriculture in the EU is now the system of direct income payments, until quite recently the most important agricultural policy programmes in the EU, both in terms of public money spent and in terms of their degree of influence on the size and shape of farming and the countryside, were those which supported the prices which farmers received. Their legacy is still with us, elements of them remain in some parts of the CAP, and provision exists for their use in exceptional circumstances, so they deserve study.

Several types of instrument have been used to intervene in markets for farm products under the EU's CAP. One was **intervention buying**, where the EU decided to buy from the market to keep prices up. This was, a major instrument used by the CAP in the 1970s and 1980s in which agencies set up by national governments (but using EU funds) purchased farm commodities. An economic analysis of support buying is shown in Fig. 10.2. In effect, EU purchasing was creating a new demand curve which was infinitely elastic at the price at which it decided to intervene. Under this mechanism, however much farmers produced, the EU had to buy it. Farmers respond to the higher price by expanding output, moving up their supply curves. Technological advances in agriculture shifted the industry supply curve to the right, increasing the amount that had to be bought with public funds. As the diagram shows, the higher price choked back the demand from other purchasers, so the amount of public buying was probably more than had been originally intended. This reduction in quantity demanded depended on the steepness of the market demand curve.

Benefit went to individual producers in the form of higher revenues from their sales, so that most benefit went to the farmers that produced the most. The competitive pressure for greater efficiency was somewhat relaxed, as product prices did not fall as much as would have been the case without intervention buying. Consumers were thought to have to pay higher prices for food, though the relationship between farm-gate price and prices in shops is not a simple one. Nevertheless anything that raised food prices was likely to be felt more keenly by low-income consumers than higher-income ones, because poor people tend to spend a higher proportion of their income on food than do richer people.

One obvious problem of support buying is that the EU found itself with a quantity of farm produce on its hands. It could store it, and the CAP at times built up large amounts of commodities in public storage facilities (and in private storage paid for out of public funds). There was much political comment about grain, beef and butter 'mountains' and the wine 'lake' created through support buying. But stocks could not continue accumulating forever. Storage is expensive, especially for frozen meat and dairy products.

Ultimately commodities that had been bought and stored needed to be dispersed. Various possibilities arose. Butter stocks could be destroyed or turned into animal feed or used for industrial purposes (**denatured**). Wine could be turned into industrial alcohol. Many of these methods proved to be politically unacceptable – voters do not like to see good food destroyed.

Some could be given away as food aid to poor countries. But the way most often used was to sell it to exporters who were given a subsidy to enable them to sell it at a loss on the world market. This also could cause problems, as for some commodities this subsidized produce pushed down the world price, disrupting the pattern of international trade and leading to claims of dumping by countries who found their exports no longer competitive (see Chapter 9). To prevent this sort of ill feeling, sometimes special deals were arranged with specific countries. But these could also cause political problems at home: at a time when the former USSR was out of international favour because of its treatment of civil rights, it was embarrassing to find that Russia was acquiring subsidized butter from the CAP at a small fraction of the price that domestic consumers were having to pay. In practice it seems

---

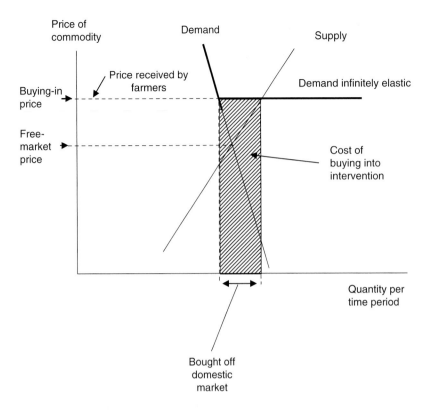

Notes:

1. When an agency of the EU buys a commodity from the market, a new 'kinked' demand curve is created which is infinitely elastic at the buying-in price (i.e. any quantity, large or small, will be bought in at the intervention price).
2. The supported price being higher than the free-market price, output from domestic farmers increases as they respond to the incentive. However, the amount taken from the market by consumers/ultizers of he commodity (other than the support buying agency) will contract.
3. Accumulating stocks of the commodity in the hands of the intervention agency have to be disposed of. Selling them on the world market will involve subsidizing their export, but this may be the least politically objectionable action; alternatives are burning, letting it rot, giving away, discretionary sales, etc.
4. Support buying will only be effective if imports are controlled, usually by import levies.
5. Consumers of food find themselves paying higher prices, a regressive change because lower-income consumers are affected proportionally greater than higher-income consumers. Higher food prices are inflationary.
6. Domestic farmers find the pressure on them to become more efficient is relaxed.

Fig. 10.2. The effect of support buying.

that agricultural policy has been shaped primarily by what governments wished to achieve for their home agricultures, without thinking too much about the international consequences.

A second main way to increase prices for farmers has been to **tax imports**, though, for this system to work, the country has to be a net importer of the commodity. This was the case in the early days of the CAP when the EU comprised only the original six Member States. The basic economics of such a tax is shown in Fig. 10.3. The effect is to raise the supply curve for imports from the rest of the world by the extent of the tax. As with support buying, this results in consumers facing higher prices in the

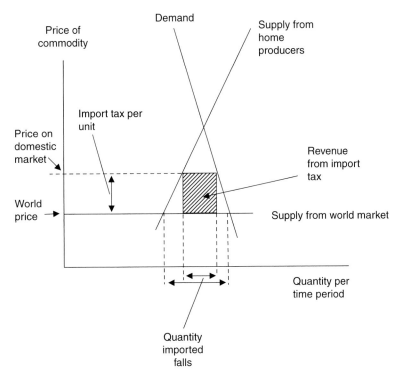

Notes:

1. A tax on imports can only work to raise domestic prices if there are some imports to tax. The free-market situation (before a tax is imposed) is one in which the EU is an importer of the commodity. The price in the EU is the same as the world market price, because no one will pay more for domestic production than for imports.
2. A tax on imports has the effect of raising the supply curve for the commodity from the world market by the extent of the levy, thereby raising the price that EU farmers receive.
3. As a result, home farmers expand their supply, but the amount demanded will shrink. In consequence, imports of the commodity are reduced. This will have an effect on the demand for foreign currency, tending to bid up the international value of the home currency.
4. An import tax will raise money for the budget of the EU.
5. Domestic prices can only be raised to the point at which no imports occur (where the domestic D and S curves intersect). To raise prices further, some other instrument will need to be used, such as support buying.
6. Food prices to consumers are raised, a regressive step which may be inflationary.
7. Pressure on domestic farmers to become more efficient is relaxed.
8. A tax is relatively simple to administer.

Fig. 10.3. The effect of a tax on imports.

home market for their food. Home farmers expand their supply, encouraged by higher prices, and imports are reduced, but part of this reduction comes from a fall in demand caused by the higher prices. Governments, rather than having to fund support, find themselves receiving revenue from the taxes on imports. But the burden of the system falls disproportionately on the poorer consumers for

whom food spending is important and who face the same price rises as richer consumers.

Taxing imports can only raise the prices that domestic farmers receive if there are some imports to tax. At higher levels of taxation imports are progressively reduced. The greatest price that can be generated this way is when the tax reduces imports to zero; then the domestic price would solely be determined by the interaction of demand with domestic supply. To support prices above this level, intervention buying or other mechanisms will have to be used. The EU's expansion to include more Member States, coupled with increases in productivity and the higher prices that encouraged domestic production, eliminated imports of many commodities and, instead, created surpluses. Under such circumstances, all an import tax can do is to prevent imports coming in to the EU market and undermining attempts to support by other means the prices farmers receive.

A further type of intervention that has been considered is the **quota** – a restriction on the quantity entering the market. Quotas can be placed on imports to keep prices up, but the most notable use in agricultural policy today is on domestic supplies, all imports being prevented by other means. Quotas on milk sales by farmers are the best-known EU example (these are due to be removed in 2015). Here we are concerned with the market impact of quotas rather than that at the farm level. The effect of a quota is shown in Fig. 10.4. Whatever the market price, the amount which farmers can supply to the market is limited to the quota amount. This creates a new supply curve. Consumers, faced with this smaller quantity, have to pay a higher price. As with the import tax, this higher price poses a bigger burden to the poor consumer than the higher-income one. Farmers as a group get a higher total amount of money from their sales, even though the quantity is less, because prices are forced up proportionally more.

How this benefit is distributed among farmers will depend on how the quota is allocated among them, but if cuts in output are applied on a percentage basis, the high-output farmer will benefit in absolute terms more than the lower-output producer, as with any other price-raising mechanism.

Quotas only work if demand is inelastic, and the steeper the demand curve the bigger the price rise which a given quota restriction will cause. As well as the impact on consumers and the lessening of the drive for efficient production and the slow down of structural adjustment caused by higher prices, quotas have additional problems. They need policing, since it will be in the interest of individual farmers to produce more at the higher prices than their quotas allow. Pressures for greater efficiency are further diminished by the knowledge that the efficient producer, however successful he becomes, cannot compete with less-efficient farmers and

expand beyond his quota limit. Quotas, unless made transferable between producers by sale or other means, are said to make rigid (or **ossify**, from the word for bone) the pattern of production of the commodity to which they apply.

Other mechanisms used to manipulate the returns received by farmers have included deficiency payments (in which the government makes up the difference between the price the farmer receives from the market and some predetermined target price), and payments per unit of production (e.g. per head of beef or sheep on the farm, or per hectare or per tonne of crop). These lower the supply curve by the extent of the payments, as producers will be willing to supply each quantity at a market price that is lower than before, knowing that the subsidy will make up the difference. Compulsory set-aside regulations, which took a proportion of land out of production, moved the supply curve to the left, thus raising the equilibrium price between supply and demand as less production came on to the market. Though compulsory set-aside is no longer used in the CAP as a supply control mechanism, one of the conditions for receiving the 'green' element of direct payments in the period after 2014 is that farmers have to put at least 5% of their arable land into an 'ecological focus area', which may have a similar effect.

Other policy instruments in use

Parallel with these direct payments and the interventions in the markets for outputs and inputs (that now comprise Pillar 1), throughout the history of the CAP there has been a group of instruments that have taken a longer view of the agricultural industry and have encouraged its structure to change and become more compatible with the economic and technical conditions of today. Put briefly, this means encouraging farm families who wish to remain in market-orientated agriculture to become competitive by increasing their farm size to a point at which they become viable and to give them the necessary technical and business skills, while at the same time helping those who wish to leave the industry to retire or (if younger) to take jobs in other industries. This **structural policy** has always been given a lower priority by governments, but it offers the more fundamental long-term solution to the problems that farming as an industry faces. The main mechanisms have been financial incentives, such as special pensions and

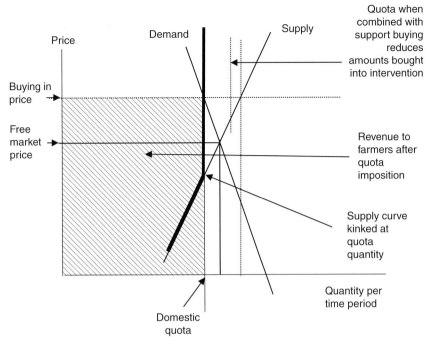

Notes:

1. By imposing a quota on the maximum amount that farmers are permitted to produce at a quantity that corresponds to less than the free-market situation, a new supply curve is created, kinked at the quota quantity (diagram in solid lines).
2. By reducing supply, the market price is raised to the point at which the quota-restricted supply curve intersects the demand curve. Where the demand curve is steep (price-inelastic demand) the smaller volume of output will be more than compensated by the higher prices, so revenue will increase. Costs of production will fall with a lower output, so incomes should improve.
3. Consumers are faced with higher prices on the market, which is regressive and potentially inflationary.
4. Pressure for producers to become more efficient is relaxed, both because of higher product prices and, if quotas are applied at the farm level, because the more efficient cannot expand unless they can obtain more quota.
5. Quota systems have to be administered and policed. Questions arise, such as: (i) at what level should quotas be imposed (e.g. national, regional or farm level)?; (ii) how should quotas be allocated, and should they be given to farmers (which is tantamount to giving the farmers a capital asset for free) or sold to them?; (iii) should they be tradable and, if so, should this be possible across national boundaries?; (iv) how should new entrants be treated for the system to be equitable?; and (v) what should happen if quotas are exceeded?
6. The dotted lines illustrate the situation with the quota on milk production in the EU. The quota quantity corresponds to a volume greater than that at which demand and supply would be at equilibrium in a free market. The imposition of a quota system was largely to contain the rapidly expanding level of output and thus to contain the amount which had to be bought into intervention and the cost of doing so. Similarly, in the beef market there are, in effect, regional quotas for the 'special premium' payable on young male animals; the main form of support, however, is intervention buying of beef.

Fig. 10.4. The effect of a quota on domestic production.

lump sums for farmers exiting the sector and, for those remaining, grants to enable training to be undertaken or payments to organizations providing the training.

From 2000 structural policy has been joined by important additional strands within the CAP. First, and of substantial importance in terms of the amount of public money used, are the attempts to use agriculture (and forestry) to deliver environmental aims (things such as increasing biodiversity, protection of landscape features, mitigation of climate change, promotion of renewable energy, etc.). Second is the aim of promoting economic and social development in rural areas in non-agricultural ways, such as by assisting the growth of tourism and craft industries to broaden the range of activities taking place and other steps to improve the quality of village life. Together these several strands (agricultural structure, environment and quality of life) form a bundle of actions that comprises Pillar 2 of the CAP (often referred to as the 'rural development' part of the CAP); currently Pillar 2 is much slimmer than Pillar 1, less than one-third of the latter's size in terms of planned expenditure of EU funds in the period 2014–2020. Implementation takes the form of 7-year Rural Development Programmes, drawn up at national or regional level, and selecting actions from a list (menu) that has been agreed at EU level; this gives authorities considerable freedom to choose what is appropriate for the particular problems of their rural areas (an example of the principle of **subsidiarity** at work, where policy decisions are made at the lowest practical level). The current set of Programmes runs from 2014 to 2020 (though some only start in 2015 because of delays in setting budgets) and are financed jointly (in various proportions) from the EU budget and by national governments.

Agri-environmental policy instruments

There is not room here to review all the instruments used to implement the increasingly important environmental role that agriculture now plays, though reference must be made to some basic concepts that lie behind them.

In Chapter 7 the notion of an 'optimum level of pollution' was introduced. This illustrates well the approach of an efficient policy, and the lessons learned from it can be applied widely. Assuming that pollution already exists, the costs of reducing it have to be set against the benefits gained by society. And because the further pollution is reduced the more costly it becomes to reduce it even more, the marginal cost of avoiding pollution (**abating** it) increases at lower pollution levels. On the other hand, as a generality, the cost to society of small amounts of pollution is low but rises as pollution levels rise. The economically efficient policy is one in which the marginal costs of abating pollution just offsets the marginal damage it causes (see Fig. 10.5). Attempting to reduce pollution more than this would result in a net loss to society. Similarly, any policy that is economically efficient will balance the marginal costs of putting it into operation against the marginal benefits this policy brings.

Assuming that information exists by which the optimum level of pollution could be established, it could be achieved by various instruments. For example, standards could be set for the quantity of

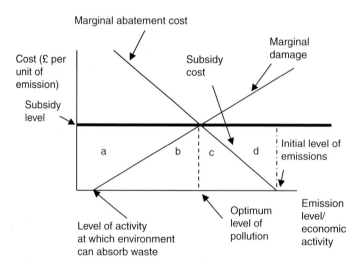

Fig. 10.5. Marginal costs and damage caused by emissions, and a subsidy not to pollute. The area of 'a' plus 'b' represents taxes, and area 'c' plus 'd' represents subsidies from the government.

nitrogen that farmers could be permitted to emit (discharge) into watercourses, with fines or other sanctions imposed if these limits were exceeded (standards are a **legal** or **regulatory** instrument). Because regulating emission levels of pollutants is technically difficult to implement (measurement of water quality might not be feasible), in practice the standards might apply to the maximum amount of nitrogen input that could be applied to fields, which in turn would result in the optimum level of pollution. Standards might also be applied to outputs (the amount of crop that could be produced) or the process by which production takes place (arable land counting more significantly than grassland that can absorb nitrogen more readily), but these again would be proxies for what really matters – the amount of pollutant emitted. Enforcing standards will incur costs, and the same standard is not necessarily appropriate for the circumstances of all producers equally. They also tend to be inflexible in the face of technical changes that should lead to them being recalculated.

Financial incentives can achieve the same end. Farmers can be paid to not do something, such as not applying high levels of fertilizer. Figure 10.5 also shows the level of subsidy that would encourage the producer not to pollute beyond the 'optimum' level. At higher levels of emissions the subsidy for not polluting is greater than the costs of taking action to stop it (e.g. the profit forgone by using less fertilizer), so the rational producer reduces emissions. As emissions are progressively reduced the cost of achieving this rises, and there is no point in reducing where the cost of doing so exceeds the subsidy for not polluting. At the optimum level of pollution the producer receives a total of c + d in subsidy from the government, but the costs of delivering this level of reduced emissions is only c (the accumulated costs of each marginal reduction of emissions). The producer receives a net gain of d.

Rather surprisingly, a tax on pollution can achieve the same result, assuming that information is available on the marginal damage and the marginal costs of abating it. If instead of a subsidy *not* to damage the environment, a tax was imposed on emissions, producers would find it rational to cut back high levels of emissions, as the tax would be greater than the cost of avoiding it. Figure 10.5 can be used to illustrate the point, where the level of tax would be identical to the level of subsidy shown. Producers would not cut back further as the costs of abating the emissions would be greater than tax due. In this case the government raises revenue by the amount corresponding to a + b. Producers lose income with a tax (they have to pay a + b) and so are understandably less keen on this particular instrument than on subsidies.

Which instruments are used to encourage agriculture to perform its environmental role will depend not only on economic rationale but also on political acceptability and administrative feasibility. While standards are applied to some major environmental threats (water and air quality) and special legal protection is given to some designated features (e.g. areas containing habitats of particular sensitivity), financial incentives are the preferred instrument for conserving biodiversity and landscape quality.

The Future Nature of Agricultural and Rural Development Policy

The policy process was shown earlier as being concerned with the setting of objectives and the design of programmes to achieve them. Over time these objectives evolve, and in the last three decades environmental and social issues have risen in relative importance to policies for agriculture and rural areas. In the past objectives have rarely been defined precisely and the ways in which the various aims might conflict with or complement each other have not been thought through. Individual policy instruments have usually been introduced to meet isolated problems without fully appreciating the ways that they could affect other programmes.

In the second decade of the 21st century some of the more obvious failings of agricultural and rural policy are likely to be made good, at least partially. Among the changes that can be expected are the following:

1. The further confirmation of the free market for agricultural commodities as the main determinant of prices that farmers receive for their products and on which they base their decisions of what to produce, how much to produce and how to produce it. It is now generally agreed that instruments that distort markets bring about inefficiencies and undesirable side effects. While a few examples of production-related support mechanisms remain (including milk quotas and sugar quotas), many are due to be phased out during the next period of the CAP (planned to run from 2014 to 2020), though history has shown that some are remarkably durable in the face of attempts to eliminate them.

2. As mentioned above, from 1992 the EU started on a process of changing its main form of support to a system of 'decoupled' direct payments (from 2014 called the Basic Payment and supplemented by a 'greened' payment which is conditional on farmers adopting certain practices designed to deliver environmental outcomes). Steps are being taken to reduce the variation between countries and types of farm in the size of these payments per hectare, but they are still primarily related to the size of the farm. In the period leading up to 2020 questions are likely to be raised about what the public gets in return for making such direct income payments to farmers. Basic Payments to farms will come under greater scrutiny and political criticism, if, as seems likely: (i) there is very little obvious benefit to society; (ii) they are hard to justify on welfare grounds (in that farmers as a group do not seem to be particularly poor, and the payments go largely to the bigger farmers who tend to have relatively high incomes and high wealth); and (iii) there are few if any arguments for these payments of a political economy nature (e.g. that they are needed to enable reform elsewhere in the CAP to take place). There are likely to be calls for changes so that the payments are targeted at those farm households in the greatest income need, in line with the stated aims of the CAP. However, this would represent a major redistribution of benefit and be opposed by the larger farmers.

3. A recognition that agriculture is 'multifunctional', in the sense that, in addition to the crops, animals and animal products it supplies to the market, it is capable of generating public goods such as attractive landscapes, biodiversity and space for recreation. A good economic case can be made for support to be given to farmers to generate such economic externalities in that, without this support, the quantities provided would be suboptimal from the point of view of society. It can be expected, therefore, that support will be linked increasingly to farmers providing these services in exchange; some of the funds currently used for Basic Payments are likely to be diverted to these environmental incentives or subsidies. Already, under existing EU Regulations, England switches 12% of its budget for payments under Pillar 1 to Pillar 2 (termed **modulation**), and this proportion can be expected to rise under political pressure.

4. A recognition that agriculture is only one part of the rural economy, and in many industrialized countries only a very small one. Often people who reside in the countryside commute to work in towns. The main providers of jobs for people living in rural areas are likely to be the service industries, the public sector, and manufacturing. Overall, the state of the rural economy is much more likely to reflect what is happening in the non-agricultural industries than on the prosperity of farming. Thus policy for agriculture is likely to be seen as just part of the broader policy concerning the use of rural resources and of the well-being of people living in rural areas. Rural development policy is therefore likely to be increasingly integrated with activities funded by the European Regional Development Fund (regional policy, including infrastructure and communications) and the European Social Fund (e.g. training for young people to combat youth unemployment).

5. A recognition that, especially in those policies where agriculture has been traditionally used as an instrument to achieve an end, alternative ways of reaching the objective may now be preferable. For example, the imposition of planning control on agricultural land use may be a more effective way of securing the preservation of areas of unique wildlife than the use of voluntary management agreements (with financial inducements) to farmers who own these sites. To take another example, though support for agriculture in hill areas of the UK is given in order, among other things, to keep people living there, more efficient ways of doing this may be available, such as by supporting rural factories or hotels. This will require the estimation of income and employment multipliers (see Chapter 8) for each type of public spending and in each type of location.

6. In line with the above, a recognition that the environmental and social characteristics of rural areas are not necessarily dependent on the existing structure of the farming industry. Land use may be more determined by the farming system than the size of farms, and the vitality of rural areas seems to depend more on demographic factors (the numbers and types of people living in communities, and migration) than on whether agriculture is their main source of livelihood. This points to a policy that is more based on the diverse needs of rural areas rather than one directed primarily at the agricultural sector.

7. A recognition that policies designed with domestic aims in mind, such as support for agriculture, can have substantial consequences for international

trade, not only in food products but, because of the reciprocal nature of trade, in non-food products as well. Hence there will be a search for ways of achieving internal objectives which do not have harmful trade implications, and for the scaling down of those which currently lead to trade problems.

All these changes involve risks, costs and disturbance of various forms. Bureaucracies (government departments, the European Commission and so on) will often prefer to keep doing what they are already doing because this is the easier option. Reform of policy has to overcome administrative inertia and vested interests.

The speed with which change takes place is in large part determined by the costs (financial and political) of keeping on with the present arrangements. Crisis speeds reform, and it is noticeable that in the EU reform of agricultural policy has taken place when budgetary crises loomed, or where major trade agreements were impeded (with negative implications for other industries), or when enlargement of the EU threatened to make an unreformed policy highly difficult to operate, threatening the political ambition of adding new Member States. Without this spur of necessity, change is less likely; there are many people and sectors, including the large farmers, some administrators and some Member States, whose interest would be harmed by reform and who oppose change which, overall, benefit the economy and society.

The improvements to agricultural policy envisaged above illustrate the sort of questioning that should be going on all the time about the purposes for which government is involved with the agricultural industry and rural areas and the most effective way of using public funds. They indicate a greater awareness of the principles of economics. They reflect a concern that resources are scarce, that they require allocating between alternative uses, and that to do so involves thinking about the objectives behind their allocation. Thus the study of agricultural policy fits well into the general definition of economics which emerged in Chapter 1.

Exercise on Material in Chapter 10

Answer the questions in the manner indicated. Answers and explanations are given in Appendix 2.

10.1. Place the following steps of the policy process in the correct order. Add numbers (1–4) in the brackets.

Policy effect (.....)
Policy implementation (.....)
Problem formulation (.....)
Policy formulation (.....)

10.2. Describe what is meant by each of the following:

(a) Evaluation

(b) Monitoring

(c) Baseline study

(d) Side effects

10.3. The government has the aim of reducing the amount of contamination of rivers by silage effluent. Describe three types of policy intervention that might be considered.

10.4. Explain what is meant when it is said that agriculture is being used in an instrumental role to achieve a policy objective.

10.5. Give five reasons why a government may wish to have an agricultural policy.

10.6. Give four reasons why the support of incomes of farmers by governments raising the prices that farmers receive for their output might be resisted.

10.7. If a rural area is having difficulties in maintaining the numbers of people living there, and policy makers decide that achieving a stable population size should be an objective, what alternatives to supporting agriculture might be considered for achieving this?

Appendix 1   Essay Questions

Chapter 1

a. What is an economic problem and how does it arise?

b. 'Even in a world of plenty all commodities are scarce.' Comment on this statement.

c. What do you understand by 'opportunity cost'? Illustrate your answer using choices individuals have to make in allocating their resources, and the way in which the State spends our money.

d. Describe the process of theory formation, testing and application, with examples drawn from economics.

e. Discuss the proposition that economics is concerned with problems of rational choice between alternative possibilities.

f. Economics has been described as being a social science. By comparison with natural sciences, how appropriate is this description?

g. What is meant by the term 'path dependency'? Illustrate your answer using examples of decisions taken by an organization with which you are familiar, such as your college or a family business.

Chapter 2

a. 'A consumer is "in equilibrium" when he is getting most for his money.' Comment on the accuracy of this statement as a definition of consumer behaviour.

b. Illustrate, with reference to indifference curves, that economics is concerned with problems of rational choice between alternative possibilities.

c. If a person was offered, as part of a pay package, a free case of wine each month, worth £100 at retail prices, or an extra £100 on the monthly salary, which should the person choose, and why? Assume that the individual pays no income tax.

Chapter 3

a. Why is the concept of demand elasticity an essential part of economic theory?

b. What explanations might be offered if more of a commodity is brought after an increase in price?

c. 'The success of modern agriculture in OECD countries has benefited everyone but the farmer.' Explain the background of this quotation in the light of demand and supply theory.

d. This notice appeared in a college handout to postgraduate students. 'Postgraduates are advised to apply early if they require one of the College's furnished flats. The demand for these flats far exceeds the number available.' Comment on that establishment's method of allocating accommodation.

e. Explain what is meant by the income elasticity of demand for a commodity and why it tends to differ for different commodities and different income levels. What is the significance to producers and consumers of the income elasticity of demand being: (i) greater than one; and (ii) negative?

f. What are the consequences for agriculture of: (i) the low price elasticity of demand; and (ii) the low income elasticity of demand for most agricultural products?

g. Discuss the implications of Engel's Law for the agricultural sector of a growing economy.

h. What changes would you expect to see in the characteristics of the demand for food, as reflected in elasticity coefficients, as a country develops and the income of its households rise? Explain the causes of these changes.

Chapter 4

a. Explain why fluctuations occur in market prices. How do fluctuations reflect the degree of competition in an industry?

b. In an industry that is virtually perfectly competitive, such as agriculture in the older Member States of the EU, increases in output by individual farmers in their quest for greater profits may be against the interests of the industry as a whole. Explain how this conflict arises, and what measures the industry can adopt to protect itself.

c. How does monopoly power arise? Is the existence of a monopoly necessarily something that is to be deplored?

d. What do you understand by 'price discrimination'? In what circumstances is it: (i) practicable; and (ii) profitable?

e. Discuss the factors which a monopolist takes into account when maximizing his or her profits. Is monopoly necessarily a bad thing?

f. Some politicians advocate price control of essential consumption goods such as energy and public transport. Comment on: (i) the desirability; and (ii) the practicality of applying price controls to basic foods such as milk and bread.

Chapter 5

a. Discuss the importance of the concept of the 'margin' in production economics.

b. What are the important production relationships which an entrepreneur must optimize if he or she wishes to generate the highest profit? What difficulties might he or she be expected to come across in aiming for this goal?

c. Economics often assume that the motivation of firms is profit maximization. Is this assumption borne out by the behaviour of firms? What other motives might bear upon their actions?

d. Describe the economies that can be achieved through large-scale production. In view of these economies, how is it that in agriculture, small firms can still flourish side by side with large ones?

e. What is the relevance of time to production?

f. Define the Law of Diminishing Marginal Returns. Discuss whether a producer should cease to expand production once his marginal returns start to diminish. Does this law apply particularly to farmers?

g. In agriculture, farms are often passed down between generations. What arguments for and against this practice can be put forward?

Chapter 6

a. 'Profits tend to equality.' Point out the ambiguities in this statement and submit a clear statement about the relations between profits in different industries and enterprises.

b. Compare the concept of transfer earnings and economic rent with that of normal and pure profits.

c. Discuss the role of the entrepreneur in the process of production.

d. Define the following terms and consider the relations between them: Wealth, Capital, Land.

e. How is factor mobility related to economic rent?

f. Examine the proposition that economic rent is likely to constitute part of the reward of *any* factor of production.

g. Robinson Crusoe's fishing would be far more effective if he were to use a net. Discuss his decision to construct a net in relation to the economic theory associated with the factors of production.

h. What is meant by 'sustainable' as applied to an economic activity? What characteristics would you look for in 'sustainable agriculture'?

Chapter 7

a. 'The market mechanism would work very well if governments would leave it alone.' Comment on the validity of this statement.

b. Describe what is meant by 'externalities'. By reference to agriculture, illustrate why they should be taken into account in decisions aimed at improving the way that the economy is organized.

c. What is a 'pure public good'? Why would a laissez-faire economic system be expected to have a smaller quantity of public goods than everyone would agree is desirable?

d. 'Governments should pay for the repair of ancient farm buildings and country churches. They are part of the national heritage and are public goods.' Comment on this statement.

e. Discuss the alternative ways in which the external costs of silage effluent can be 'internalized'. Put forward an argued case for the alternative you prefer.

f. What economic arguments could be put for compelling school children to stay in full-time education to the age of 16 years when many would wish to leave at 14 years? Imperfect knowledge and externalities should find a place in your answer.

g. What advantages and disadvantages would be expected to flow from replacing a system of state-provided education at school level with one of vouchers that parents could use to buy education at privately or publicly owned schools of their choice?

Chapter 8

a. Discuss the proposition that saving is a private virtue but a public vice.

b. In what ways can the government of a developed country influence the working of its general economy?

c. 'The multiplier is a two-edge sword. It cuts for you or against you.' Explain and discuss this statement.

d. Discuss the chief factors which will cause a rise in general prices. How are such price changes brought under control?

e. Discuss the relative merits of economic growth.

f. Outline the problems encountered when a government of a Western developed economy attempts to achieve simultaneously low employment, stable prices and economic growth.

g. Outline the factors on which a nation's standard of living depends. To what extent does a rising GNP represent increases in the nation's quality of life?

Chapter 9

a. 'It is comparative advantage, not absolute advantage, which determines the pattern of trade between countries.' Elucidate and discuss.

b. Although the law of comparative advantage shows that, given certain assumptions, free trade maximizes welfare, most nations in fact impose tariffs. What reasons are usually advanced in support of this policy, and to what extent are they justified?

c. Should the UK import some goods which it could produce at home?

d. Is it economically sound to appeal to the inhabitants of the UK to 'Buy British'?

e. 'Most arguments for protection for a home industry against competition from abroad are simply rationalizations for special benefits to particular pressure groups.' Discuss.

f. At one time the import of fresh milk into the UK for either liquid consumption or for manufacture into dairy products was, in effect, banned. What arguments might be put forward for and against such a constraint on trade?

g. What economic effects might be expected to flow from the purchase by foreigners of farms in the UK?

h. Why should a government be interested in the international exchange rate of its country's currency?

i. What are the relative advantages and disadvantages of participating in Monetary Union with a single currency?

Chapter 10

a. What is meant by the 'policy process'? Illustrate your answer with examples drawn from EU agricultural policy.

b. What are the chief economic characteristics of agriculture in the EU? What problems do they pose for government?

c. What were the objectives of the Common Agricultural Policy of the EU, as set out in 1957 and repeated in 2010? To what extent is the new set agreed in the late 1990s as part of *Agenda 2000* still appropriate for the present time?

d. Why is public policy needed to address environmental issues in industrialized economies?

e. Explain the economic concept of the 'optimum level of pollution'. What difficulties do you anticipate might be encountered in the practical use of this concept within government policy on the environment?

f. Compare and contrast the characteristics and potential use of any two of the following policy instruments: (i) standards; (ii) a tax; and (iii) a subsidy.

Appendix 2 Suggested Answers and Explanations for the Exercises following each Chapter

Chapter 1

1.1. The three missing essentials are:
   (a) Resources available to man are *scarce*.
   (b) Their allocation between alternative uses requires *choice*.
   (c) This allocation is designed to achieve certain *objectives*.
A better definition would be 'Economics is the study of how individuals and society distribute scarce resources between alternative uses in pursuit of given objectives'.

1.2. The resources of land, labour, capital and entrepreneurship available to man for the production of goods and services to satisfy his wants are limited in supply. Because they are of limited quantity man has to choose how he allocates them between alternative uses. Any good which is limited in supply relative to the wants for it (i.e. where choice of use is involved) is economically scarce. Any commodity which is not limited in supply is not economically scarce. No problems of allocation arise; sufficient quantities are available to satisfy all wants completely. Such commodities are termed 'free goods'. Example:
   Air: Only in exceptional circumstances is air limited in supply and hence is 'scarce' (e.g. in mines, submarines).
   Water: Water is normally a scarce commodity. We therefore encourage people to turn off taps rather than leave them running. Prices are used here; householders pay for water through meters and water charges, or flat charges dependent on the value of the house. The indiscriminate use of water would be reflected back eventually in higher costs. In times of drought, particular care has to be exercised in the uses to which water is put; it becomes a more scarce commodity and other methods for its allocation may need to be used (e.g. water rationing). To a man adrift on a raft on a freshwater lake water is not a scarce commodity; the problem of allocation between different uses does not arise. There is enough for all the uses the man can devise for water.
   Sand: To the UK economy sand is a scarce resource – it is limited in supply in comparison to the requirements it could serve. Prices are used as a method of allocating it. A building firm in a sandy desert could use as much as it wished.

1.3. (a) Positive.
   (b) Positive statements concern what is, was or will be, and normative statements concern what ought to be. Disagreements over positive statements can be settled by an appeal to the facts, while disagreements over normative statements cannot be settled in this way. Normative statements depend on value judgements.

1.4. Opportunity cost is the cost of making a choice in terms of forgone alternatives. More specifically the opportunity cost of using something in a particular way is the benefit forgone by not using it in its *best alternative use*. Given that the State can spend money on a hospital *or* a road, the opportunity cost of building the road is the hospital.

1.5. Which is forgone in each case?
   (a) his evening of study
   (b) his evening of table-tennis.

1.6. (a) the £10,000/year job plus convenience
   (b) the £11,000/year job.
Alternatively
   (a) the £10,000/year job
   (b) the £11,000/year job plus necessary travel.
It helps if alternative choices are thought of in terms of the satisfaction which they can give. In this example satisfaction comes not only from monetary reward but also from convenience.

1.7. In terms of forgone alternatives, the opportunity cost of growing wheat is the net revenue from potatoes. If the opportunity cost of growing wheat were greater than the net revenue produced by growing wheat, the farmer, if motivated by profit, should switch from growing wheat to growing potatoes.

1.8. The sequence of steps is as follows:

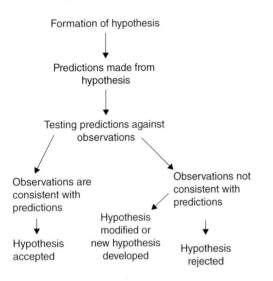

1.9. (a)
    (i) Form a hypothesis: 'Farmers on dairy farms of 100–150 ha in Devon who have invested in buildings over the last 10 years get lower net incomes now than those who did not.'
    (ii) Make a prediction: 'If farms were visited, one would find lower net incomes on farms which had invested than on those that had not.'
    (iii) Test the prediction against reality by mounting a survey of farms. Analyse the results.
    (iv) If the survey's findings agreed with the predictions, the hypothesis could be accepted.
(b) No cause-and-effect relationship has been proved, just an association shown.

1.10. Comments on the three scenarios could be as follows:
    (a) Though there may be some relationship between crime rates and the intensity of policing, there are many other factors that influence the level of crime, and changes in these are likely to have dominated the picture. For example, the likelihood of being caught on camera has increased in most areas, with a greater potential for conviction. Politicians would be advised to avoid proposing simple solutions to phenomena that have complex causes.

    (b) If there is no causal relationship between the amount spent on packaging and the number of people who smoke, it might be instructive to ask: 'Why not use the simplest forms?' Does attractive packaging influence particular groups, such as by encouraging the younger smoker to take up the habit, while not impacting on those who are well-established smokers?

    (c) Is there scientific evidence that badgers can carry TB and it is technically possible for transmission of TB from badgers to cattle? This should not be difficult to establish from simple science. On the assumption that there is scientific evidence, the issue becomes one of assessing the degree of culling that is required. A controlled experiment would be appropriate. Maybe in some areas a cull could be used and in other areas not, with a review of the results. But the issue is unlikely to be simple. What proportion of badgers need to be killed to make a significant difference to the incidence of TB in cattle? This suggests an experiment with various levels of culling is needed. And how does this compare in effectiveness and costs with alternatives that may exist, such as inoculation of badgers so that they do not develop the disease? Real-life policy choices are rarely simple.

Chapter 2

2.1. The man has a want – for his hunger to be satisfied. Goods which satisfy this want generate utility.

Utility is generated by the ability of a commodity to satisfy a want. Demand is the desire for a commodity coupled with the ability to pay for what is desired.

(i) Bread gives utility (it can satisfy the want) but we do not know if the man has the ability to pay for the loaf – demand is not established.

(ii) Bread gives utility but there is no demand because the man has no purchasing power.

(iii) Bread gives utility and there is demand – the man can afford to acquire the loaf.

(iv) Bread cannot satisfy the man's want – he is averse to it. Bread therefore generates no utility. Demand is therefore automatically absent.

(v) No utility or demand.

(vi) To a man who is fully satisfied a loaf of bread will give no additional satisfaction (i.e. it will generate no utility). Demand cannot exist if there is no desire for a commodity.

	Situation	Utility	Demand
(i)	A starving man wanting a loaf of bread	Yes	Don't know
(ii)	A starving penniless man wanting a loaf costing £10	Yes	No
(iii)	A starving wealthy man wanting a loaf costing £10	Yes	Yes
(iv)	A starving man who has an aversion to bread (the aversion is total)	No	No
(v)	A starving wealthy man who has an aversion to bread (again the aversion is total; resale is banned)	No	No
(vi)	A man whose hunger has already been fully satisfied	No	No

2.2. The total and marginal utility schedules are:

Units of X consumed per time period	Total utility	Marginal utility
1	9	9
2	21	12
3	35	14
4	50	15
5	65	15
6	79	14
7	91	12
8	100	9
9	105	5
10	105	0

When plotted they look like Figs 2.2 and 2.3 (Chapter 2) except that there is an initial rise in marginal utility before the fall sets in.

2.3. The 'Law of Diminishing Marginal Utility' states that 'the utility of additional units of a commodity to any consumer decreases as the quantity of that commodity he is already consuming increases'.

2.4. Yes, the law should really start 'Once a consumer's stock of commodity has reached a certain size, the utility of additional units, etc.'. The point at which diminishing marginal utility units sets in is called the 'origin'. In the example above the origin is 5 units.

2.5. Car wheels to a motorist. Dining chairs which form a set? Another interesting example is noise. In a totally silent environment a person alone at home all day can feel isolated and lonely. Some sound from neighbours or passing traffic can be a source of satisfaction if it relieves the feelings of being cut off. More sound in an initially very quiet environment could give increasing marginal utility but soon diminishing marginal utility would set in. As the decibels built up, a level would be reached beyond which further increases would cause a reduction in satisfaction – sound would have become noise pollution and the marginal utility of sound would have fallen to negative quantities.

2.6. Assuming that the Law of Diminishing Marginal Utility applies to both goods A and B, for maximum satisfaction this consumer should buy *more* of A and *less* of B. This adjustment of the purchasing pattern will make the ratio of *marginal utilities closer to* the ratio of *prices*.

2.7. (a) The indifference curves should be similar to Fig. 2.5 in Chapter 2.

(b) The marginal rate of substitution (MRS) is the number of units of Y which have to be given up for an increase of 1 unit of X to maintain a constant level of satisfaction. The MRS on:

I_1 between X5 and X10 = 2.0
I_2 between X15 and X20 = 1.2
I_3 between X20 and X25 = 1.2

(c) Budget lines are shown in Figs 2.7 and 2.8 in Chapter 2. The different levels of income correspond with three parallel budget lines, the one for lowest income nearest the origin and with higher incomes further away from it. The change in relative prices (the last three options shown) alters the slope of the budget line. Reading off the points of tangency gives the following quantities.

Income per week (£)	Price of goods (each)		Quantity purchased per week	
	X	Y	X	Y
60	£2	£2	10	20
88	£2	£2	20	24
110	£2	£2	25	30
110	£2	£3	25	20
110	£2	£6	25	10

2.8. From the final three rows of the table shown in 2.7(c) the demand schedule for commodity Y is

Price of Y	Quantity demanded
£2	30
£3	20
£6	10

2.9. A consumer is said to be in equilibrium when the *total utility* from his purchases is *maximized*. This also implies that the satisfaction he derives from his purchases is *maximized*. He allocates his spending, probably subconsciously, to this end by practising the Principle of *Equimarginal Returns*. The consumer's equilibrium is termed *stable* since any departure from this position caused by external forces will call into play forces which tend to restore the equilibrium position.

Chapter 3

3.1. The total market demand for onions is found by adding together the demands of the three income groups at each price:

Price per kilogram (£)	Demand (thousand kg/week)			
	Rich group	Middle group	Poor group	Total market demand
1.0	4	10	10	24
0.9	5	12	12	29
0.8	6	15	16	37
0.7	6	18	23	47
0.6	6	22	30	58
0.5	5	26	40	71
0.4	5	29	52	86

3.2. If the price elasticity of demand of a good is equal to minus 3, then a 1% fall will *raise* the quantity *demanded* by 3%. The total revenue of the sellers of the good will *increase* and the total expenditure of the buyers of the good will *increase*.

3.3. Income elasticity of demand of a good (see equation at bottom of page).

The percentage change in expenditure on housing is given (20%).

The percentage change in income is $400/2000 \times 100 = 20\%$

$E_{Dy} = 20/20 = 1$

3.4. Engel's Law states that 'the proportion of personal expenditure devoted to necessities decreases as income rises'. Statements (b) and (c) are compatible with this law. Statement (a) is not because, in absolute terms, the same or more may be spent on essentials, although their relative proportion of total spending will decline.

3.5. The elasticity of supply with respect to product price is estimated as the percentage change in quantity supplied per time period *divided* by the percentage change in price. The elasticity of supply is generally *lower* in the short run than in the long run, and this implies that the long-run supply curve is generally *less* steep than the short-run curve. In addition, the elasticity of supply will be *lower* if the producer has no alternative lines of production open to him; if the production cycle is short the elasticity of supply will be *greater* than if it is long.

$$\text{Income elasticity of demand } (E_{Dy}) = \frac{\text{Percentage change in quantity demanded}}{\text{Percentage change in income}}$$

3.6. **(a)** Any three of the following:

- Change in consumer incomes.
- Change in prices of complementary or competitive goods.
- Change in tastes of consumers.
- Change in population size.

(b) Any three of the following:

- Change in the prices of other goods the firms could produce.
- Change in the cost of factors of production.
- Change in price of joint products.
- Change in the state of technology.
- Change in the goals of firms.

3.7. **(a)** Plotting the schedules produces a 'scissors' graph.

(b) The equilibrium price is 5p.

(c) **(i)** At a market price of 6p demand would not be sufficient to clear the quantity of carrots supplied to the market. With time this surplus would build up until such time as the government decided either to drop the price or to remove the surplus by physical destruction or the operation of a discriminating monopoly (including dumping abroad).

(ii) A market price of 4p would cause demand at that price to exceed supply. Methods other than the price system would arise to distribute the carrots, e.g. queuing, rationing, and a black-market might spring up.

(d) **(i)** The government will need to purchase 53m kg/week.

(ii) Its expenditure (price × quantity) will be £3.18m.

(iii) The demand curve shows that consumers will demand this quantity only if the price is 4p/kg.

(iv) The government will receive from selling carrots (price × quantity) £2.12m.

(v) The government's loss is expenditure minus revenue = £0.06m loss.

(e) If demand increases by 4 million kg at all prices, the demand curve will move uniformly to the right by 4 million kg. As demand has changed without any shift in the supply curve, the supply curve is cut at a higher price. The new equilibrium price would be 5.5p and the quantity sold 51 million kg.

(f) The conventional method of showing the effect of such tax is to shift vertically the supply curve by the extent of the tax. This assumes that producers pay the tax and add it to the price at which they sell. If producers were willing to supply 60 million kg at 8p before the imposition of a tax of 1p/kg, then after the imposition of the tax the price which the market will need to pay to cause farmers still to produce 60 million kg will be 8p + 1p = 9p. This is repeated at all levels of output. Thus a new supply curve is created above the original. This new supply curve cuts the original demand curve at 5.5p, when 47 million kg will be bought.

NB: When the market price is 5.5p, the government will raise 1p × 47 million kg = 47 million p revenue from the tax. Producers will receive 4.5p of the market price. This is 0.5p less than they were receiving before the tax imposition. Consumers are paying 0.5p more than they were before the tax (5.5p as opposed to 5p). The *incidence* of the tax in this case is said to fall equally on the producer and the consumer, as both suffer a similarly disadvantageous price change of 0.5p/kg.

(g) Farmers given a grant of 2p/kg will be willing to supply to the market at 7p that quantity they were previously willing to supply at 9p and so on. A new market supply curve will be created vertically below the original one. This will cut the demand curve at a market price of 4p, when 53 million kg will be bought.

(h) **(i)** If the demand for carrots has a negative income elasticity of demand, an increase in consumers' incomes will cause demand for carrots to contract (i.e. the demand curve will shift to the left). From the scissors graph it can be seen that such a shift will cause prices to fall.

(ii) The argument is the reverse of (i).

3.8. **(a)** Slope of this straight line is –1.

(b) Price elasticity of demand (see equation at bottom of page).

$$\text{Price elasticity of demand } (E_{Dp}) = \frac{\text{Percentage change in quantity demanded}}{\text{Percentage change in price}}$$

The prices are given. The quantities demanded at these prices must be read from the graph.

  (i) –11.4
  (ii) –2.6
  (iii) –0.45

From i, ii and iii we can say: 'The price elasticity of a straight-line demand curve is *not* constant along its whole length' (i.e. the statement in the question is false).

  (c) Using the formula given, which measures the elasticity at single points, the following results are obtained:

  (i) at price 2 units EDp is –0.25
  (ii) at price 6 units EDp is –1.5
  (iii) at price 8 units EDp is –4.0

Where two curves cross (i.e. price and quantity demanded are the same for both curves) the price elasticity of demand of each curve is inversely proportional to the slope of the curve. The curve with the steeper slope has the elasticity coefficient nearer to zero (i.e. is the less elastic). Hence the second sentence of the statement should read thus: 'At the point of intersection the steeper curve possesses the *lower* price elasticity of demand, and is therefore said to be the *less* elastic of the two'.

3.9. Stable prices imply that supply has to expand at the same rate as does demand. We therefore have to estimate the change in quantity that would be demanded assuming prices remain the same. At the start of the 10-year period the average consumption of the 10 million people is 0.1 t of cereals/year. At the end of the period there are 11 million people, but they will consume more at given prices because of their rise in income. With an income 30% higher, and an income elasticity of demand of 0.5, their consumption per head will be 15% higher (30×0.5). In absolute terms this is 1.15×0.1 t = 0.115 t. With the larger population the total amount demanded will be 11 million people $\times$ 0.115 t, or 1.265 million t. Thus supply has to expand by 0.265 million t (or 26.5%) to keep prices at an unchanged level.

Chapter 4

4.1. In a perfectly competitive industry:
  (a) there are *many* producers
  (b) entry to the industry is *free*
  (c) each producer *cannot* influence market price.

In a complete monopoly:
  (a) there is *one* producer
  (b) entry to the industry is *restricted*
  (c) the producer has *considerable* influence on price.

To a producer in a perfectly competitive industry:
  (a) marginal revenue *is* constant with varying levels of output
  (b) marginal revenue *is* identical with average revenue
  (c) the demand curve which he faces is infinitely *elastic*.

To a producer in an imperfectly competitive industry:
  (a) marginal revenue *is not* constant with varying levels of output
  (b) marginal revenue is normally *less* than average revenue
  (c) the demand curve for his products is normally *down-sloping*.

A perfectly competitive industry normally faces a demand curve which is *less than infinitely* elastic.

The marginal revenue of a perfectly competitive industry normally *declines* with increasing output.

4.2. Schedules for the sale of rubber boots.

Price per pair (average revenue) (£)	Quantity demanded/week	Total revenue (£)	Marginal revenue (£)
2.75	1	2.75	2.75
2.45	2	4.90	2.15
2.20	3	6.60	1.70
1.90	4	7.60	1.00
1.65	5	8.25	0.65
1.45	6	8.70	0.45
1.20	7	8.40	–0.30
1.00	8	8.00	–0.40
0.80	9	7.20	–0.80
0.60	10	6.00	–1.20

  (a) Price per pair that the shopkeeper receives is identical with his average revenue for pairs of boots. Hence plotting average revenue and marginal revenue involves the plotting of the first and fourth columns above.
  (b) The price of boots when marginal revenue is £1.00 is £1.90 and the quantity sold at this price is four pairs. This answer can be obtained either directly from the table, or from the graph. The marginal cost of boots is constant in this case at £1, so that

| National income | Agriculture | | Other industries | |
	Receives in exchange for its products	Percentage of national resources receiving this	Receive in exchange for their products	Percentage of national resources receiving this
£100m	£50m	50%	£50m	50%
£110m	£52m	50%	£58m	50%

the shopkeeper's marginal cost curve is a horizontal straight line at £1 on the vertical axis. This line will cut the MR curve, and the quantity sold and average revenue (price) can be read.

The AR curve is the same as the demand curve faced by the producer (shopkeeper). It shows the relationship between price and quantity sold. The slope of the demand curve at price £1.90 is about –0.22. (The exact measurement of the slope is unimportant.) Using the formula (EDp = 1/slope × P/Q) to estimate the price elasticity of demand gives a result of about –2.2. Hence the price elasticity of demand at this most profitable level of sales is greater than (–) 1. This agrees with the prediction that the most profitable level of output for a monopolist occurs when he has restricted his output where the demand for his produce is elastic (i.e. price elasticity of demand coefficient greater than 1). His MR at this point will be positive. If the agricultural *industry* is like a monopolist in the supply of food, and if the price elasticity of demand for its products is between –1 and 0 can you suggest what it could do to increase its profit?

4.3. Distribution of national income by spenders can be found in the table at the top of the page.

A 10% rise in national income is synonymous with a 10% rise in the income of consumers making up the nation. This will cause an increase in demand, the size of the increase for each product being reflected in the particular income elasticity of demand for that product (see equation at bottom of page).

By rearranging the formula above we can calculate the resulting change in demands for agricultural and other products.

Agriculture:
Percentage change in demand = 0.4 × 10% = 4%
Other industries:

Percentage change in demand = 1.6 × 10% = 16%
Expansion in expenditure:
Agriculture: a 4% increase on £50m is £2m (**a**).
Other industries: a 16% increase on £50m is £8m (**b**).

Thus the final distribution of national income by spenders is £52m on agriculture (**c**) and £58m on the other industries (**d**).

After the increase in national income the half of the nation's resources used in agriculture is attracting a lower share of the national expenditure than the half in the other industries. In the initial state the resources inside and outside agriculture were rewarded to the same extent (i.e. there was parity of reward between the two sectors of this simplified economy) whereas after the rise in national income this parity is destroyed, with the 50% of national resources used in 'other industries' attracting more than 50% of total expenditure. This is reflected in higher wages, interest rates, rents and profits in 'other industries'. One would expect resources (land, labour, capital, management) to be attracted from agriculture by these higher returns. However, resources used in agriculture are notoriously slow to move to other forms of production, resulting in the rewards in other industries being persistently higher and leading to demands for income support to agriculture. Thus the statement in 4.3 should read:

> After the rise in national income the half of the nation's resources in agriculture is earning *less* than half in the other industries. Resources should be transferred *from* agriculture *to* the other industries if the returns to the nation's resources are to be equated in each sector.

4.4. If a firm in perfect competition is attempting to maximize profits it should produce that level of output where the addition to total costs caused by producing the last unit of output (called *marginal* cost of production) just equals *marginal* revenue. In perfect

$$\text{Income elasticity of demand } (E_{Dy}) = \frac{\text{Percentage change in quantity demanded}}{\text{Percentage change in income}}$$

competition this will also *equal* the price of the product. A monopolist *will* achieve maximum profits by pursuing the same policy and in his case the marginal revenue at his optimum level of output will be *less than* the price which he charges for his product.

4.5. Compared with a non-discriminating monopolist, a monopolist engaged in price discrimination will generally have a *greater* revenue. This is because the discriminating monopolist equates the *marginal* revenue for his product in each of the markets between which he discriminates.

4.6. (a) The two conditions are: (i) two or more markets for the monopolist's product which have different demand characteristics; and (ii) the ability to prevent 'leaks' or 'seepage' between the markets by legal, time, distance or other constraints.

(i) 30p (you will need to read this from the graph you drew)
(ii) 4 gallons
(iii) 22p
(iv) 8 gallons
(v) £2.96

Chapter 5

5.1. (a) Refer back to Fig. 5.1 in Chapter 5 for the relevant set of curves.
(b) (i) At the point of inflection of the production function *marginal* product is at its maximum.
(ii) At the point where a line from the origin is tangential to the production function *marginal* product and *average* product are equal.
(iii) When total product is maximal *marginal* product is zero.
(iv) When total product is declining with increasing quantities of input *marginal* product is negative and *average* product is positive and *declining*.
(v) The point where marginal product is maximal is also called the point of *diminishing (marginal) returns*.
(vi) Where total product has been reduced to zero by using a very high level of input. Average product is also zero. An example might be where extremely high levels of nitrogen fertilizer are used on cereals. Lodging of the crop,

(b)

Liquid consumption				Cheese making			
Quantity sold (gal/day)	Average revenue (£)	Total revenue (£)	Marginal revenue (£)	Quantity sold (gal/day)	Average revenue (£)	Total revenue (£)	Marginal revenue (£)
1	0.45	0.45	0.45	1	0.29	0.29	0.29
2	0.40	0.80	0.35	2	0.28	0.56	0.27
3	0.35	1.05	0.25	3	0.27	0.83	0.25
4	0.30	1.20	0.15	4	0.26	1.04	0.23
5	0.25	1.25	0.05	5	0.25	1.25	0.21
6	0.20	1.20	−0.05	6	0.24	1.44	0.19
7	0.15	1.05	−0.15	7	0.23	1.61	0.17
8	0.10	0.80	−0.25	8	0.22	1.76	0.15
9	0.5	0.45	−0.35	9	0.21	1.89	0.13
				10	0.20	2.00	0.11
				11	0.19	2.09	0.09
				12	0.18	2.16	0.07
				13	0.17	2.21	0.05
				14	0.16	2.24	0.03
				15	0.15	2.25	0.01
				16	0.14	2.24	−0.01
				17	0.13	2.21	−0.03
				18	0.12	2.16	−0.05
				19	0.11	2.09	−0.07
				20	0.10	2.00	−0.09

disease and killing by excessive nitrogen may reduce the harvestable crop to nothing.

5.2. The completed table can be found at the bottom of the page.

  (a) The firm is producing under perfect competition as its MR does not fail with increasing output.

  (b) Fixed costs are those which it has to bear whether or not it produces anything, and so are its costs when output is zero. Total costs when output is zero are all fixed costs, and are £110.

  (c) The graph will appear similar to Fig. 5.22 in Chapter 5. Some allowance must be made in interpreting the results shown below for the slightly different ways in which curves may be drawn by hand.

  (d) A firm is in equilibrium when *marginal* cost equals *marginal* revenue. (Marginal cost is the increase in total cost caused by the last unit of output and marginal revenue is the increase in total revenue caused by the last unit of output.)

  (e) 8 units of output.

  (f) (i) Total revenue (£400) minus total costs (£269) = £131.

   (ii) Average revenue (£50) minus average total costs (£33.6) multiplied by the number of units of output (8). (£50 − £32.6) × 8 = £131.2) This is the maximum profit that can be earned by the firm. Try subtracting total costs from total revenue at other levels of output. Lower profits will always be found.

  (g) The price of an article is synonymous with its average revenue. In the long run a firm will need to cover its fixed costs (e.g. capital depreciation, rent and rates, regular labour force, interest charges on borrowed capital), and if the average price it receives for its product is not sufficient to at least cover the fixed and variable costs, it will eventually go out of production. From your graph it may be seen that if the price of its product is about £31, then the firm will just cover its fixed plus variable costs if it produces 7 units of output. If the price were below £31, at no level of output would its average total costs be covered. At higher prices the firm will have a range of outputs where price (average revenue) exceeds average total costs, and within this range is situated the most profitable level of output. At this most profitable level, average total cost will not be at its lowest point, but on the rise. So the answer in the long run (i) is: minimum long run price £31; quantity produced 7 units. (ii) In the short term production will continue as long as variable costs are covered. This will happen as long as the price is above £14, at which 6 units will be produced.

5.3. (a) The isoquants should be drawn. It will be similar to Fig. 5.10 of Chapter 5.

  (b) An iso-cost line must be constructed. This connects the quantities of oats and barley which can be bought for the same expenditure. If the prices of the cereals are equal at £30/t, then say £120 will buy 4 t of each, or any combination of the two lying along the iso-cost line connecting 4 t of barley to 4 t of oats. Iso-cost lines further away from the origin represent higher levels of cost, but the slopes will not alter as long as the relative

Output (units)	Total revenue	Marginal revenue	Total costs	Fixed cost	Variable cost	Marginal cost	Average variable cost	Average total cost
0			110	110	0			
1	50	50	140	110	30	30	30	140
2	100	50	162	110	52	22	26	81
3	150	50	175	110	65	13	21.7	58.3
4	200	50	180	110	70	5	17.5	45
5	250	50	185	110	75	5	15	37
6	300	50	194	110	84	9	14	32.3
7	350	80	219	110	109	25	15.6	31.3
8	400	50	269	110	159	50	19.9	33.6
9	450	80	349	110	239	80	26.5	38.8

Answers for the Exercises

prices of oats and barley are constant. Hence, once one iso-cost line has been drawn it may be slid parallel until it is tangential to an isoquant, and the optimum combination of the two cereals read off. This is the combination which produces the quantity of eggs represented by the isoquant at the lowest cost. Other combinations *may* be used, but these will correspond with iso-cost lines which are further from the origin (i.e. more costly) and which *cut* the isoquant.

Optimum quantity to feed to achieve 230 eggs/bird/year is:

oats 3.6 t; barley 2.6 t

(c) Relative proportions: Oats/barley = 3.6/2.6 = 1/0.7

(d) Oats: 3.6 × £30 = £108. Barley: 2.6 × £30 = £78. Total cost = £186

(e) Slide the original iso-cost line until it is tangential to the second isoquant. The inputs at the point of tangency are; oats 4.2 t, barley 3.6 t. The proportions have changed in this higher production diet as well as the absolute quantities. Oats/barley is now 1/0.86. This change is caused by difference between the isoquants. It is unlikely that the proportions will remain the same as production is increased.

(f) A new iso-cost line must be constructed. The new optimum is; oats 5 t, barley 1.6 t.

(g) As might be expected, a fall in the price of oats has caused more of them to be used. (Compare the quantities in (f) with those in (b) which produce the same number of eggs.) If barley is still at £30/t, then the total cost of the diet will have fallen. The optimum inputs are now: oats 5 t × £15/ton

= £75, barley 1.6 t × £30 = £48. Total costs £123 (as opposed to £186 previously).

5.4. (a) The production possibility curve you plot will be similar to Fig. 5.16.

(b) (i) The MRS of products at one level of resource use is the change in the number of units of product on the vertical axis which occurs when production of the product on the horizontal axis is increased by 1 unit. With products which compete for the firm's resources, an increase in production of one product is accompanied by a fall in the other, and the MRS is negative (i.e. a negative change divided by a positive change). Over the range quoted in this part of the question hay and wheat are complementary, an increase in hay production also producing an increase in wheat production (see first equation at bottom of page).

(ii) Over this range the products are competitive (see second equation at bottom of page).

(c) To find the most profitable combination of products an iso-revenue line must be constructed. It links the quantities of hay and wheat which bring in the same revenue. All combinations of hay and wheat lying on this line will also bring in that same revenue. Iso-revenue lines further from the origin represent higher (and therefore more desirable) levels of revenue. The optimum combination of products is where the production possibility curve touches an iso-revenue line furthest from the origin.

If wheat and hay are the same price per tonne, then an iso-revenue line would connect (say) 200 t

$$\text{Marginal rate of substitution of products} = \frac{\text{Change in } y}{\text{Change in } x}$$

$$= \frac{(+)50\,t}{(-)50\,t}$$

$$= (-)1$$

$$\text{Marginal rate of substitution of products} = \frac{\text{Change in } y}{\text{Change in } x}$$

$$= \frac{(-)60\,t}{(+)50\,t}$$

$$= (-)1.2$$

of hay with 200 t of wheat, as both these quantities would bring in the same revenue. This line may be slid parallel away from the origin to represent higher iso-revenue lines. Where the production possibility curve just touches an iso-revenue line is the optimum combination of products. Note that it is the *relative* prices of the products which determines the slope of the iso-revenue line and hence the optimum combination.

(i) Wheat 260 t; hay 100 t

(ii) A new iso-revenue line is required from (say) 100 t wheat to 200 t of hay – both of these will bring in the same revenue. Proceed as above. Wheat 260 t; hay 100 t

(iii) As (ii) above; wheat 260 t; hay 100 t

(d) If hay produces no revenue than an infinite amount of it will be necessary to produce the same revenue as, say, 200 t wheat. In this case, therefore, the iso-revenue line is horizontal and can be slid up and down to find the optimum combination of the two products, which will be 260 t of wheat and 100 t of hay. Even when hay produces no revenue it will be desirable to produce 100 t, as this increases wheat production.

For convenience, this example has been constructed with a sharp angle to the production possibility curve; this produces identical answers for the various price levels quoted. In practice a smoother curve is likely and the optimum combinations are likely to be more sensitive to changes in relative prices.

5.5. Option c is the best description of a fixed cost.

5.6. (a) If marginal cost is rising with increasing output, average total cost *may be rising or falling*. Average fixed cost will be *falling*. The minimum price which the entrepreneur will be able to accept in the short run will correspond to the lowest point on the *average variable cost* curve, but in the long run price must be at least the lowest point on the *average total cost* curve. The rising marginal cost curve cuts both the *average variable cost* curve and the *average total cost* curve at their lowest points. To make maximum profits an entrepreneur will select that level of output at which *marginal revenue* equates with *marginal cost*.

For sources of economies (**b**) and diseconomies (**c**) of scale, refer to the text to check your answers.

5.7. (a) The essential nature of a technological advance is that it improves the relationship between inputs and outputs so that, at the level of the individual firm, the margin between costs and revenues is greater than under the old system.

(b) A technological advance:
(i) shifts the production functions upwards (giving a greater volume of output at each level of variable input) and lowers average and marginal cost curves.
(ii) shifts the iso-quant nearer to the origin.
(iii) shifts the production possibility boundary out further from the origin.

(c) Factors that affect the speed at which new technology is taken up by the operators of firms include their capital requirements, management requirements, risk impacts, and scale of operation requirements.

(d) Characteristics of entrepreneurs that tend to make them faster adopters of new technology include attitude to risk, ability to withstand risk (which will in turn reflect size of business and capital base), age, education, and good contact with change agents (disseminators of information such as consultants and technical advisers).

(e) The correct components of this description of the 'treadmill' of technological advance, as expounded by Cochrane, are:

The *average* farmer is on a treadmill with respect to technological advance. In the quest for *increased returns*, or the minimization of losses ... he runs *faster and faster* on the treadmill. But by running *faster* he does not reach the goal of *increased returns*; the treadmill simply turns over *faster*.

(f) Likely implications of the treadmill of technological advance for the structure of the agricultural industry include: reduction in the numbers of small businesses and the increase in the number of large ones, concentration of commodity production among large producers, increase of diversification into non-farm types of production, and shedding of labour from agriculture.

Chapter 6

6.1. The demand for a factor of production is a derived demand as the factor is not wanted for its own sake but for the contribution it makes to the production of goods which *are* wanted for the satisfaction of wants. The demand for factors is derived from the demand for these want-satisfying goods. Anything which affects the demand for these goods will affect the demand for the factors producing them. The products of one firm (e.g. tractors) may be factors of production of other firms (e.g. farmers). The demand for steel for tractor making is derived from the demand for tractors which in turn is derived from the demand for agricultural products.

6.2. Five different goods from which timber derives its demand: (**i**) furniture; (**ii**) houses; (**iii**) boats; (**iv**) garden sheds; and (**v**) magazines (via paper).

6.3. The price elasticity of demand for a factor depends on:

- the price elasticity of demand for the product;
- the cost of the factor relative to total costs of production; and
- the presence (or absence) of good substitutes for the factor.

> (**a**) Rises in the price of glass will cause prices of glasshouses to rise. A price-elastic demand for glasshouses will result in a considerable fall-off in the quantity of glasshouses sold, which will be reflected as a fall in the quantity of glass demanded. The more elastic the demand for glasshouses, the more elastic will be the demand for glass.
>
> (**b**) If the cost of water represents only a very small part of the total costs of bread making, the price of water can vary a lot before it significantly alters the price of bread and the quantity of bread demanded and subsequently is reflected back in a reduced demand for water. The demand curve for water is hence very steep and the demand price inelastic.
>
> (**c**) The presence of good substitutes means that producers will switch between factors as the prices of competing fertilizer brands change. Small changes in the price of one brand cause large changes in the quantity demanded – demand is price elastic. For fertilizer in general, however, there is no good substitute and the general level of price can fluctuate widely with relatively small changes occurring in the quantity demanded.

6.4. An entrepreneur with a given quantity of labour available to distribute between the various enterprises of his business, must so allocate it that the *last*

hour used in each enterprise yields *the same* values of output. This will of course *not mean* that the same amount of labour must be devoted to each enterprise.

This is an example of the Principle of Equimarginal Returns. If labour were not allocated in this way the farmer would gain by moving labour from where it was earning a low marginal revenue product to where it was earning a higher marginal revenue product.

6.5. To use the Principle of Equimarginal Returns one must derive the marginal products of each day of cultivation on each field from the total product figures (yield) given. The marginal product of, say, the third day of cultivation is the yield from 3 days' work minus the yield from 2 days' work and so on. Marginal products of labour:

Days of cultivation	Potatoes (t)		
	Field A	Field B	Field C
0			
1	15	20	10
2	10	10	7
3	8	9	6
4	4	6	1

To derive the best use of his 8 days cultivation time he must obviously use all 8 days. The Table readily shows what the time allocation would be if he had only 5 days, as the marginal product of a day's cultivation is 10 t of potatoes on each field if he spends 2 days on A, 2 days on B and 1 day on C. The best use of the 6th day would be on B, as its marginal product is greatest (9 t). The 7th day would be best spent on giving field A a third day of cultivation (marginal product 8 t) and the 8th day best spent on C (marginal product 7 t). The best allocation of 8 days of cultivation is thus: field A, 3 days; field B, 3 days; and field C, 2 days. The total yield is 96 t from A, 79 t from B and 82 t from C, totalling 257 t. Any other time allocation will give a lower overall yield.

6.6. (**a**) Farmer A is geographically *mobile*, as he is willing to move around the country, but occupationally *immobile* as he is unwilling to leave sheep farming.

(**b**) His labour is specific – he cannot be switched easily to other industries (or even within farming it appears).

(**c**) Specific labour is often geographically mobile – e.g. the brain drain to USA, opera singers jet-hop around the world's opera houses.

(d) Non-specific labour is usually the more occupationally mobile – e.g. a labourer can work in many industries.

(e) The more specific are: (i) road roller; (ii) moorland; (iii) older worker. They are specific because they cannot be easily switched to uses other than those in which they are normally found.

6.7. (a) Economic rent is the surplus received by a factor above its transfer earnings.

(b) A £100,000/year actor who has a best-paid alternative employment yielding £20,000/year has earnings which consist of (i) £20,000/year transfer earnings, and (ii) £80,000/year economic rent.

(c) The actor receives economic rent in the payment for his services because: (i) his services are economically scarce (they are limited in supply relative to the desire for them); and (ii) his services are inelastic in supply. If demand for his services increases, the price of his services will rise, and an even larger proportion than before will consist of economic rent.

(d) Other people will be attracted into the theatrical profession in the long run by higher economic rents if these contain an element of pure profit. Incomes will tend to be pushed down and so economic rents in the profession as a whole will decline. However, for individual good actors there are no perfect substitutes as their services are always scarce (simply because there is demand for their unique talents).

6.8. (a) By definition, risks are possibilities of future loss which can be formally insured against. Entrepreneurs thus can remove risk by means of insurance policies.

(b) Two sources of risk in agriculture, both insurable, are loss by fire and loss by personal accident.

(c) Sources of uncertainty include:
   (i) Changes in price and/or costs between when production is initiated and when its products are ready for sale. The longer the production period the greater the uncertainty, *ceteris paribus*.

(ii) Failure of physical production (yields, etc.) caused by weather, disease, etc.

(iii) Changes in government policy.

(iv) Changes in the other sectors with which farmers trade directly or indirectly. The invention of new man-made fibres may reduce the demand for wool. Vegetable, milk and cream substitutes may cut back the demand for the natural products, etc.

(v) New production processes may be developed in agriculture which increase production, lower the price and place those farmers not adopting the new techniques in untenable positions (e.g. the widespread use of yard and parlour systems cf. cows housed in milking sheds; bulk handling of grain made bigger combines uneconomic to operate).

(vi) The personal circumstances if the farmer and family have a great influence on the farm business (e.g. death of a father may necessitate sale of part of the farm to pay capital taxes).

(d) Reduction of uncertainty is brought about by:
   (i) Choice of enterprises which are relatively safe.

(ii) Diversification within the farm business (i.e. having more than one enterprise on the farm).

(iii) Production flexibility including: (i) cost flexibility (e.g. the choice of a system of production with low fixed costs enables an entrepreneur to scrap an enterprise more easily); (ii) product flexibility (processes which can switch products are less subject to uncertainty); and (iii) time flexibility (enterprises with short production cycles tend to be safer than those with long cycles, *ceteris paribus*).

(iv) Informal insurances (e.g. potato blight can be controlled by preventative spraying).

(v) Contracting and vertical integration: contracts both for inputs and products can limit cost and price uncertainty, and ensure supplies and outlets for products.

(vi) The use of economic and statistical data and research: any producer who is not aware of the economic characteristics of his product – price cycles, long-term trends, etc. – is in a worse position to react to changes than his informed fellow-producers.

(vii) Hedging: dealing in futures.

6.9. (a) When prices and costs are stable an industry in perfect competition will so arrange itself that its entrepreneurs are earning *normal* profits. If the beef-producing industry is in such an equilibrium, and the demand for beef suddenly increases because of an increase in consumers' incomes, in the short term beef producers will earn *surplus* profits. New entrepreneurs will be attracted into the industry and force *down surplus* profits until a level is restored where entrepreneurs are again earning *normal* profits.

(b) Normal profit is the payment necessary to keep an entrepreneur in a particular line of production.

(c) Normal profit is a return to entrepreneurship and so interest on capital will have already been removed.

(d) He or she will leave the industry if they persistently earn less than normal profits.

(e) Entrepreneurs will require a high normal profit in industries where the possibility of a loss is high. The factors of production put into making a record which flops are lost. Differences in uncertainty determine differences in normal profits between industries and enterprises within industries.

Chapter 7

7.1. Reasons why governments may wish to intervene in markets to change the allocation of resources and outcomes include: (i) market failure (correcting for externalities and lack of public goods, or consumer lack of information, or monopoly power); (ii) reasons of equity (fairness); or (iii) to regulate the economy (assuming its self-correcting forces need assistance).

7.2. The market mechanism is best suited to handling *private* good and services. The pure private good displays *rivalness* and *excludability*. In contrast the pure public good displays *non-rivalness* and *non-excludability*.

7.3. Defence and lighthouses are examples of public goods. The others are private goods. Despite the label, seats/places on public buses are both rival and excludable. Similarly, dental treatment is a private good; the market mechanism could work perfectly well to supply dental services. Governments may choose to make payments through the tax system and the national health service rather than have people pay directly, but rivalness and excludability both apply.

7.4. The presence of externalities to a particular production process means that, in order to achieve an optimal level of production from the point of view of society, governments will need to *either reduce or increase production depending on circumstances*.

7.5. Taking into account the effect of external costs of private transport to society will *raise* the supply curve based on private costs of production. If there are external benefits from consumption of education, the demand curve for educational services from society's viewpoint will be *above* that from individuals and families.

7.6. (a) All except the cost of the car can be regarded as transactions costs.

(b) A single currency operating in both the UK and Germany will remove costs of currency exchange and reduce the risk associated with the price that has to be paid for the imported good (expressed in home currency).

(c) The likely outcome on the demand for German cars in the UK of provision on the Internet of information on procedures for importing vehicles will be to increase it, because transaction costs (in this case the time spent on collecting information on how to do it) will fall.

(d) (i) International trade will expand and (ii) overall the welfare of people in both countries as a whole will increase. This does not deny that some individual may be made less well off, at least in the short term, but overall society can be expected to gain (following the logic of international specialization and exchange).

Chapter 8

8.1. In the circular flow of income, the following are injections (I), withdrawals (W) or neither (N):

(a) Saving of wages by workers – W

(b) Wages paid to civil servants by the government – I

(c) Spending by foreigners on our exports – I
(d) Income tax – W
(e) Investment by a firm in new machinery – I
(f) A gift of £5 from a father out of his spending money to his son – N (This is simply a transfer of spending power, though if the son saves it, a withdrawal will occur.)
(g) Spending of wages by workers – N

8.2. The marginal propensity to withdraw is the *proportion* of an increase in *total* income which is withdrawn by saving, taxation, purchase of imports, etc.

8.3. The relationship between the marginal propensity to save (MPS) and the multiplier is thus: Multiplier = 1/MPS (in this case 1/0.25). This means that the effect of £1 extra spending in the economy by investment or any other form of injection will cause total spending to rise by £4.

8.4. National income will be in equilibrium when the intended level of injections is *equal to* the intended level of withdrawals. This equilibrium level *need not* correspond to one at which all the nation's productive resources are fully employed.

8.5. (a) We are given some important pieces of information about the economy illustrated in the diagram. The first is that the national income is in equilibrium; this means that savings (S) and investment (I) must be equal because injections and withdrawals must be the same (other forms of injections and withdrawals are excluded in this simplified economy). If consumption spending (C) is £600,000 out of a national income of £1m, then £400,000 must be saved (S). Because national income is in equilibrium, this means that I also is £400,000.

(b) Of the £1m national income, £600,000 is spent on consumption, i.e. 0.6 of the total. Average propensity to consume (APC) is therefore 0.6. As the question states that APC = MPC, MPC is therefore 0.6.

(c) With a national income of £1m and an MPC of 0.8 where MPC = APC, £800,000 will be spent on consumption (C) and £200,000 saved (S). An equilibrium in national income will mean that I will also need to be £200,000.

(d) For equilibrium S must equal I, i.e. S must be £400,000. What level of income will give this level of S? If MPC is 0.8, MPS (marginal propensity to save) must be 0.2, i.e. 0.2 of each £1 of income will be saved. The level of national income (Y) to give £400,000 of saving is:

Y = £400,000/0.2 = £2m

Consumption spending (C) will be:

Y – S = £2m – £400,000 = £1.6m

8.6. (a) True – the C–C line rises as national income (Y) increases.
(b) True – at levels of Y below L.
(c) False – when consumption (C) is greater than (Y), the difference must be made up by drawing on reserves, i.e. dis-saving or negative saving.
(d) False – the proportion (the average propensity to consume or APC) falls as income rises.
(e) False – in this case the C–C line is straight. If you were to measure how much consumption increased with successive small rises in income you would find that the same proportions of the extra income was always consumed and saved (i.e. MPC and MPS are constant).
(f) True.
(g) True.

8.7. (a) Raise – relaxation of the controls on borrowing encourages consumer spending; an enlarged C will raise all the other curves.
(b) Lower – C will contract and maybe I too, taking the other curves with it.
(c) Raise – the G component of C + I + G will expand.
(d) Lower – C will contract.
(e) Lower – C and I will both contract because of the higher cost of personal and business loans.
(f) Raise – for the opposite reason to (e).
(g) Raise – C expands.

8.8. (a) The deflationary gap is the shortfall in aggregate demand, measured at the full employment level of national income by which demand would need to be expanded to achieve full employment; RS in the diagram.
(b) Conversely to (a), the inflationary gap is QR. It is the amount by which aggregate demand should be reduced to produce an

equilibrium national income at the full employment level.

(c) Prices will rise due to demand exceeding output at the initially prevailing prices.

8.9. The most promising alternative is (b) as it will lower the level of aggregate demand. (a) will increase aggregate demand and the inflationary gap. (c) will also increase aggregate demand; some of the tax payment will be financed by reducing saving so that the amount of *consumer* expenditure curtailed by the taxes will be *less* than the extra *government* expenditure. (d) Poor people have a lower propensity to save than the rich, so the transfer will cause overall savings to fall and consumption spending to rise, raising aggregate demand.

8.10. $M \times V = P \times T$

8.11. Demand-pull inflation and cost-push inflation.

8.12. In the Phillips curve the *annual percentage change in the level of prices* is shown as being *inversely related to* the level of unemployment, so that the higher the level of unemployment, the *lower* the level of inflation.

8.13. UK gross national product (GNP) is a measure of *only some* of the production activity under the control of UK nationals. A rise in GNP *often* implies a greater level of well-being of the population.

Chapter 9

9.1. The commodity forgone – 5 t of wheat.

9.2. A country is said to possess *comparative* advantage vis-à-vis another country in the production of a good 'X' if the expansion of production of good 'X' by 1 unit involves forgoing the *production* of a *lesser* quantity of other goods in the first country than in the second.

9.3. No. Trade depends on comparative advantage, not on the absolute levels of efficiency. Other countries may have a comparative advantage over the UK in food production despite the high absolute efficiency of UK agriculture.

9.4. (a) Maxitry has the comparative advantage in peach production; its opportunity cost of expansion of peach production is lower in terms of coconuts forgone than Minitry's.

(b) Minitry.

(c) Minitry, since it would expand coconut production and so have a surplus for export in exchange for peaches.

(d) Both.

9.5. Specialization and trade should be in such a direction that Country A exports wheat and Country B exports maize. (This arrangement follows as A has the comparative advantage in wheat production. One extra tonne involves the loss of 1 t of maize, whereas in B it involves the loss of 6/4 = 1.5 t of maize.)

9.6. Under perfect competition the relative prices of goods indicate the relative quantities of resources going into their production. For example, in Country C, 1 kg of beef takes three times the resources of 1 kg of butter. Every extra kilogram of beef produced then would involve the loss of 3 kg of butter. In Country D an extra 1 kg of beef would involve the loss of only 2 kg of butter. Country D thus has a comparative advantage in beef production, since it has the lower opportunity cost. It follows that the answers are thus: (a) Country D; (b) Country C.

9.7. Yes, we can predict the relative prices of butter and beef, but with the information given we cannot predict a precise figure, only a range within which it will lie. It will be somewhere between the two original internal price ratios, i.e. between 1: 3 and 1: 2.

9.8. Any three of the following:
- To protect infant industries.
- For defence purposes.
- To prevent dumping.
- To allow time for factors of production to be reallocated.
- For the protection of society against undesirable imports (drugs, illegal firearms, etc.).

9.9. The Terms of Trade are said to 'improve' when the *index of export prices rises* relative to the *index of import prices*. Such an 'improvement' will have an effect on the Balance of Payments which *can be either beneficial or deleterious depending on circumstances*.

9.10. a, b, c, d, f, h.

9.11. a, d.

9.12. b, c, f, h.

9.13. e (g is usually treated as a monetary movement).

9.14. b, c, e. Each of these increase the demand for foreign currency and increase the supply of pounds sterling.

Chapter 10

10.1. The correct order of the policy process is: policy effect (4); policy implementation (3); problem formulation (1); policy formulation (2).

10.2. The meanings of the terms are:

(a) Evaluation: judging the outcome in relation to objectives and resources used.

(b) Monitoring: collecting data during a policy programme for use in evaluation.

(c) Baseline study: a look at the situation before a programme starts.

(d) Side effects: effects felt outside the programme.

10.3. The following types of policy intervention might be considered by governments to reduce the amount of contamination of rivers by silage effluent.

- Legislation on water course standards, setting a maximum of permitted pollution, with fines for exceeding these levels.
- Taxes on pollution above certain levels.
- Subsidies to producers to enable them to invest in systems of silage storage that did not create pollution.
- Education and training schemes funded by the government.
- Government purchase of land in the catchment area of rivers, so that what goes on there can be controlled directly.

10.4. When agriculture is described as being used in an instrumental role, the policy aim is one not immediately concerned with agricultural production, rather for example in creating rural jobs, or in producing a cared-for countryside.

10.5. Reasons why a government may wish to have an agricultural policy include (there are others):

- the promotion of food security;
- economic growth;
- rural jobs and a vital rural economy;

- farm incomes;
- labour mobility;
- balance of payments; and
- environmental appearance.

10.6. Reasons why the support of incomes of farmers by governments raising the prices that farmers receive for their output might be resisted include:

- The distortion in price signals they give to farmers, so that farmers produce more than a competitive market for farm outputs would call forth, indicating an inefficient use of resources at economy level.
- The extra supply is likely to have an impact on international markets, so that the pattern of trade is distorted, to the detriment of specialization and exchange.
- There may be harmful effects on the environment where agriculture generates negative externalities (as can happen with more intensive land use).
- Most of the benefits of higher incomes goes to the larger producers and little to small producers who probably face the most severe income shortages.
- Pressure for farming to become more efficient is relaxed.
- Consumers find food prices rising, which impacts disproportionately heavily on the poorest members of society.

10.7. If a rural area is having difficulties in maintaining the numbers of people living there, alternatives to supporting agriculture that might be considered for achieving this include:

- Encouraging the expansion or establishment of non-farm businesses located in the countryside, such as tourism or services.
- Improving communications technology and transport links, so that people can live in rural areas but work elsewhere.
- Supporting the provision of basic services, such as health care and education.
- Building human capital (education and training) and social capital (the ability of communities to help themselves by networking, collective action, etc.).

The important point is that basic agriculture is unlikely to act as a growth point for incomes and jobs in most developed countries, so other means have to be found.

References

Cochrane, W.W. (1958) The agricultural treadmill. In: *Farm Prices, Myth and Reality*. University of Minnesota Press, Minneapolis, Minnesota, pp. 85–107.

Department for Environment, Food and Rural Affairs (Defra) (2013) *Family Food 2011*. Defra, London. Available at: https://www.gov.uk/government/collections/family-food-statistics (accessed 16 March 2013).

European Parliament (1998) *Health Care Systems in the EU – a Comparative Study*. Directorate General for Research Working Paper. European Parliament, London.

Malthus, T. (1798) *An Essay on the Principle of Population, as it Affects the Future Improvement of Society with remarks on the speculations of Mr. Godwin, M. Condorcet, and other writers*. Published anonymously.

Ministry of Agriculture, Fisheries and Food (MAFF) (1988) *National Food Survey, 1986*. Annual Report of the National Food Survey Committee, MAFF. HMSO, London.

Muhammad, A., Seale, J.L. Jr, Meade, B. and Regmi, A. (2011) *International Evidence on Food Consumption Patterns: an Update Using 2005 International Comparison Program Data*. USDA Technical Bulletin No. TB-1929. March 2011. United States Department of Agriculture (USDA), Washington, DC.

Smith, A. (1776) *An Inquiry into the Nature and Causes of the Wealth of Nations*. W. Strahan and T. Cadell, London.

Thomas, E. (1949) *An Introduction to Agricultural Economics*. Thomas Nelson & Sons, London.

Thünen, von J.H. (1826) *Der Isolierte Staat in Beziehung auf Landschaft und Nationalökonomie*. Translated by C.M. Wartenberg (1966) *Von Thünen's Isolated State*. Pergamon Press, Oxford.

Suggested Further Reading

The titles are listed chronologically, with newest titles first.

General texts to supplement all chapters of this book

Lipsey, R.G. and Chrystal, A. (2011) *Economics*, 12th edn. Oxford University Press, Oxford.

Samuelson, P.A. and Nordhaus, W.D. (2009) *Economics*. McGraw-Hill, New York.

Begg, D., Fischer, S. and Dornbusch, R. (2008) *Economics*. McGraw-Hill, Maidenhead, UK.

Bannock, G., Baxter, R.E. and Rees, R. (1994) *The Penguin Dictionary of Economics*, 3rd edn. Penguin, New York.

Texts specific to the economics of agriculture

Drummond, H.E. and Goodwin, J.W. (2011) *Agricultural Economics*, 3rd edn. Prentice Hall, Eaglewood Cliffs, New Jersey.

Penson, J.B. Jr, Capps, O.T., Rosson, C.P. III and Woodward, R.T. (2009) *Introduction to Agricultural Economics*, 5th edn. Prentice Hall, Eaglewood Cliffs, New Jersey.

Andreosso-O'Callaghan, B. (2003) *The Economics of European Agriculture*. Palgrave Macmillan, Basingstoke, UK.

Brassley, P. (1997) *Agricultural Economics and the CAP: an Introduction*. Blackwell Scientific, Oxford.

Ritson, C. (1997) *Agricultural Economics: Principles and Policy*. Granada, London.

Hodge, I. (1995) *Environmental Economics*. Macmillan, Basingstoke, UK.

Colman, D. and Young, T. (1989) *Principles of Agricultural Economics*. Cambridge University Press, Cambridge.

Hill, B. and Ray, D. (1987) *Economics for Agriculture: Food, Farming and the Rural Economy*. Macmillan, Basingstoke, UK.

Hill, B.E. and Ingersent, K.A. (1982) *An Economic Analysis of Agriculture*, 2nd edn. Heinemann Educational Books, London.

Farm production economics

Cramer, G.L., Jensen, C.W. and Southgate, D.D. Jr (2001) *Agricultural Economics and Agribusiness*, 8th edn. Wiley, New York.

Casavant, K.L., Infanger, L. and Bridges, D.E. (1998) *Agricultural Economics and Management*. Prentice Hall, Eaglewood Cliffs, New Jersey.

Little, R.D. (1997) *Economics: Applications to Agriculture and Agribusiness*, 4th edn. Prentice Hall, Eaglewood Cliffs, New Jersey.

Barnard, C.S. and Nix, J. (1979) *Farm Planning and Control*, 2nd edn. Cambridge University Press, Cambridge.

Bishop, C.E. and Toussaint, W.D. (1958) *Introduction to Agricultural Economic Analysis*. Wiley, New York.

The EU's Common Agricultural Policy and public decision making

Hill, B. (2012) *Understanding the Common Agricultural Policy*. Earthscan Food and Agriculture. Routledge, London.

Organisation for Economic Co-operation and Development (OECD) (2010) *Agricultural Policies in OECD Countries: At a Glance*. OECD, Paris.

Oskam, A., Meester, G. and Silvis, H. (2010) *EU Policy for Agriculture, Food and Rural Areas*. Wageningen Academic Publishers, Wageningen, The Netherlands.

Ackrill, R. (2000) *The Common Agricultural Policy*. Sheffield Academic Press, Sheffield, UK.

Economic Research Service (ERS), United States Department of Agriculture (USDA) (1999) *The*

EU's CAP – Pressures for Change. ERS/USDA, Washington, DC.

Grant, W. (1997) *The Common Agricultural Policy.* St Martin's Press, New York.

Ritson, C. and Harvey, D.R. (1997) *The Common Agricultural Policy.* CAB International, Wallingford, UK.

The policy process

Hill, M. (2006) *The Public Policy Process*, 4th edn. Pearson-Education, Harlow, UK.

Wallace, H. and Wallace, W. (2005) *Policy-Making in the European Union.* Oxford University Press, Oxford.

Hogwood, B.W. and Gunn, I.A. (1984) *Policy Analysis for the Real World.* Oxford University Press, Oxford.

Recent information on the workings of the EU institutions and the process of decision making within the Common Agricultural Policy is available on the website of the institutions (www. europa.eu.int). The views from the Commission and Parliament are sometimes somewhat contrasting.

Addresses of relevance include the following (all have the prefix http://):

consilium.europa.eu/ (Council)
ec.europa.eu/ (Commission)
europarl.europa.eu/ (European Parliament)
curia.europa.eu (Court of Justice)
eca.europa.eu (Court of Auditors)
eesc.europa.eu (Economic and Social Committee)
cor.europa.eu (Committee of the Regions)

Index

absolute advantage 174–177, 188
additionality 196
adoption (technical advances) 46
aggregate demand 155–159, 161, 162, 180, 199
average capital:output ratios (ACOR) 166
average cost (AC) 46, 53, 55, 57–59, 62, 81,
    92, 94, 96, 98, 140
average fixed cost (AF$_x$C) 92, 93
average product (AP) 74–76, 165
average propensity to consume (APC) 155
average propensity to save (APS) 155
average propensity to withdraw *see* average propensity
    to save (APS)
average revenue (AR) 54, 55, 57–59, 62, 93
average value product (AVP) 76, 77
average variable cost (AVC) 79–81, 92, 93

balance of payments
    on capital account 185
    correcting measures 188
    on current account 185, 187
    total 185
balance of trade 183, 185
baseline study 196
basic payment (to farms) 212
Bennet's Law 37
bilateral monopoly 63
bilateral trade 176
budget line 17–19, 21, 83
    see also iso-expenditure line
business form 72

capital
    consumption of 164
    definition of 118, 131, 219
    diminishing returns to 116
    economic growth and 131, 163, 167, 183
    as factor of production 116–122
    international trade and 172, 173, 177, 180,
        181, 183, 185, 187
capital account 185
capital flows and international trade 185
capital goods 117–120, 154, 155, 164, 180
capital market 120, 121
'cartel' 62, 64
cause and effect 8

centrally planned economies 4, 131, 146, 166
ceteris paribus 6, 20, 21, 28, 182
circular flow of income 150–155
Coase's Theorem 145
Cobweb Theorem 67
Common Agricultural Policy of the European
    Economic Community 7, 102
comparative advantage 173–178, 180, 181, 191
competitive goods (in consumption) 37–38
competitive products (in supply) 45
complementary products (in supply) 45
    see also joint products
consumer goods 51, 117, 120, 132, 154, 155, 163,
    164, 167
    see also capital goods
consumer spending and the level of national
    income 161
consumer surplus 20–21
contracts
    and price cycles 67, 68
    and reduction in uncertainty 125, 126
    and vertical integration 127
controlled experiments 6–8
controls 7
cost flexibility 126, 127
'cross-compliance' 204
cross elasticity of demand 38
cross elasticity of supply 45, 46

deficiency payment 208
deflationary domestic policy and balance of payments
deflationary gap 158
demand
    curve as faced by the individual farmer 32, 33
    curve as faced by the industry 32
    elasticity *see* cross elasticity of demand;
        price elasticity of demand
    for factors of production 46, 103, 132–133
        see also derived demand
    household 28, 29
    market 28, 39, 47, 65, 148, 157, 159, 177, 178,
        185–187, 191, 205
    schedule 20, 26, 27, 47, 54
derived demand for factors 132
differentials
    dynamic 133
    equilibrium 133, 134

Diminishing Marginal Utility, Law of 13, 14, 74
Diminishing Returns, Law of 21, 73, 74, 116
direct (income) payments 204, 205, 212
discriminating monopoly 61–62
'disutility' 13
diversification 125, 126, 205
division of labour 95, 97, 119
dumping 181–182, 205

economic characteristics of agriculture 198, 200, 202
economic growth
    and agriculture 163, 167, 199, 200
    balance of payments 160, 183
    costs of 167–169
    definition 163, 164
    factors related to the rate of 163
economic problem 2–3, 6–8, 14, 26, 146, 150,
    191, 198, 201
economics, definitions 1, 3, 131
economies of scale and size 94
education, as human capital 122, 165
elasticity see cross elasticity of demand; cross elasticity
    of supply; income elasticity of demand; price
    elasticity of demand; price elasticity of supply
emissions see pollution
Engel's Law 34–35, 47
entrepreneurship, as a factor of production 116, 118,
    123–127
equilibrium
    consumer 21, 23
    in differentials 133, 134
    equilibrium price 21, 23, 26, 27, 55, 208
    of a firm 72
    level of national income 152–154, 156–158
    stable and unstable 21–23
European Central Bank 179, 186, 187
evaluation 7, 196–198
exchange rates between currencies 150
excludability 148
expansion path 86
external benefits and external economies 143, 144
external costs and external diseconomies 143
externalities
    and economic growth 144

factor–factor relationship 73, 82–88, 91, 99
factor–product relationship 73–81, 88, 91, 99
factors of production
    classification 116, 127, 129
    demand and supply of 132–133
    non-specific and specific 44, 127
    see also mobility of factors
    prices and supply 66, 98, 102
fixed costs 41, 60, 79, 91–93, 126, 140
fixed proportions (inputs) 84, 87

flexible exchange rate 178–179
floating exchange rate 186
Foreign Exchange Market 177, 185
free goods 2, 13–14, 219
free market and economic system 4
'free riders' 147
Friedman, M. 159
'functional relationship' 6

GATT (General Agreement on Tariffs and Trade) 174,
    181, 182, 187
global market 51
government spending and the level
    of national income 152, 157
gross national product (GNP) 163
guaranteed prices 68, 120

hedging (with 'futures') 127
horizontal integration 97, 127
horizontal summation (of supply curves) 55
household demand 28
hypothesis 4–6, 8, 123

imperfect competition 53–54, 57–61, 64, 140–141, 192
imperfections in the market 53, 191
imperfect knowledge (by consumers) 148
import tax 208
    see also tariffs
income
    distribution and redistribution of 139, 140
    effect 18, 19
    and inflation 160, 161
    and trade 181
income elasticity of demand
    significance to agriculture 35–37
income support (in agriculture) 203
incremental capital:output ratio (ICOR) 165, 166
indifference
    curves 16–18, 21, 83, 176
    theory 14–20
infant industries, and protection 180
inferior goods 33, 35
inflation
    cost-push 162
    demand-pull 161, 162
    effects on an economy 150
    and exchange rates 150, 160
inflationary gap 158, 161
injections, to circular flow of income 151, 152
innovators 104, 105, 107, 109, 202
instrumental role (for agriculture) 198, 201, 208, 212
insurance
    formal 124
    informal 126–127

internalization (of external costs and
         benefits) 145
International Monetary Fund (IMF) 186
international trade
         and economic growth 183
invention and innovation 164, 167
         see also technical advance
investment *see* capital; economic growth; injections,
         to circular flow of income
'invisible hand' 138–139, 150
invisibles (in trade) 183–185
iso-cost line 83–86
iso-expenditure line 17, 18
         see also budget line
isoquant 82–86, 99
iso-resource curve 88–91, 99
iso-revenue line 89, 90

joint demand goods 38
joint products (in supply) 45

Keynes, J.M. 157, 159

labour
         in agriculture 8, 65, 97, 129, 202
         and economic growth 163, 200
         as factor of production 44, 73, 115, 116,
                  122–123
         quality 97, 122
         quantity 86, 122, 156, 165
         supply curve of 31, 133
         wages and unionization 43, 46, 51, 91, 103,
                  107, 132, 151, 153, 162
laggards 109
'laissez-faire' 138, 150
land, as a factor of production 44, 116–117
Law of Diminishing Marginal Product/Law
         of Diminishing Returns *see* Diminishing
         Returns, Law of
Law of Large Numbers 7
long-term trend (in agricultural prices) 64, 65
Luddites 103, 130
'lumpy' commodities 21, 22
'lumpy' inputs 86, 87

macroeconomics 130, 148–169, 179
MacSharry reforms 204
Malthus, T. 116, 165, 166
margin 12–13, 41, 43, 93, 99, 103, 120, 124,
         125, 133, 164, 204
marginal abatement cost (of pollution) 147
marginal approach to optimizing the factor–product
         relationship 79–80

marginal cost (MC) 54, 55, 57, 58, 61, 62, 79–81, 88,
         92–94, 99, 140, 143, 146, 210, 211
marginal damage (of pollution) 210, 211
marginal factor cost (MFC) 76–78
marginal product (MP) 73–76, 131, 172
marginal propensity to consume (MPC)
marginal propensity to save (MPS) *see* marginal
         propensity to withdraw (MPW)
marginal propensity to withdraw (MPW) 154
marginal rate of substitution
         between inputs 82, 83
         between products 88
marginal rate of substitution (MRS) 16, 62, 82–84,
         88–91, 175
marginal rate of transformation (MRT) 88
marginal revenue (MR) 54, 55, 57, 58, 61, 80, 88, 93,
         99, 143
marginal utility 13–15, 20–21, 74
marginal value product (MVP) 76–78, 200
market demand 28, 39–40, 65
medium-period booms or slumps 65
medium-period cyclical price movements 66
         see also price cycles
Milk Marketing Board 53, 61, 125
mobility
         of factors 65, 127–130
         geographical 127–130
         and international trade 173
         occupational 127–130
models, use of 3, 51
modulation 212
'monetarists' 161
monetary movements 183, 185, 186
money supply 122, 160–163
monitoring 196
monopoly 53, 54, 57–62, 64, 97, 134–135, 140–141,
         147, 192, 203
monopoly profit 59, 97, 134–135
monopsony 54
multifunctional/mutifunctionality 202, 212
multilateral trade 176, 177
multiplier 153–155, 160, 196, 212

National Health Service (NHS) 8, 141, 152
national income 4, 103, 151–154, 156–159, 161–163,
         180, 181, 202
         see also equilibrium
nationalized industries 135, 141, 159
neo-Keynesians 161
new varieties of crops with higher yields 31
         see also technological advances
non-excludability 146
non-rivalness 146
normal profits 46, 52, 57–59, 63, 134–135
normative statements 5, 9, 219
         see also value judgements

objectives (and goals)
    of agricultural policy 8, 47, 104, 204
    of firms 3, 40, 47
    of government policy 105, 191–213
    of individuals 2, 3
    of society 2, 3
oligopoly 53, 54, 64, 140
oligopsony 53, 64
opportunity cost 1–2, 21, 44, 173–176
optimum level of pollution 147, 211

Pareto optimum 139
path dependency 7–8, 192, 198
perfect competition
    and individual producer 54–56
    and industry 54–60
    and social welfare 139
perfect knowledge 51, 52, 140, 141, 148
perfect market 52, 53, 62
perfect substitutes (inputs) 83, 84, 86
Phillips, A.W. 162
Phillips Curve 162
planned economic system *see* centrally
        planned economies
policy
    for agriculture's problems 191, 198, 202–203
    process 192, 194, 198, 211
    for rural areas 191–213
pollution 3, 53, 138, 144, 145, 167, 168, 191,
        195, 204, 210, 211
population, size of
    and demand 39, 40
    and economic growth 164–166
positive statements 5, 6
price cycles 66–68, 124, 202
price effect 18–20
price elasticity of demand
    for factors of production 132
    significance to agriculture 30–33
    and tariffs 182
price elasticity of supply 41
price movements in agriculture
    booms or slumps 65–66
    cyclical 66
    daily 64
    long-term trend 64–66
    medium-period cyclical 66
price system
    functions of 52–53
    imperfections in 53, 138
Principle of Equimarginal Returns 15, 61, 90,
        131, 172, 201
private benefits 143, 145, 146
private costs 142–146
privatization 135, 141, 159, 167
problem formulation 192–195

product flexibility 126
production cycle, length of 43
production flexibility 126
production function 78, 81, 82, 131, 135, 165
production possibility boundary (or curve) 88
product–product relationship 73, 88–89, 91, 99
profit, functions of 135
protectionism 183
    see also restrictions on trade
public goods 53, 146–149, 169, 191, 202, 212

quantitative easing 161
Quantity Theory of Money 160–163
quasi-rent 133
quota 14, 32, 61, 138, 179, 181–183, 186,
        208, 209, 211

rate of technical transformation (RTT) 88
relative income 36, 188
reserve currency 185
reserves, of gold and foreign currency 185
restrictions on trade, arguments for 181, 183
retaliation in tariff war 182
Ricardo, D. 116
risk and uncertainty 124–127
rivalness 147, 148
    see also non-rivalness
rural development
    future of 211–213
    policies for 201, 211, 212

savings (private) 161
    see also withdrawals (from circular flow of income)
scarcity, economic 2, 3
scientific approach 4–6, 8, 110
scissors graph 27
self sufficiency 51, 115, 172, 173, 200
    see also restrictions on trade
side effects 196, 211
single payment scheme/single farm payment 204
Smith, A. 138, 139, 150
social benefit 123, 143–147, 149, 191
social goods *see* public goods
social welfare
    social welfare frontier 193
specialization (and exchange) 51, 115, 167, 172, 175,
        176, 179–181, 191
specific and non-specific factors
    of production 44, 127, 128
stabilization of markets 203
standard
    as applied to pollution 144
    of living 3, 51, 120, 150, 155, 156, 160,
        165, 167, 173, 179, 181, 204

sterling balances 185
structural adjustments 203, 208
sub-marginal producer 55
subsidiarity 210
supply
      cross elasticity of 45, 46
      curves 27, 31, 46, 47, 55, 69, 132, 200, 205
      of factors of production 40, 46, 132–134
      lag 43, 124
      price elasticity 41
support buying 205, 206
supra-normal profits (or surplus or excess profits) 52
sustainable/sustainability 14, 117, 118, 162, 163

tariffs 179–183, 186
tastes of consumers 28, 38–39, 148
tax
      on imports *see* tariffs
      on pollution 211
technical advances 46, 47, 102, 128, 180, 203
technical unemployment 130
technology 40, 46–47, 98–111, 124, 130, 141,
          144, 163, 165, 168, 180, 200
terms of trade 183, 184
theory
      verification 6
Theory of Comparative Advantage 173–174, 179
      see also comparative advantage
Theory of Consumer Choice 11–23, 72, 83
Theory of Demand 28–40, 46
      see also demand
Theory of Distribution 115–135
Theory of Supply 31, 40–47
      see also supply
Theory of the Firm 72–111
Thomas, E. 118
time flexibility 126, 133
total factor cost (TFC) 76–78
total product (TP) 73, 74, 118, 153
total revenue (TR) 46, 54, 55, 57, 80
total value product (TVP) 76–78, 99

total variable cost (TVC) 79–81, 92
trade-off
      between inflation and unemployment 162
trade union 132, 148, 162, 163
transactions cost 142
transfer earnings (of factors) 44
transfers/transfer payments 151
treadmill of technological advance 47, 104

unemployment
      frictional 131
      and international trade 167, 182
      and level of national income 153, 157
      mass (or cyclical) 130
      residual 130
      seasonal 130
      structural 130
      see also equilibrium
utility 2, 11–16, 20–21, 72, 140, 148
Utility Theory of Consumer Choice 14–15

value judgements of society 139–140
      see also normative statements
variable cost (VC) 41, 79, 91–93, 126
velocity of circulation 160, 161
vertical integration 127
visibles (and) visible trade 183
von Thunen, J.H. 116, 117

'waiting' period in capital investment 120
wealth
      and capital 118, 140
'Wealth of Nations' 138
weather, effect on demand and supply 26, 31
withdrawals
      from circular flow of income 151–153
      from the market 52
World Trade Organization (WTO) 174, 181,
          182, 187